MICROSUPERCAPACITORS

Woodhead Publishing Series in Electronic and Optical Materials

MICROSUPERCAPACITORS

Edited by

KAZUFUMI KOBASHI

KAROLINA LASZCZYK

ELSEVIER

WP
WOODHEAD PUBLISHING
An imprint of Elsevier

Woodhead Publishing is an imprint of Elsevier
The Officers' Mess Business Centre, Royston Road, Duxford, CB22 4QH, United Kingdom
50 Hampshire Street, 5th Floor, Cambridge, MA 02139, United States
The Boulevard, Langford Lane, Kidlington, OX5 1GB, United Kingdom

Notices
Knowledge and best practice in this field are constantly changing. As new research and experience broaden our understanding, changes in research methods, professional practices, or medical treatment may become necessary.

Practitioners and researchers must always rely on their own experience and knowledge in evaluating and using any information, methods, compounds, or experiments described herein. In using such information or methods they should be mindful of their own safety and the safety of others, including parties for whom they have a professional responsibility.

To the fullest extent of the law, neither the Publisher nor the authors, contributors, or editors, assume any liability for any injury and/or damage to persons or property as a matter of products liability, negligence or otherwise, or from any use or operation of any methods, products, instructions, or ideas contained in the material herein.

Library of Congress Cataloging-in-Publication Data
A catalog record for this book is available from the Library of Congress

British Library Cataloguing-in-Publication Data
A catalogue record for this book is available from the British Library

ISBN: 978-0-08-102888-9 (print)
ISBN: 978-0-08-102889-6 (online)

For information on all Woodhead publications
visit our website at https://www.elsevier.com/books-and-journals

Publisher: Matthew Deans
Acquisitions Editor: Kayla Dos Santos
Editorial Project Manager: Rachel Pomery
Production Project Manager: Anitha Sivaraj
Cover Designer: Vicky Pearson

Typeset by STRAIVE, India

Contents

Contributors

Bihag Anothumakkool
TNO Holst Centre, Dutch National Institute for Applied Scientific Research, Eindhoven, The Netherlands

Yuan Chen
School of Chemical and Biomolecular Engineering, The University of Sydney, Sydney, NSW, Australia

Meena Ghosh
Academy of Scientific and Innovative Research (AcSIR), Ghaziabad, Uttar Pradesh; Physical and Materials Chemistry Division, CSIR-National Chemical Laboratory, Pune, India

Małgorzata Graś
Institute of Chemistry and Technical Electrochemistry, Poznan University of Technology, Poznań, Poland

Mihir Kumar Jha
Department of Chemistry, Indian Institute of Technology Bombay, Mumbai, Maharashtra, India

Kazufumi Kobashi
CNT-Application Research Center, National Institute of Advanced Industrial Science and Technology, Ibaraki, Japan

Łukasz Kolanowski
Łukasiewicz Research Network, Institute of Non-Ferrous Metals Division in Poznan Central Laboratory of Batteries and Cells, Poznań, Poland

Sreekumar Kurungot
Physical and Materials Chemistry Division, CSIR-National Chemical Laboratory, Pune, India

Karolina Urszula Laszczyk
Faculty of Microsystem Electronics and Photonics, Wroclaw University of Science and Technology, Wrocław, Poland

Zhi Li
Department of Chemical and Materials Engineering, University of Alberta, Edmonton, AB, Canada

Grzegorz Lota
Institute of Chemistry and Technical Electrochemistry, Poznan University of Technology; Łukasiewicz Research Network, Institute of Non-Ferrous Metals Division in Poznan Central Laboratory of Batteries and Cells, Poznań, Poland

Katarzyna Lota
Łukasiewicz Research Network, Institute of Non-Ferrous Metals Division in Poznan Central Laboratory of Batteries and Cells, Poznań, Poland

Jijeesh Ravi Nair
Helmholtz Institute Münster, IEK-12, Forschungszentrum Jülich GmbH, Münster, Germany

Piotr Śliwiński
Faculty of Microsystem Electronics and Photonics, Wroclaw University of Science and Technology, Wrocław, Poland

Ziyou Song
Department of Electrical Engineering and Computer Science, University of Michigan, Ann Arbor, MI, United States

Roby Soni
Electrochemical Innovation Lab, Department of Chemical Engineering, University College London, London, United Kingdom

Chandramouli Subramaniam
Department of Chemistry, Indian Institute of Technology Bombay, Mumbai, Maharashtra, India

Vidyanand Vijayakumar
Helmholtz Institute Münster, IEK-12, Forschungszentrum Jülich GmbH, Münster, Germany; Academy of Scientific and Innovative Research (AcSIR), Ghaziabad, Uttar Pradesh; Physical and Materials Chemistry Division, CSIR-National Chemical Laboratory, Pune, India

Martin Winter
Helmholtz Institute Münster, IEK-12, Forschungszentrum Jülich GmbH; Institute of Physical Chemistry, University of Münster; MEET Battery Research Center, Münster, Germany

Shengli Zhai
School of Chemical and Biomolecular Engineering, The University of Sydney, Sydney, NSW, Australia; Department of Chemical and Materials Engineering, University of Alberta, Edmonton, AB, Canada

Preface

Historically, the first supercapacitor was developed by Becker et al. who revealed his invention in a patent application, but under the title "Low voltage electrolytic capacitor"[a] in 1954 [1]. Since then, although parallel semiconducting technology, i.e., integrated circuits, was developed [2], mobile energy storage devices played a minor role, until the boom in smartphones, especially in the first decade of the 2000s of the 21st century [3]. The rapid jump in the available internet-related technologies via mobile phones opened the route for the new applications to be used right now and at the place [4–7]. Nowadays, the concept of connecting living organisms (humans, animals), plants, and objects to a unified network for continuous monitoring is commonly known around the globe and applied for the research [8–11]. Although a networking has its opponents considering personal security and right to privacy [12–14], for many of us it gives hope to resolve a lot of issues. Especially those pointed as the most critical for the human being, such as environmental pollution, food- and drinkable-water-deficit diseases, especially in the context of the latest pandemic of COVID-19, but also so-called civilization diseases (cancer, diabetes, depression, and many others), society aging, and labor shortage. On the other hand, the space exploration and possible settlement of humans at the other planets rise the demand for urgent solutions to supply with the basics products to enable the settles to survive [15]. This boom on the Internet of Things (IoT), which encompasses the point-of-care [16], smart home management [17], green revolution [18], and many others, definitely needs the energy to run. While the group of researchers focuses on harvesting energy [19–21], the point is how to store the overdosage of the energy to have an energy supply, when it cannot be deployed by renewable energy sources like wind and solar.

Therefore miniaturized energy storage devices, especially for mobile sensing and actuating devices, medical implants, or medical examination, are the rapidly growing topic of interest for a wide diversity of brands. Parallel to microbatteries, we can find hundreds of scientific papers every year that discuss new findings on the storage mechanism, efficient materials, and applications, that enable novel functionality, to realize the concept of

[a] Today, we know well, the electorolytic capacitors are distinguished from the supercapacitors through the governing mechanism of the energy storage. Thus the title does not describe adequately the invention.

networking, and support the investigations toward solving the critical problems. In order to bring the recent development closer, we would like to present you the collection of chapters written by various scientists around the world discussing in a straightforward way on major aspects of the miniaturized supercapacitor also named microsupercapacitors. The book is intended to be a form of a guide through the basics terms, starting from the applicable materials, applications, technology, and recent developments. We hope it would help the students, engineers, and everybody interested in the multidisciplinary fields to discover the potential of the microsupercapacitors based on the knowledge gathered in this book and make a step forward in sustainable development.

We would like to thank here all coauthors who sacrificed their time and share their knowledge in one of the fields required to build and examine the microsupercapacitors. We would like to thank the managers who guided us through the book process preparing and our families who supported us with their patience and indulgence.

Karolina Laszczyk[a] and Kazufumi Kobashi[b]
[a]Faculty of Microsystem Electronics and Photonics, Wroclaw University of Technology, Wrocław, Poland
[b]CNT-Application Research Center, National Institute of Advanced Industrial Science and Technology, Ibaraki, Japan

References

[1] H. J. Becker, and V. Ferry, Low voltage electrolytic capacitor. US Patent Office, no. 2,800,616, Patented July 23, 1957.
[2] S. Kumar, N. Krenner, Review of the semiconductor industry and technology roadmap, J. Sci. Educ. Technol. 11 (2002) 229–236.
[3] G. Cecere, N. Corrocher, R.D. Battaglia, Innovation and competition in the smartphone industry: is there a dominant design? Telecommun. Policy 39 (2015) 162–175.
[4] H. Jee, Review of researches on smartphone applications for physical activity promotion in healthy adults, J. Exerc. Rehabil. 13 (2017) 3–11.
[5] C. Pronello, P. Kumawat, Smartphone applications developed to collect mobility data: a review and SWOT analysis, Adv. Intell. Syst. Comput. 1251 (2021) 449–467.
[6] J. Torous, A.C. Powell, Current research and trends in the use of smartphone applications for mood disorders, Intern. Interv. 2 (2015) 169–173.
[7] S. Lee, J. Suh, Y. Choi, Review of smartphone applications for geoscience: current status, limitations, and future perspectives, Earth Sci Inform. 11 (2018) 463–486.
[8] K.D. Chaitanya, R.V. Adiraju, S. Pasupuleti, D. Nandan, A review of smart greenhouse farming by using sensor network technology, Adv. Intell. Syst. Comput. 1245 (2021) 849–856.

[9] K.S. Adu-Manu, C. Tapparello, W. Heinzelman, F.A. Katsriku, J.D. Abdulai, Water quality monitoring using wireless sensor networks: current trends and future research directions, ACM Trans. Sens. Netw. 13 (2017) 4.

[10] L.M. Gómez Peláez, J.M. Santos, T.T. de Almeida Albuquerque, N.C. Reis, W.L. Andreão, M. de Fátima Andrade, Air quality status and trends over large cities in South America, Environ. Sci. Policy 114 (2020) 422–435.

[11] M. Takahashi, Z. Feng, T.A. Mikhailova, O.V. Kalugina, O.V. Shergina, L.V. Afanasieva, R.K.J. Heng, N.M.A. Majid, H. Sase, Air pollution monitoring and tree and forest decline in East Asia: a review, Sci. Total Environ. 742 (2020) 140288.

[12] L. Khandare, D.K. Sreekantha, K. Sairam, A study on encryption techniques to protect the patient privacy in health care systems, in: Innovations in Power and Advanced Computing Technologies (i-PACT), Vellore, India, 2019, 2019, pp. 1–5.

[13] X. He, X. Yang, W. Yu, J. Lin, Q. Yang, Towards an iterated game model with multiple adversaries in smart-world systems, Sensors 18 (2018) 674.

[14] C. Naseeb, "Future" children—can IoT devices help save the world? IT Profess. 22 (2020) 13–17.

[15] K.A. Farley, K.H. Williford, K.M. Stack, Mars 2020 mission overview, Space Sci. Rev. 216 (2020) 142.

[16] S. Nayak, N.R. Blumenfeld, T. Laksanasopin, S.K. Sia, Point-of-care diagnostics: recent developments in a connected age, Anal. Chem. 89 (2017) 102–123.

[17] G. Lobaccaro, S. Carlucci, E. Löfström, A review of systems and technologies for smart homes and smart grids, Energies 9 (2016) 348.

[18] M. Tabaa, F. Monteiro, H. Bensag, A. Dandache, Green industrial internet of things from a smart industry perspectives, Energy Rep. 6 (2020) 430–446.

[19] J. Singh, R. Kaur, D. Singh, Energy harvesting in wireless sensor networks: a taxonomic survey, Int. J. Energy Res. 45 (2021) 118–140.

[20] S.K. Karan, S. Maiti, J.H. Lee, Y.K. Mishra, B.B. Khatua, J.K. Kim, Recent advances in self-powered tribo-/piezoelectric energy harvesters: all-in-one package for future smart technologies, Adv. Funct. Mater. 30 (2020) 2004446.

[21] Y. Tian, X. Liu, F. Chen, Y. Zheng, Harvesting energy from sun, outer space, and soil, Sci. Rep. 10 (2020) 20903.

PART I

Materials and Defining Performance for Microsupercapacitors

CHAPTER 1

Materials under research: Nanomaterials, aerogels, biomaterials, composites, inks

Grzegorz Lota[a,b], Katarzyna Lota[b], Łukasz Kolanowski[b], and Małgorzata Graś[a]

[a]Institute of Chemistry and Technical Electrochemistry, Poznan University of Technology, Poznań, Poland
[b]Łukasiewicz Research Network, Institute of Non-Ferrous Metals Division in Poznan Central Laboratory of Batteries and Cells, Poznań, Poland

1.1 Introduction

Due to the diminishing of fossil fuels, carbon dioxide emissions, and climatic changes, many countries are forced to increase the energy efficiency not only in the industry but also in transportation and domestic heating or cooling systems. Energy consumption in our life should always be related to environmental protection. Due to the increasing use of portable electronics and high interest in electric vehicles, the power sources industry has developed noticeably. Many electronic devices require the use of rechargeable batteries (or fuel cells). This strong demand for efficient power sources contributed to rapid research and development of new electrode materials for electrochemical cells. There are few main electrochemical power sources that could be used to store and convert the energy, mainly rechargeable batteries (accumulators), fuel cells, electrochemical capacitors, and conventional (dielectric) capacitors. Each system is characterized by a different type of electrochemical properties and hence offers different energy supply characteristics. Accumulators are able to store high energy at low power density, but electrochemical capacitors can deliver power over 20 kW kg^{-1} despite their low energy storage properties [1–5].

Electrochemical capacitors are systems that accumulate energy by charging/discharging the electric double layer due to electrostatic adsorption of ions on the electrode surface. In comparison with batteries, they are characterized by high power density, linear discharge profile, and superior cyclability [6–9]. In the case of symmetric (carbon materials) systems, in which the formation of electrical double layer occurs, there is only

https://doi.org/10.1016/B978-0-08-102888-9.00001-X

electrostatic attraction between electrolyte ions and electrode materials. In electrostatic attraction, there are no structural changes in electrode materials which ensures high cycling stability (>100,000 cycles). Electrostatic attraction should not provide the changes in electrode materials therefore the decrease in capacitance during the cycling should be negligible [10]. Currently, scientists are trying to increase the capacitance by modifying the electrode materials, using transition metal-based materials, or conducting polymers and their composites. These materials store more energy than carbon-based systems due to the presence of faradaic processes. Unfortunately, such materials are characterized by lower charge propagation and cycling stability. This is mainly due to slow redox reaction kinetics and lower electrical conductivity in the comparison with carbon materials. Moreover, the crystal structure of materials could collapse during charging/discharging processes. In the case of transition metal-based materials, the dissolution significantly reduces the cycling stability and should not be neglected [11]. For conducting polymers, the potential range and electrolyte ions significantly influence the cycling stability. During the charging/discharging process the conducting polymers undergo swelling and shrinking which affect disruption in conducting mechanisms. Another important factor is the adherence of the polymer to the current collector. It has been stated that deposition of polypyrrole on exfoliated graphite surface causes 97% capacitance retention during 10,000 cycles (6 A g^{-1}) [12]. Therefore, apart from the electrode materials properties, the current collector/electrode and electrode/electrolyte interfaces possess significant factors in cycling stability. It should be noted that transportation kinetics of electrons/ions, electrolyte decomposition, the loss of electrical contact between all components and ion transfer between interfaces influence the capacitor damage and thus cycling stability [10, 11].

Significant problems of electrochemical capacitors include high self-discharge and relatively low energy density of the system. The increase of capacitance could be realized by faradaic processes. Faradaic reactions occur at the electrode/electrolyte interface in the presence of heteroatom functional groups [13–17], transition metal oxides [18], and conducting polymers [19]. It should be noted that faradaic reactions could be derived from the electrolyte. For example, the aqueous solution of KI is able to provide electrochemical reactions due to the various oxidation states of iodine [20].

Due to the high interest in portable devices, wearable electronics, sensors, and wireless communications, the size of electronic systems becomes smaller. The miniaturization of electrical circuits and other systems resulted in intensified research and development of microelectromechanical systems (MEMS). MEMS are associated with all devices obtained from microfabrication except integrated circuits (ICs) [21]. Microelectromechanical systems can contribute to a change in world economy and our life due to numerous opportunities for the commercialization of new applications. For example, electrochemical capacitors can be charged by piezoelectric transducers in "smart" clothes which could also be used for military purposes [22]. Therefore, there is a strong demand for efficient micro-power systems to meet the requirements of MEMS. Thus, microsupercapacitors (MSC) are a very important class of advanced electrochemical capacitors (also called supercapacitors or ultracapacitors) as power sources for future microelectronic devices. MSC refers to a device with a footprint area at a millimeter/centimeter scale and thickness lower than 10 μm [23]. In this chapter, different types of electrode materials have been reviewed and the classes considered are presented in Fig. 1.1.

1.2 Principles of working and background

Microsupercapacitors consist of two electrodes immersed in the electrolyte and work mainly due to charging/discharging of the electric double layer. This mechanism occurs due to the adsorption of cations to the negative

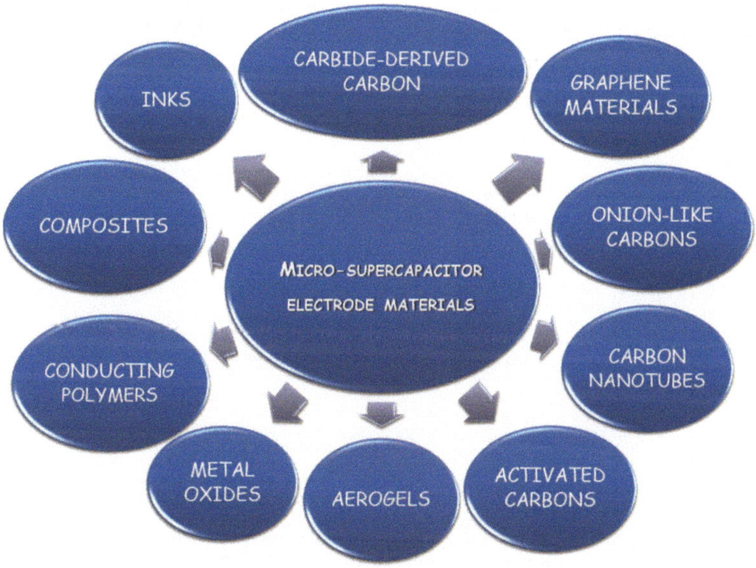

Fig. 1.1 Electrode materials of microsupercapacitors described in this chapter.

electrode and anions to the positive one. During discharging, ions will release into the electrolyte solution. The total capacitance is determined by the electrode with the smallest value of capacitance according to Eq. (1.1):

$$\frac{1}{C}=\frac{1}{C_1}+\frac{1}{C_2} \tag{1.1}$$

where C_1 and C_2 are the capacitance values of positive and negative electrodes. The specific capacitance resulting from accumulated ions can be described by the following Eq. (1.2):

$$C=\frac{\varepsilon_r\varepsilon_0 A}{d} \tag{1.2}$$

where ε_r is the electrolyte dielectric constant and ε_0 is the vacuum dielectric constant, d is the thickness of electric double layer and A is the surface area of the electrode which is accessible to the electrolyte ions. The energy of the system is related to the capacitance and depends on the voltage as given by Eq. (1.3):

$$E=\frac{1}{2}CU^2 \tag{1.3}$$

It should be noted that the energy of the system is also limited by the electrodes. High specific surface area should result in high specific capacitance. The relationship between BET surface area and capacitance of the system is more complicated. It should be noted that the access of electrolyte to the micropores is crucial. In case when the pores in the electrode are too small, the material will not contribute to the total capacitance for bigger solvated ions.

During the charging/discharging the charge passed through the electrodes should be the same (for symmetric and asymmetric systems), as follows: $q_+=q_-$. Therefore, in the case of an asymmetric system, the potential range of each electrode should be determined in order to optimize the mass ratio of electrodes. The charge stored by electrodes depends on the capacitance C, potential range ΔE, and electrode mass [24–26], according to Eq. (1.4):

$$q=C\cdot\Delta E\cdot m \tag{1.4}$$

Apart from low energy density, the other important drawback of microsupercapacitors is their relatively high self-discharge in the comparison with batteries. Self-discharge is a quick voltage drop after the charging process which results in lower energy and power that can be stored. Several types of self-discharge mechanisms have been proposed, such as ohmic leakage, parasitic reactions, and charge redistribution. From the practical and commercial point of view, such processes should be taken into consideration during the design of microsupercapacitors with desirable parameters. The rate of self-discharge strongly depends on the charging/discharging conditions, type of electrode used, electrolyte, supercapacitor construction, and the presence of any species (contaminants) that undergo oxidation/reduction [27–29]. Moreover, the impact of the current collector and its corrosion should not be neglected [30].

Microsupercapacitors use the same materials and could base on similar processes that can be used for electrochemical capacitors. Very often, electrode materials are deposited/synthesized on interdigitated current collectors on insulating substrates. The cost of the MSC mainly depends on the fabrication method used. The deposition/synthesis and transfer of the electrode material on the current collector is a very important task during the fabrication process because the distance between interdigitated electrodes ranges from 500 to 5 μm [31, 32]. Fabrication methods can be divided into two main groups. The first one is based on the transfer of synthesized electrode material in the form of dispersion of its particles in a solvent into the current collector. The electrode material can be deposited by different printing techniques or electrophoresis [33–35]. The second one applies to direct synthesis of the electrode material during fabrication, for example, laser scribing, electrochemical deposition, and polymerization. The second approach enables the use of electrode materials without binders and conductive additives. Moreover, such electrodes are characterized by high adhesion to the current collectors [36, 37].

1.3 Nanomaterials, aerogels, and biomaterials

Carbon materials are mainly used for the construction of electrochemical capacitors and microsupercapacitors. Due to their relatively high conductivity (in the range of 10^3 S m^{-1} for amorphous carbon to 10^5 S m^{-1} for graphite [38]) and chemical stability, carbon materials ensure good reversibility and cycling stability. Such materials are usually produced as a powder with designed pore size distribution in order to provide high capacitance (which exceeds 100 F g^{-1} [6, 7] by proper ion diffusion and formation of the electric double layer. Carbon materials can be obtained by thermal or hydrothermal treatments and halogenation. Etching metals from carbides by chlorine at high temperatures is a very common method. This class of materials is called carbide-derived carbon (CDC) and is characterized by excellent performance because their microstructure can be adjusted by synthesis parameters. Moreover, the initial structure of the carbide influences the final properties of CDC. Pore size distribution is strongly associated with chlorination temperature and the structure of carbides. For example, the formation of micropores depends on the initial carbide density. CDC films could be used in aqueous and organic electrolytes. It was stated that the volumetric capacitance decreases with increasing coating films. It should be noted that the capacitance of a 2 μm film can reach 180 F cm^{-3} in $TEABF_4$ but in 160 F cm^{-3} 1 M H_2SO_4 [39]. One of the most commonly used carbides is TiC, which can be deposited by chemical vapor deposition (CVD) or physical vapor deposition (PVD). After that, carbon materials are synthesized by chlorination at 250–500 °C, according to the following reaction [40]:

$$TiC + 2Cl_2 \rightarrow TiCl_4 + C \tag{1.5}$$

The research conducted by Heon et al. [41] showed that a CDC film synthesized by reactive magnetron sputtering of TiC and chlorination at 400 °C reached 180 F cm^{-3} in a 1.5 M $TEABF_4$/acetonitrile electrolyte. Volumetric capacitance of microsupercapacitors based on CDC film decreases with decreasing chlorination temperature [41]. CDC films can be obtained using different substrates, such as Si wafers, glassy carbon, highly ordered pyrolytic graphite (HOPG), or Al_2O_3 [41, 42].

Special attention is also paid to graphene materials. These materials are characterized by unique properties mainly due to their electric conductivity (~10^8 S m^{-1} [38]), high surface area (~1500 m^2 g^{-1}), and ability to work in the broad electrochemical window. Graphene materials for MSC application can be obtained on any substrates by using a LightScribe DVD burner. In this process, graphene oxide (GO) absorbs light and is converted into graphene. This material is usually called laser-scribed graphene (LSG) and it can be obtained in less than 30 min (more than 100 microsupercapacitors in one step). This type of microsupercapacitors is characterized by high power density (200 W cm^{-3}) [43]. It should be noted that the resulting LSG material reached high conductivity (1738 S m^{-1}) and surface area up to 1520 m^2 g^{-1}. This material can be used as an electrode without binders and other conductive additives. Moreover, this process allows for the production of LSG with an open network structure which prevents the agglomeration of graphene layers [44]. The process of CO_2 laser writing and the resulting films are presented in Fig. 1.2 [45]. Graphene oxide can be

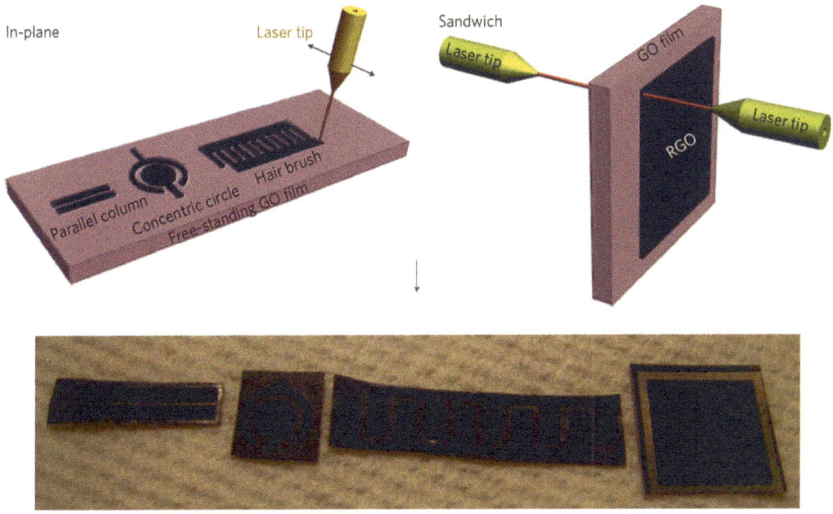

Fig. 1.2 Scheme of CO_2 direct laser writing of microsupercapacitors. *Reprinted with permission from W. Gao, N. Singh, L. Song, Z. Liu, A.L.M. Reddy, L. Ci, R. Vajtai, Q. Zhang, B. Wei, P.M. Ajayan, Direct laser writing of micro-supercapacitors on hydrated graphite oxide, Nat. Nanotechnol. 6 (2011) 496–500.*

obtained on rigid (Si wafers) or flexible (PET) substrates in order to fabricate in-plane microsupercapacitors through micropatterning of graphene films with a thickness of 6–100 nm. Graphene oxide can be subjected to CH_4 plasma reduction which can ensure an additional carbon source (C^+; CH_n^+ or CH_n, $n = 1$–3) that may be incorporated into the defects of GO. In such a case, C/O ratio is increased which is associated with high conductivity of this material (345 S cm^{-1}). Therefore, the fabricated microsupercapacitor can store an area capacitance of 80 μF cm^2 in H_2SO_4/PVA gel electrolyte [46]. The fabrication of microsupercapacitors concerns expensive methods to construct much thinner and flexible microelectronics. One of the promising and cheap methods is graphene-based in-planar supercapacitor obtained by in-situ reduction of GO. GO can be deposited by interfacial gelation of GO water suspension at 80 °C. The thickness of the deposit can be adjusted by immersion time. After that, the reduction of GO can be provided in-situ in a 0.1 M ascorbic acid at 90 °C for 2 h. Moreover, due to the further deposition of Ni (as a current collector) on the rGO layer, the prepared microsupercapacitor is characterized by lower contact resistance with the areal capacitance of 12.5 mF cm^{-2} at 5 mV s^{-1} in PVA/H_3PO_4 gel electrolyte [47]. Gel electrolytes are desirable for wearable electronics because they ensure good mechanical flexibility and facilitate the production process without the need for complicated sealing of components. Wu et al. [48] proposed self-assembled graphene oxide and its reduction on a Cu/Ag layer at room temperature. Cu, due to lower redox potential, is able to reduce GO and it is dissolved in this process, while the Au layer ensures contact with rGO and acts as current collector. The resulting MSC based on this material is characterized by 0.95 mF cm^{-2} at 0.43 mA cm^{-2} (PVA/H_2SO_4 gel electrolyte). The process of graphene synthesis provides its unique properties and thus the final parameters of MSC. Chemical synthesis of graphene is usually achieved by oxidation of graphite which should be exfoliated to give single sheets of graphene oxide (GO). In order to overcome the interaction between layers (van der Waals forces), it is very important to separate the layers and to produce graphene. It should be noted that chemical oxidation, intercalation, and then exfoliation are one of the most promising ways to produce graphene oxide at a large production scale. In such a case, graphene oxide can be obtained at low processing costs with the possibility of chemical functionalization [49]. Graphene materials are characterized by higher sp^2 sites than sp^3 in their structure. High-quality graphene materials as electrode materials do not require the use of conductivity additives and binders [50]. Special attention in graphene chemistry is also paid to chemically reduced graphene oxide, also called chemically converted graphene (CCG) [51]. In this process, the carbon to oxygen atomic ratio (C/O) is one of the important factors to produce high-quality graphene. The initial C/O ratio of GO is in the range of 4:1 to 2:1. After the reduction process, the C/O ratio increases, and graphene aromatic structure is retrieved. This structure of the graphene layer ensures high conductivity of this material and thus high performance of the MSC. The chemical reduction of GO can be provided by numerous chemical reagents. Hydrazine is one of the most commonly used reductants [52, 53]. Other reducing agents include ascorbic acid [54], sulfur compounds (for example, $NaHSO_3$, SO_2, $SOCl_2$) [50, 55], sodium borohydride [56], inorganic and organic bases [51, 57]. Yang et al. [58] proposed CCG hydrogel films which were compressed by capillary pressure in order to increase the density by removal of volatile solvent. The resulting supercapacitors based on this graphene material were characterized by exceptionally high performance with volumetric energy up to 60 Wh dm^{-3}.

One of the most important issues in the design of MSC is the pore size of the porous carbon. The transport of electrolyte ions affects the performances of the microsupercapacitors. Moreover, the surface of carbon materials should be available for solvated ion adsorption. The so-called carbon onions or onion-like carbons (OLC) are promising materials which are characterized by moderate specific area. OLCs offer very interesting physicochemical properties due to the absence of porous network inside the particles. Therefore, the microsupercapacitors based on this type of materials are able to deliver energy at high current density in comparison with other carbon materials with a comparable surface area. Onion-like carbon can be produced by vacuum annealing of detonation nanodiamonds (NDs) at high temperature. The temperature between 1327 and 1527 °C strongly influences the ratio of sp^2 and sp^3 sites. OLC can be deposited onto gold current collectors on silicon wafers. The prepared microdevice is able to work up to 3 V in a 1 M tetraethylammonium tetrafluoroborate in propylene carbonate electrolyte. Measurements indicate that, despite the higher capacitance of microsupercapacitor based on the activated carbon (AC) with higher surface area, OLC-based MSC is characterized by lower relaxation time constant $\tau_0 = 26$ ms (700 ms for AC-based MSC) and maximum power density close to 1 kW cm^{-3}. Better charge propagation of onion-like carbon in comparison with activated carbon can be explained by the limitation of ion transfer inside the pores of AC [59–62]. OLC-based MSC can be also obtained by electrophoretic deposition (EPD) combined with eutectic mixture of ionic liquids (IL). Huang et al. [62] proposed a microsupercapacitor which operated in a 1:1 IL electrolyte mixture of *N*-methyl-*N*-propylpiperidiniumbis [fluorosulfonyl]imide (PIP13FSI) and *N*-butyl-*N*-methylpyrrolidiniumbis [fluorosulfonyl]imide (PYR14FSI). Despite the high leakage current, this microdevice is able to store 1.1 mF cm^{-2} at 200 mV s^{-1} within 3.7 V. Moreover, the MSC based on eutectic electrolyte can work from −50 to 80 °C.

Carbon nanotubes are other carbon materials that have attracted interest as electrode materials in electrochemical capacitors as well as microsupercapacitors. Carbon nanotubes are characterized by higher conductivity than activated carbon (10^4–10^7 S m^{-1} [38]), low mass density (1.3–1.8 g cm^{-3} [63, 64]), and high chemical and electrochemical stability. Due to the relatively low surface area resulting from poor microporosity, the capacitance value is in the range of 10–40 F g^{-1}. This capacitance could be enhanced by changing the structures of carbon nanotubes (activation process or production of carbon nanotubes with open central canal) and surface functionalization by other heteroatoms (nitrogen, oxygen, sulfur) [65, 66].

Low areal energy density of microsupercapacitors is one of the main drawbacks which could be solved by graphene/carbon nanotube composites. Carbon nanotubes allow to store energy and can increase the conductivity of materials. Moreover, CNTs are able to prevent graphene aggregation. Wang et al. [67] synthesized graphene/CNT ink for writing of interdigitated electrodes. The authors stated that a 5% addition of CNTs to the composites increases the performance of MSC. The capacitance of MSC based on these materials is equal to 9.81 mF cm^{-2} at 0.05 mA cm^{-2}.

Reduced graphene oxide and carbon nanotube composites can also be obtained by electrostatic spray deposition (ESD) combined with photolithography lift-off methods. This approach allows obtaining rGO without the need for additional thermal or chemical reduction of GO. In this method, it is possible to obtain binder-free MSC, the capacitance of which reached 6.1 mF cm^{-2} at 0.01 V s^{-1}. It was reported that this microsupercapacitor is characterized by high power response and an RC time constant of 4.8 ms. It should be noted that the number of CNTs higher than 20% increased the power characteristics but decreased the capacitance [68].

Carbonaceous hydrogels and aerogels are another large class of carbon materials for supercapacitor purposes. Hydrogels are cross-linked materials with adsorbed liquids. This type of material is characterized by high water absorption and can be used as electrode materials for energy storage sensors and adsorbents [69]. Soborio et al. [70] obtained flexible electrodes based on carboxymethylcellulose hydrogels. This type of material could also be obtained as a composite with conducting polymers in order to improve the stability of the device. Yin et al. [71] proposed a hydrogel supercapacitor based on a conducting polymer in order to avoid delamination of electroactive materials due to external force. Authors synthesized hydrogel-based supercapacitors with a polypyrrole-polyvinyl alcohol/sulfuric acid sandwiched device with the energy density of 6.94 μWh cm^{-2}. On the other hand, aerogels possess similar properties to hydrogels. Liquids in hydrogels are replaced with air without damaging the interconnected microstructure. Aerogels are characterized by very low density (0.2–1.0 g cm^{-3} [72]) and exhibit a foam-type structure. Such materials are obtained by carbonization of organic aerogels obtained by polycondensation of monomers (resorcinol, melamine, formaldehyde) to form the sol. The sol is then subjected to curing to form a "wet gel." Before the carbonization process, the wet gel is dried under supercritical conditions. This process is expensive and complicated. In addition, the structure of carbon aerogels is responsible for the high internal resistance of these materials [73, 74]. The increase in conductivity can be achieved by introducing carbon nanotubes into the carbon aerogels structure at the stage of their synthesis. The appropriate dispersion of carbon nanotubes in the carbon aerogel matrix provides a modification of the porosity, morphology of the resulting composites, and thus affects the physical and electrochemical properties [75]. Zhu et al. [76] obtained 3D hierarchical graphene aerogels for a supercapacitor operating in 3 M KOH.

The synthesis process is presented in Fig. 1.3. In this fabrication process, the composite ink was prepared by combining the GO precursor, silica powder, and resorcinol-formaldehyde solution with sodium carbonate. The composite ink was extruded by a micro-nozzle to pattern 3D architectures. Supercapacitors based on this material deliver power density of 4.079 kW kg^{-1} at the energy density of 0.26 Wh kg^{-1}.

Due to the limited thickness and volume of electrodes, high surface area is desirable for the construction of MSC. Therefore, activated carbons can be also used for the construction of MSC electrodes. Activated carbons can be obtained from organic substances which are rich in elemental carbon. In such a case, biomass (especially biomass waste materials) such as cellulose-lignin composites can be used as a precursor for the production of activated carbons. The most common activated carbon precursors include wood, coconut shell, or coal. The first step in activated carbon production is the carbonization process, which is the thermal decomposition in an inert atmosphere at elevated temperature. The formation of sp^2 sites occurs at this stage, which is very important in order to increase the conductivity of the carbon material. The second step is the activation process, which can be provided by physical or chemical means. Physical activation uses CO_2 or steam, whereas chemical activation is based on the use of acid (H_3PO_4), alkalis (NaOH, KOH), and salts ($FeCl_3$, $ZnCl_2$, and K_2CO_3) [77–80]. The type of the activated precursor has high influence on the final pore size distribution. The microstructures of bio-inspired carbons developed by different plant tissues are presented in Fig. 1.4 [81].

An interesting idea is to use bio-template synthesis for ecological and cheap production of MSC electrode material. In this case, biotemplates are extracted and electrode materials are deposited. Several types of biomaterials are considered as templates, such as microorganisms, mineral-based templates, and plant organ-based templates [77, 81–84]. Shim et al. [85] obtained 3D-hierarchical Co_3O_4 superstructures from bacteria-templating synthesis (Fig. 1.5). For this purpose, a spherical, Gram-positive bacterium (*Micrococcus lylae*) was used as a biotemplate. The supercapacitor based on this material is characterized by the capacitance of 2.04 F cm^{-2} at 19.02 mA cm^{-2}.

The use of nature and biotechnology to fabricate different types of nanomaterials with controlled microstructure is attracting much interest. Nature is abundant in natural biomaterials which allow to fabricate green and facile electrode materials for microsupercapacitors [77]. Most of the technologists and scientists are looking for cheap and suitable electrode materials which could be obtained in an uncomplicated and time-efficient manner. A major challenge is to increase the specific capacitance and energy density without a

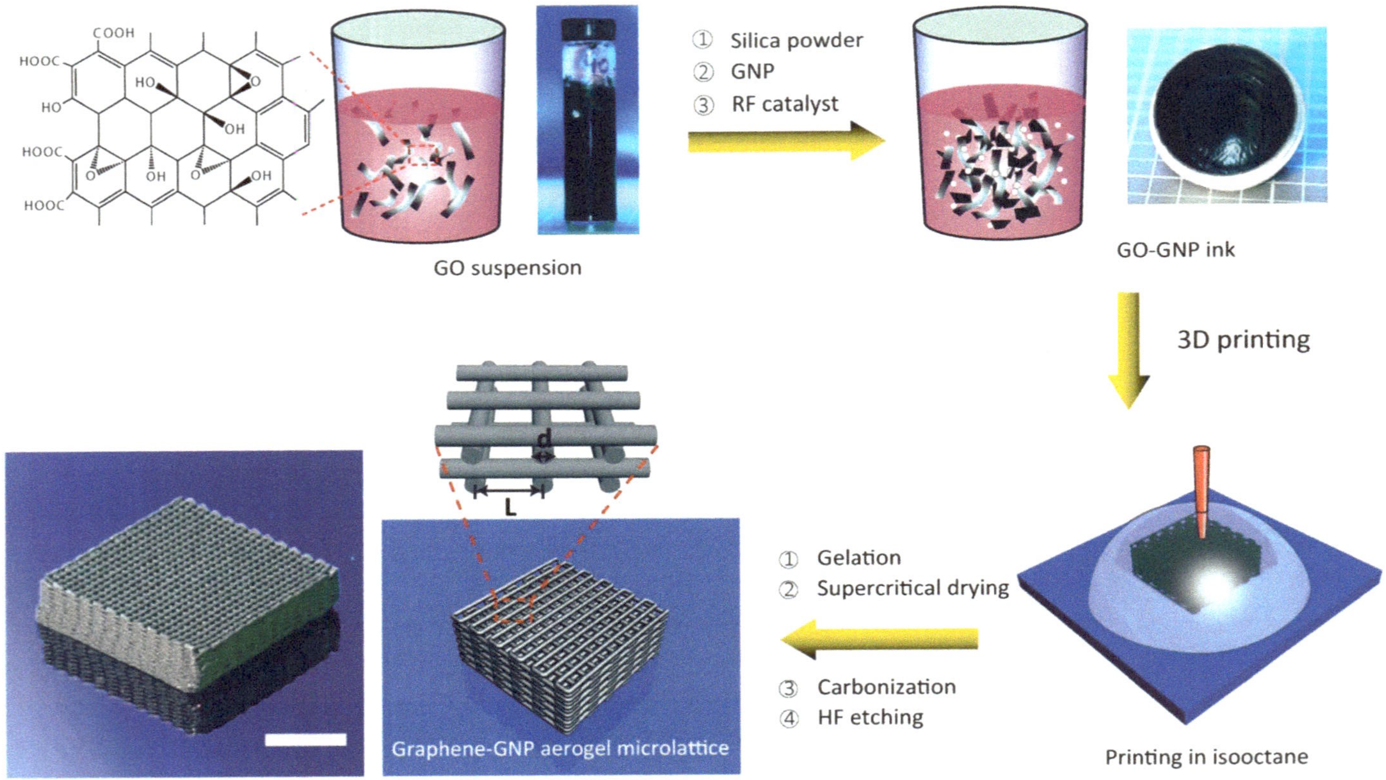

Fig. 1.3 Schematic representation of the fabrication process of graphene-GNP aerogel microelectrode. *Reprinted with permission from C. Zhu, T. Liu, F. Qian, T. Yong-Jin Han, E.B. Duoss, J.D. Kuntz, C.M. Spadaccini, M.A. Worsley, Y. Li, Supercapacitors based on three-dimensional hierarchical graphene aerogels with periodic macropores, Nano Lett. 16 (2016) 3448–3456.*

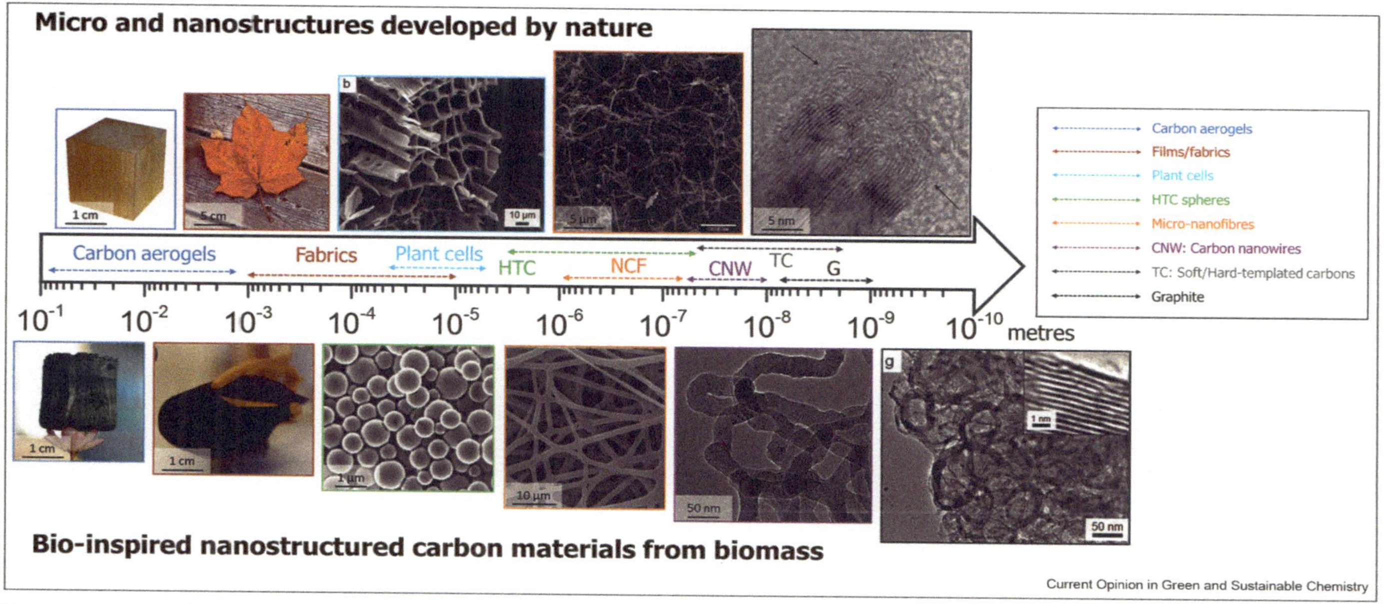

Fig. 1.4 Micro- and nanostructured carbon materials derived from biomass. *Reprinted from S. Herou, P. Schlee, A.B. Jorge, M. Titirici, Bio-mass-derived electrodes for flexible supercapacitors, Curr. Opin. Green Sustain. Chem. 9 (2018) 18–24.*

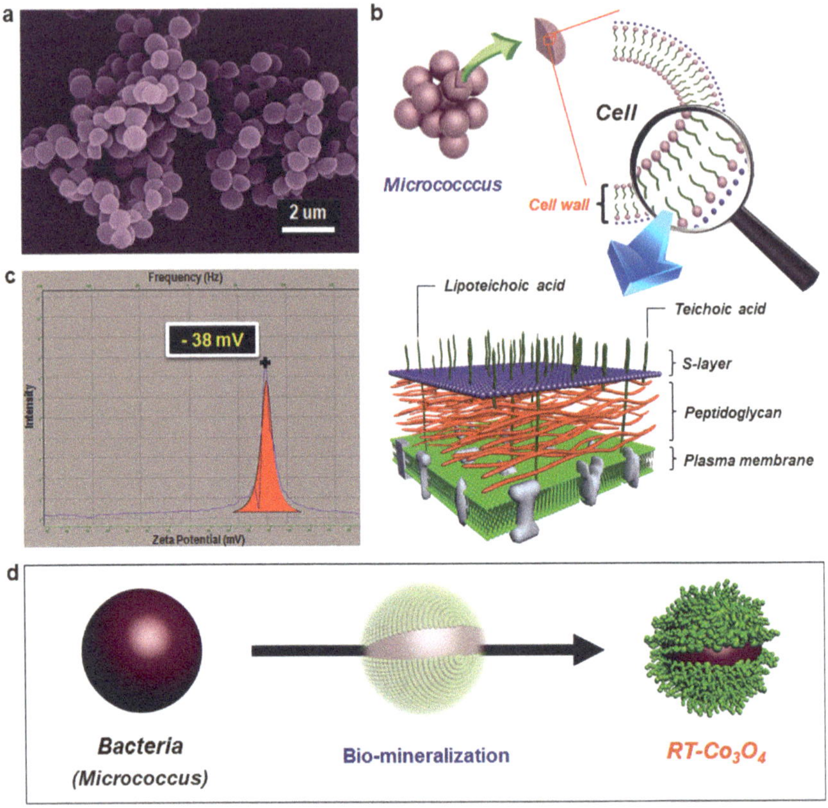

Fig. 1.5 Scheme of the 3D-hierarchical Co_3O_4 superstructures obtained from bacteria-templating synthesis. FE-SEM image of *Micrococcus lylae* (A), the structure of the cell wall of Gram-positive bacterium (B), Zeta potential measurements indicating negatively charged surface of the bacteria (C), scheme of the synthesis process (D). *Reprinted with permission from H.W. Shim, A.H. Lim, J.C. Kim, E. Jang, S.D. Seo, G.H. Lee, T.D. Kim, D.W. Kim, Scalable one-pot bacteria-templating synthesis route toward hierarchical, porous-Co3O4 superstructures for supercapacitor electrodes, Sci. Rep. 3 (2013) 1–9.*

decrease of the cycling stability. This could be obtained by proper modification of nanomaterials and their synergistic effect in composites. Therefore, in the next section, the pseudocapacitive materials, composites, and inks have been described.

1.4 Pseudocapacitive materials, composites, and inks

The working principles of EDLC are based on non-Faradaic process. Fast redox reactions which occur due to the presence of pseudocapacitive materials are able to provide more charge than EDLC based on high surface area carbon electrodes. In case of the pseudocapacitive materials, faradaic charge transfer at the electrode/electrolyte interface is responsible for higher capacitance. Different types of materials are subjected to extensive research in order to increase the energy density of the device, such as carbon materials modified by heteroatoms, transition metal oxides/hydrooxides, and conducting polymers. In case of the modified carbon materials, the surface functionalities can contribute to the pseudocapacitive effect. As it was stated for oxygen functionalities, quinone/hydroquinone redox couples are responsible for the pseudocapacitotive effect [86]. It should be noted that functional groups are associated with the conjugated graphene layers hence the electron transferred from/to oxygen functional groups are delocalized. Therefore, the cyclic voltammogram measurements do not present characteristic current peaks due to the constant flow of current [87, 88]. Moreover, excessive oxygen content in carbon materials is not desirable due to the changes/ disruption of surface conjugation and increased content of sp^3 sites. This can be confirmed by Raman spectroscopy measurements in case of which the I_d/I_g ratio decreases during oxidation of carbon resulting in conversion of sp^2 sites to sp^3 sites (according to the three-step amorphization described by Ferrari and Robertson [89]). Moreover, incorporation of other heteroatoms (such as B, N, S, P) is also possible. For example, in case of nitrogen pyrronic, pyridinic, and quaternary N sites are present after nitrogen doping [90].

Metal oxides are other types of materials that can be used for electrode construction. Metal oxides are able to provide fast redox reactions at the electrode/electrolyte interface which are also the source of the pseudocapacitive effect. In this case, transition or noble metal oxides (RuO_2, MnO_2) and nitrides (Mo_xN) such as TiN, VN, RuN are currently under research [91, 92]. Several methods have been developed to fabricate metal oxide-based electrodes for MSC. The hydrothermal method is used most commonly. In the framework of this method, it is possible to obtain different types of nanoparticles (such as nanosheets/nanorods) by changing the synthesis parameters: temperature, pressure, and reaction time. For example, this method allows obtaining of α-MnO_2 nanorods in the mixture of $KMnO_4$ and HNO_3 in a Teflon-lined autoclave [93]. Electrospinning is another effective method that allows to prepare transition metal oxide for nanotechnology purposes. This method allows to synthesize various ternary transition metal oxides $(CoMn_2)_4$, $NiCo_2O_4$. Electrospinning is based on the electric field force in order to accelerate the injection of precursor solution [94]. Other typical methods are electrodeposition [95], microwave-assisted heating (MAH) [96], and sol-gel methods [97].

RuO_2 was one of the first discovered metal oxides which exhibit high pseudocapacitive behavior. Pseudocapacitive properties are based on fast redox reactions which occur with the protonation of the material [91, 98]:

$$RuO_2 + xH^+ + xe^- \leftrightarrow RuO_{2-x}(OH)_x \tag{1.6}$$

where

$$0 < x < 2$$

Despite the fact that hydrous ruthenium oxide exhibits the highest reported specific capacitance, its high cost has hindered its development. In order to increase the performance, it is desirable to obtain composites by deposition on carbon nanowalls (CNW). The deposition of ruthenium oxide on CNWs leads to a reduction in the mass of oxide used. Moreover, microsupercapacitor does not require a high quantity of electrode materials. It should be noted that utilization of this composite causes an increase of capacitance of graphene-based microsupercapacitors (21 μF cm^{-1}) [99]. CNWs are vertically oriented graphene sheets that exhibit a 3-dimensional structure of 2D graphene domains. Such new hybrid material delivers a specific capacitance of over 1000 mF cm^{-2} at 2 mV s^{-1} (in 0.5 M sulfuric acid) [100].

On the other hand, MnO_2 is another promising metal oxide that is abundant, cheap, and environmentally friendly. The charge storage principles also occur in the presence of protons and are based on the process of intercalation from an aqueous solution as follows [91, 101]:

$$MnO_2 + xH^+ + yC^+ + (x+y)e^- \leftrightarrow MnOOH_xC_y \tag{1.7}$$

where x, y are moles of H^+ and C^+ cations (Na^+, Li^+, K^+) intercalated in manganese oxide.

Pseudocapacitive properties can also originate from conducting polymers (polypyrrole, polyaniline, polythiophene). The operating principles of conducting polymers are based on ion doping. Among the conducting polymers, p- or n-type doping materials can be distinguished. Such materials ensure a low ESR value, fast charge/discharge process and can be obtained at lower costs in comparison with noble metal oxides [102, 103].

Sun et al. [104] proposed a 3D polypyrrole electrode on a silicon substrate which operated in a 0.5 M NaCl solution. The 3D symmetric supercapacitor was fabricated by using MEMS technology. The electropolymerization of PPy was conducted in the presence of sodium p-toluenesulfonate (TOSNa) as the surfactant and supporting salt. The capacitance properties arise from p-type doping/dedoping of anions in accordance with the following reactions:

$$PPy + TOS^- + e^- \rightarrow PPy + TOS^- \quad (1.8)$$

$$PPy + Cl^- - e^- \rightarrow PPy^+ Cl^- \quad (1.9)$$

$$PPy^+ Cl^- + e^- \rightarrow PPy + Cl^- \quad (1.10)$$

Reaction (1.8) occurred only during the discharge. After the first discharge, Cl^- replaces TOS– due to smaller size and high concentration [8, 9]. This type of MSC is characterized by 0.128 F cm^{-2} at 20 mV s^{-1} [105].

PEDOT is a different type of conducting polymer that could be used for MSC construction. PEDOT is characterized by high stability for electrooxidation. The current research shows that PEDOT-based microsupercapacitors reached an areal capacitance of 9 mF cm^{-2} with volumetric capacitance of 50 F cm^{-3} in 1 M H_2SO_4 [106].

Conducting polymers and transition metal oxides provide high capacitance due to the pseudocapacitive effect and thus contribute to high energy density. This type of material enables fast redox reactions, but they are characterized by short-term cycling stability and lower structural stability. Moreover, these materials possess low rate-capacity and poor electrical conductivity due to the presence of redox reactions which are responsible for the increase in capacitance properties. In order to overcome the above-mentioned problems, much attention has been devoted to the fabrication of different types of composites, such as carbon/carbon composites, carbon materials/transition metal oxides, carbon materials/conducting polymer, and conducting polymer/metal oxides [107–109].

Carbon nanotubes open new possibilities to fabricate printable thin-film electrodes. This can be realized by preparing the suspension with water and spraying with airbrush pistol on polyethylene terephthalate. The performance of the device working in PVA/H_3PO_4 electrolyte indicates the specific capacitance up to 36 F g^{-1} (at 30 mA g^{-1}) [111]. Carbon nanotubes-based microsupercapacitors can also be obtained by means of 3D printing. In this case, the inks are uniformly extruded from the micronozzle. 3D-printed MSC based on carbon nanotubes exhibits 4.69 F cm^{-2} at 50 mV s^{-1} [112]. Flexible MSC based on single-walled carbon nanotubes (SWCNT) is presented in Fig. 1.6. SWCNT was spray-deposited on the electrode area. This type of device was characterized by 18 F cm^{-2} and low time constant (1 ms) [110]. Lin et al. [113] obtained 3-dimensional graphene/carbon nanotubes composites (3D G/CNTCs) as a hybrid material on nickel

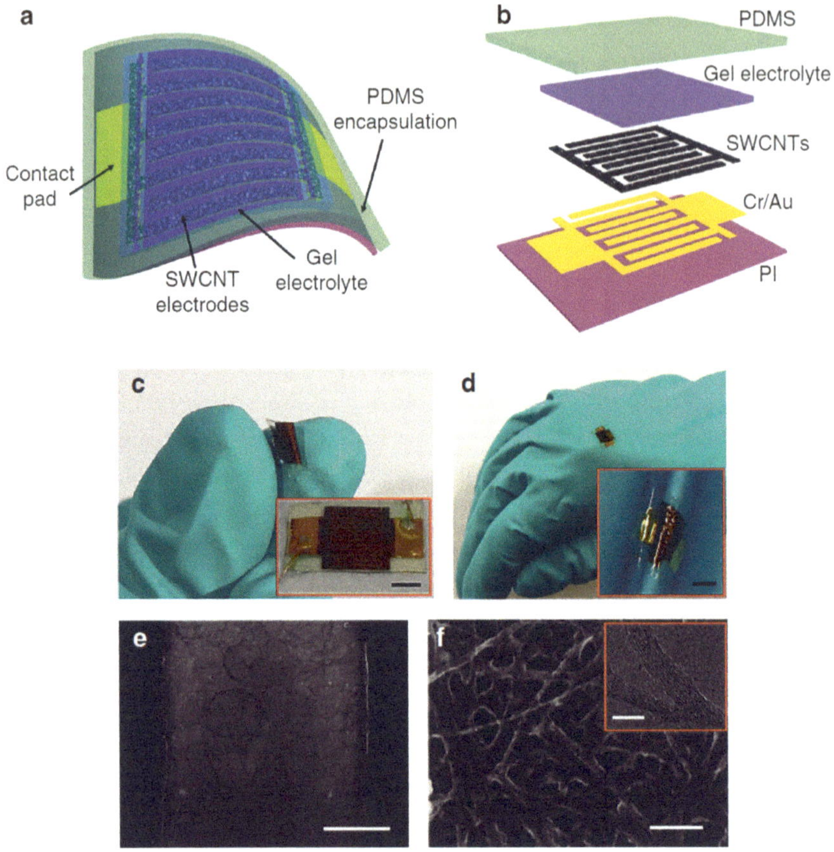

Fig. 1.6 Scheme of the flexible SWCNT-based microsupercapacitor (A and B); photograph of flexible MSC (C) and MSC mounted on a latex glove (D); SEM image of SWCNT (E and F), and inset in (F) shows a TEM image. *Reprinted with permission from J. Pu, X. Wang, R. Xu, S. Xu, K. Komvopoulos, Highly flexible, foldable, and rollable microsupercapacitors on an ultrathin polyimide substrate with high power density, Microsyst. Nanoeng. 4 (2018) 1–11.*

current collectors (Fig. 1.7). Microsupercapacitors based on this material exhibit specific capacitance of 2.16 mF cm^{-2} in an aqueous electrolyte and 3.93 mF cm^{-2} in ionic liquids. Moreover, the authors measured the electrochemical performance in comparison with different times of carbon nanotubes growth (Fig. 1.7F–H) for 1, 2.5, and 5 min. This process allows obtaining carbon nanotubes with different heights of 10, 15, and 20 μm, respectively. Moreover, the capacitance value increases with increasing time of growing. This could be probably explained by the increased surface area of the electrode. Moreover, the electrode material obtained after 5 min of growth is characterized by the lowest value of diffusion resistance. It should

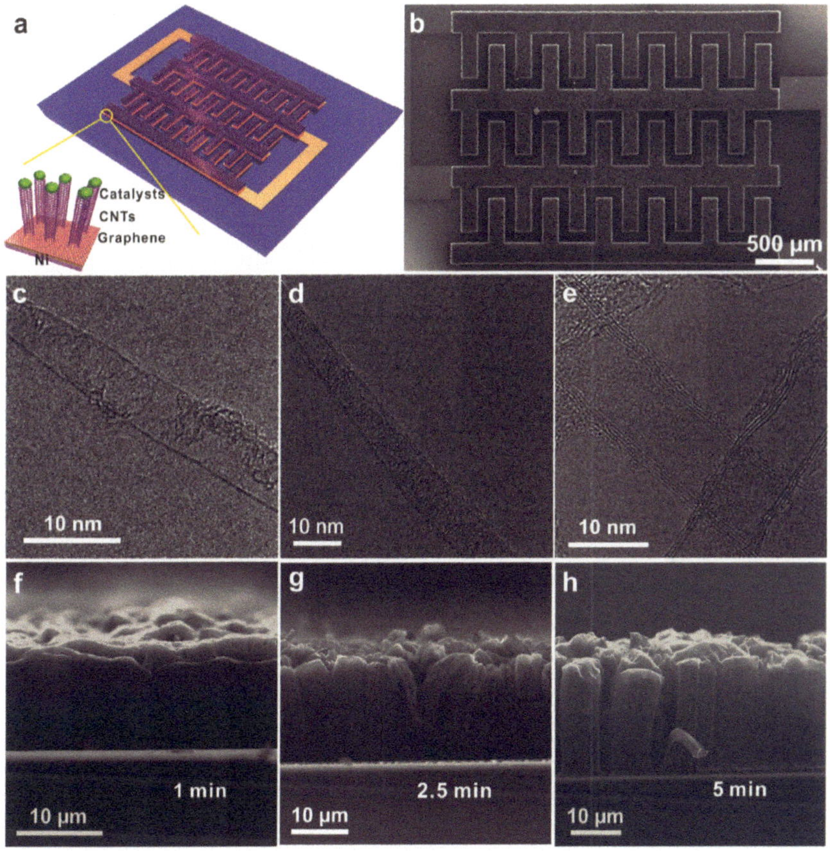

Fig. 1.7 Schematic representation of 3D G/CNTCs-based MSC (A); SEM image of interdigitated electrodes of MSC (B); SEM image of 3D G/CNTCs (C–E); SEM image of CNTCs synthesized for 1, 2.5, and 5 min, respectively (E–H). *Reprinted with permission from J. Lin, C. Zhang, Z. Yan, Y. Zhu, Z. Peng, R.H. Hauge, D. Natelson, J.M. Tour, 3-Dimensional graphene carbon nanotube carpet-based microsupercapacitors with high electrochemical performance, Nano Lett. 13 (2013) 72–78.*

be noted that this method allows obtaining high-quality carbon nanotubes confirmed by Raman spectroscopy measurements, which indicates a low band intensity ratio of I_D/I_G (1:9).

Carbon nanotubes are also desirable materials for composites with metal oxides and hydroxides. Cobalt-based oxide or hydroxide are also promising materials due to their high chemical stability and interesting capacitance (ca. 400 F g^{-1} [114, 115]). In order to improve their poor electrical conductivity, various composites with carbon materials (carbon nanotubes, graphene) have been proposed [116, 117]. Special focus is also devoted to 3D composites, for example, 3D nanocomposites of cobalt oxide (CoO) nanoflowers woven with carbon CNTs, which can be obtained using the hydrothermal method. MSC based on such materials exhibited 17.4 F cm^{-3} with an energy density of 3.48 mWh cm^{-3} [116].

There are several methods to fabricate MSC. Inkjet printing is one of the most often used procedures. Inkjet printing is a method of preparing interdigital electrodes by pushing droplets of ink on various substrates. Solubility and chemical stability are the critical factors for preparing inks which should be adjusted in order to avoid the close up of the nozzle. Therefore, the rheological properties, such as surface tension, and viscosity of inks are crucial to generate droplets during the printing [118, 119]. This method also enables the production of the asymmetric microsupercapacitors. Shen et al. [120] obtained MSC based on V_2O_5 (cathode) and graphene-vanadium nitride quantum dots (anode). Such 3D printed microsupercapacitors exhibit a high capacitance of 207.9 mF cm^{-2}. Moreover, extensive research was devoted to MXene inks. MXenes are two-dimensional carbides and nitrides of transition metals. Two types of MXene inks are presented in Fig. 1.8: organic inks for inkjet printing and for extrusion printing [121]. Extrusion printing seems to be a very promising area of research. This method also enables the fabrication of composite electrodes with conducting polymers. Extrusion printing of MSC based on GO/PANI (Fig. 1.9) interdigitated electrodes enables the fabrication of devices with outstanding areal capacitance (153.6 mF cm^{-2} at 5 mV s^{-1}) [122].

To sum up, different types of electrode materials for microsupercapacitors are presented in Table 1.1. Carbon materials are the most common

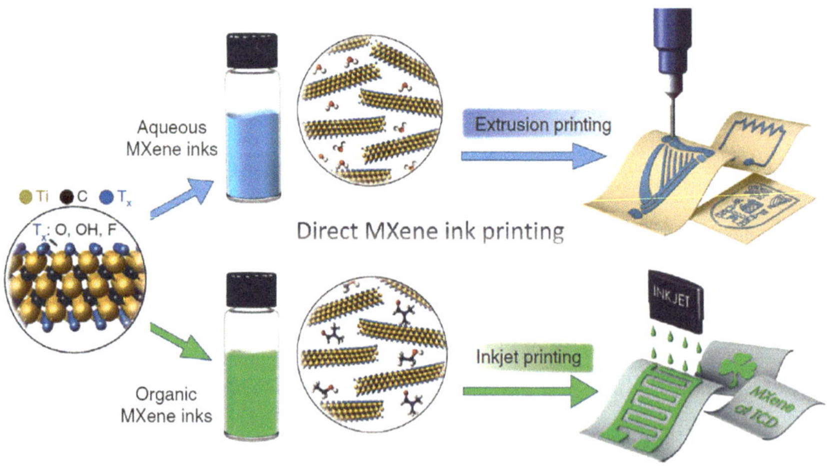

Fig. 1.8 The scheme of direct MXene ink printing. *Reprinted from C. Zhang, L. McKeon, M.P. Kremer, S.H. Park, O. Ronan, A. Seral-Ascaso, S. Barwich, C. Coileain, N. McEvoy, H.C. Nerl, B. Anasori, J.N. Coleman, Y. Gogotsi, V. Nicolosi, Additive-free MXene inks and direct printing of micro-supercapacitors, Nat. Commun. 10 (2019) 1–9.*

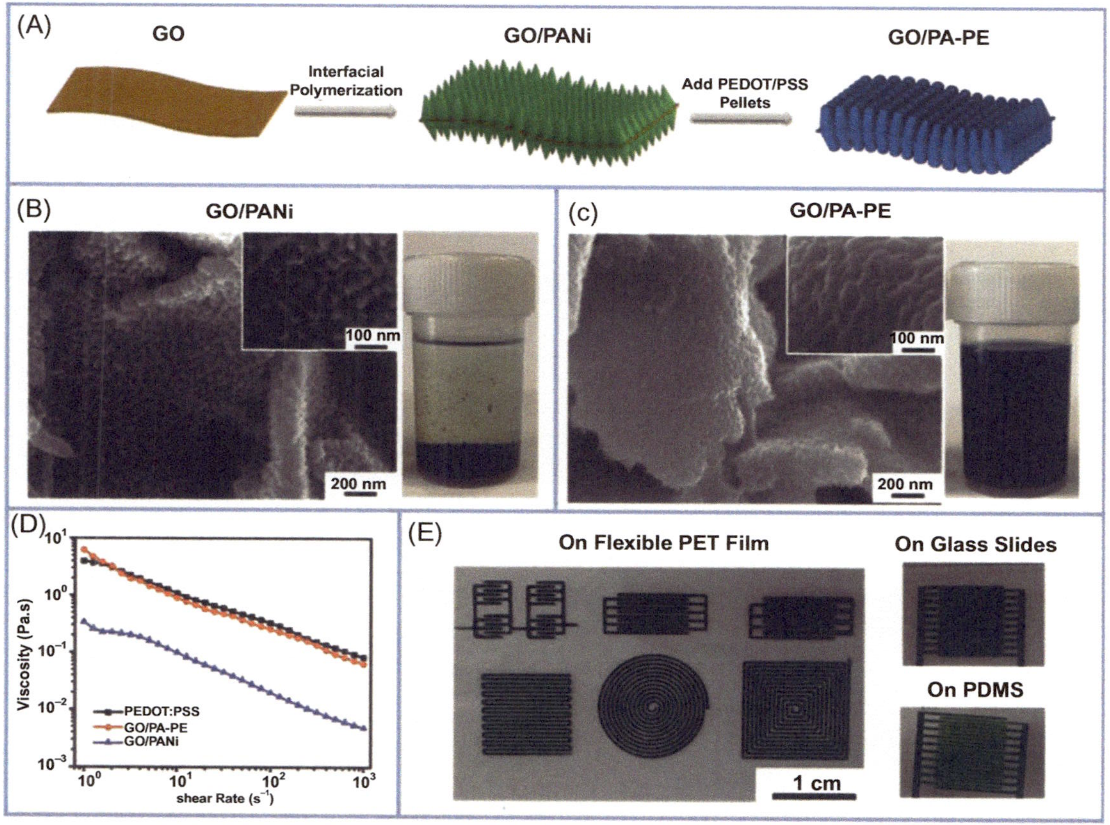

Fig. 1.9 Preparation of GO/PA-PE inks (A), SEM image and photographs of GO/PANi (B) and GO/PA-PE composites (C), viscosity vs shear rate (D), digital photographs patterns on different substrates (E). *Reprinted with permission from Y. Liu, B. Zhang, Q. Xu, Y. Hou, S. Seyedin, S. Qin, G.G. Wallace, S. Beirne, J. M. Razal, J. Chen, Development of graphene oxide/polyaniline inks for high performance flexible microsupercapacitors via extrusion printing, Adv. Funct. Mater. 28 (2018) 1–12.*

Table 1.1 Properties of typical microsupercapacitors in literature.

Electrode material	Electrolyte	Capacitance	Cycling stability	References
Activated carbon	1 M H_2SO_4	250 F cm^{-3} (45 A g^{-1})	95% after 10,000 cycles (10 A g^{-1})	[123]
TiC-CDC	1 M H_2SO_4	110 F g^{-1} (10 mV s^{-1})	Stable after 10,000 cycles (100 mA)	[124]
Onion-like carbon (OLC)	1 M Et_4NBF_4/PC	0.9 mF cm^{-2} (100 V s^{-1})	Stable after 10,000 cycles (50 V s^{-1})	[61]
rGO-CNTs	3 M KCl	2.8 mF cm^{-2} (50 V s^{-1})	–	[68]
Graphene/NiOOH/$Ni(OH)_2$	PVA-KOH	0.75 mF cm^{-2} (5 mV s^{-1})	80% after 3000 cycles (500 mV s^{-1})	[125]
Graphene/polyaniline inks	1 M H_2SO_4	82 F g^{-1} (20 mV s^{-1})	Stable after 1000 cycles (at 1.4 A g^{-1})	[122]
MXene	PVA-3 M H_2SO_4	562 F cm^{-3}	97% after 10,000 cycles (200 μA cm^{-2})	[121]

materials with stable cycling stability. On the other hand, pseudocapacitive materials are characterized by higher capacitance and thus energy density. The composites of carbon materials with pseudocapacitive effect could increase not only the capacitance but also the cycling stability. Moreover, new types of materials and their composites allow obtaining micromechanical devices for wearable and stretchable electronics.

1.5 Summary

In this chapter, we have discussed different types of materials under research in case of microsupercapacitors. The structure of materials and their physicochemical properties affect the performance of the device as well as the method of fabrication. Carbon materials, which are characterized by high chemical and electrochemical stability, high conductivity, and cycling stability, are the most widely used materials. Nevertheless, much attention is devoted to carbon-based microelectrodes, which are characterized by lower values of volumetric and areal capacitance compared to pseudocapacitive materials and their composites. It should be noted that, despite the increasing number of reports regarding the MSC materials and fabrication techniques, the technology is still undeveloped for large-scale production. Moreover, the standardization of MSC performance is needed to compare different results in terms of thickness of materials, capacitance, energy/power density, etc. In case of further development, materials with 3D architecture seem to be an interesting approach for system miniaturization.

Acknowledgment

This work was supported by funds from the National Science Centre (Poland) granted on the basis of decision number DEC-2018/31/B/ST8/01619.

References

[1] B.E. Conway, Electrochemical Supercapacitors: Scientific Fundamentals and Technological Applications, Kluwer, 1999.
[2] F. Béguin, E. Frąckowiak, Carbons for Electrochemical Energy Storage and Conversion Systems, CRC Press, Taylor & Francis Group, 2010.
[3] F. Béguin, E. Frąckowiak (Eds.), Supercapacitors: Materials, Systems and Applications, Wiley-VCH, Weinheim, Germany, 2013.
[4] S. Rusi, R. Majid, Green synthesis of in situ electrodeposited rGO/MnO_2 nanocomposite for high energy density supercapacitors, Sci. Rep. 5 (2015) 16195. 1–13.
[5] T. Kesavan, T. Partheeban, M. Vivekanantha, M. Kundu, G. Maduraiveeran, M. Sasidharan, Hierarchical nanoporous activated carbon as potential electrode materials for high performance electrochemical supercapacitor, Microporous Mesoporous Mater. 274 (2019) 236–244.
[6] A. Burke, R&D consideration for the performance and application of electrochemical capacitors, Electrochim. Acta 53 (2007) 1083–1091.
[7] R. Kötz, M. Carlen, Principles and application of electrochemical capacitors, Electrochim. Acta 45 (2000) 2483–2498.
[8] J.M. Miller, A. Burke, Electrochemical capacitors: challenges and opportunities for real-world applications, Electrochem. Soc. Interface 17 (2008) 53–57.
[9] A. Burke, Ultracapacitor technologies and application in hybrid and electric vehicles, Int. J. Energy Res. 34 (2010) 133–151.
[10] M. He, K. Fic, E. Frąckowiak, P. Novák, E.J. Berg, Towards more durable electrochemical capacitors by elucidating the ageing mechanisms under different testing procedures, ChemElectroChem 6 (2019) 566–573.
[11] T. Wang, H.C. Chen, F. Yu, X.S. Zhao, H. Wang, Boosting the cycling stability of transition metal compounds-based supercapacitors, Energy Storage Mater. 16 (2019) 545–573.
[12] A.M. Bryan, L.M. Santino, Y. Lu, S. Acharya, J.M. D'Arcy, Conducting polymers for pseudocapacitive energy storage, Chem. Mater. 28 (2016) 5989–5998.
[13] Y. Zhao, M. Liu, X. Deng, L. Miao, P.K. Tripathi, X. Ma, D. Zhu, Z. Xu, Z. Hao, L. Gan, Nitrogen-functionalized microporous carbon nanoparticles for high performance supercapacitor electrode, Electrochim. Acta 153 (2015) 448–455.
[14] D. Zhu, Y. Wang, L. Gan, M. Liu, K. Cheng, Y. Zhao, X. Deng, D. Sun, Nitrogen-containing carbon microspheres for supercapacitor electrodes, Electrochim. Acta 158 (2015) 166–174.

[15] G. Lota, J. Tyczkowski, R. Kapica, K. Lota, E. Frackowiak, Carbon materials modified by plasma treatment as electrodes for supercapacitor, J. Power Sources 195 (2010) 7535–7539.
[16] G. Lota, P. Krawczyk, K. Lota, A. Sierczyńska, Ł. Kolanowski, M. Baraniak, T. Buchwald, The application of activated carbon modified by ozone treatment for energy storage, J. Solid State Electrochem. 20 (2016) 2857–2864.
[17] Ł. Kolanowski, M. Graś, J. Wojciechowski, M. Baraniak, P. Krawczyk, G. Lota, The influence of carbon material modification on the pseudocapacitive effect, Mater. Today Proc. 6 (2019) 36–41.
[18] T. Cottineau, M. Toupin, T. Delahaye, T. Brousse, D. Bélanger, Nanostructured transition metal oxides for aqueous hybrid electrochemical supercapacitors, Appl. Phys. A 82 (2006) 599–606.
[19] E. Frackowiak, V. Khomenko, K. Jurewicz, K. Lota, F. Béguin, Supercapacitors based on conducting polymers/nanotubes composited, J. Power Sources 153 (2006) 413–418.
[20] G. Lota, E. Frackowiak, Striking capacitance of carbon/iodide interface, Electrochem. Commun. 11 (2009) 87–90.
[21] J.W. Judy, Microelectromechanical systems (MEMS): fabrication, design and applications, Smart Mater. Struct. 10 (2001) 115–1134.
[22] J.R. Miller, Valuing reversible energy storage, Science 335 (2012) 1312–1313.
[23] C. Liu, Z. Zhang, Electrochemical materials design for microsupercapacitors, in: Supercapacitor Design and Application, IntechOpen, 2016.
[24] T. Brousse, D. Bélanger, J.F. Long, To be or not to be pseudocapacitive, J. Electrochem. Soc. 162 (2015) A5185–A5189.
[25] V. Khomenko, E. Raymundo-Piñero, F. Béguin, Optimization of an asymmetric manganese oxide/activated carbon capacitor working at 2 V in aqueous medium, J. Power Sources 153 (2006) 183–190.
[26] L. Demarconnay, E. Raymundo-Piñero, F. Béguin, Adjustment of electrodes potential window in an asymmetric carbon/MnO_2 supercapacitor, J. Power Sources 196 (2011) 580–586.
[27] H.A. Andreas, J.M. Black, A.A. Oickle, Self-discharge in manganese oxide electrochemical capacitor electrodes in aqueous electrolytes with comparisons to faradaic and charge redistribution models, Electrochim. Acta 140 (2014) 116–124.
[28] H.A. Andreas, Self-discharge in electrochemical capacitors: a perspective article, J. Electrochem. Soc. 162 (2015) A5047–A5053.
[29] J. Black, H.A. Andreas, Effects of charge redistribution on self-discharge of electrochemical capacitors, Electrochim. Acta 54 (2009) 3568–3574.
[30] J. Wojciechowski, Ł. Kolanowski, A. Bund, G. Lota, The influence of current collector corrosion on the performance of electrochemical capacitors, J. Power Sources 368 (2017) 19–29.
[31] N.A. Kyeremateng, T. Brousse, D. Pech, Microsupercapacitors as miniaturized energy-storage components for on-chip electronics, Nat. Nanotechnol. 12 (2017) 7–15.
[32] H. Zhang, Y. Cao, M.O.L. Lam Chee, P. Dong, M. Ye, J. Shen, Recent advances in micro-supercapacitors, Nanoscale 11 (2019) 5807–5821.
[33] Y. Wang, Y.Z. Zhang, D. Dubbink, J.E. ten Elshof, Inkjet printing of δ-MnO2 nanosheets for flexible solid-state micro-supercapacitor, Nano Energy 49 (2018) 481–488.
[34] Y. Lin, Y. Gao, F. Fang, Z. Fan, Recent progress on printable power supply devices and systems with nanomaterials, Nano Res. 11 (2018) 3065–3087.
[35] M. Singh, H.M. Haverinen, P. Dhagat, G.E. Jabbour, Inkjet printing-process and its applications, Adv. Mater. 22 (2010) 673–685.

[36] J.B. In, B. Hsia, J.H. Yoo, S. Hyun, C. Carraro, R. Maboudian, C.P. Grigoropoulos, Facile fabrication of flexible all solid-state micro-supercapacitor by direct laser writing of porous carbon in polyimide, Carbon 83 (2015) 144–151.
[37] H.-C. Huang, C.-J. Chung, C.-T. Hsieh, P.-L. Kuo, H. Teng, Laser fabrication of all-solid-state microsupercapacitors with ultrahigh energy and power based on hierarchical pore carbon, Nano Energy 21 (2016) 90–105.
[38] I.H. Tavman, Preparation and characterization of conductive polymer nanocomposites based on ethylene-vinylacetate copolymer (EVA) reinforced with expanded and unexpanded graphite, Adv. Mater. Res. 1114 (2015) 92–99.
[39] E.N. Hoffman, G. Yushin, T. El-Raghy, Y. Gogotsi, M.W. Barsoum, Micro and mesoporosity of carbon derived from ternary and binary carbides, Microporous Mesoporous Mater. 112 (2008) 526–532.
[40] J. Chmiola, C. Largeot, P. Taberna, P. Simon, Y. Gogotsi, Monolithic carbide-derived carbon films for micro-supercapacitors, Science 328 (2010) 480–483.
[41] M. Heon, S. Lofland, J. Applegate, R. Nolte, E. Cortes, J.D. Hettinger, P. Taberna, P. Simon, P. Huang, M. Brunet, Y. Gogotsi, Continuous carbide-derived carbon films with volumetric capacitance, Energ. Environ. Sci. 4 (2011) 135–138.
[42] P. Huang, M. Heon, D. Pech, M. Brunet, P. Taberna, Y. Gogotsi, S. Lofland, J.D. Hettinger, P. Simon, Micro-supercapacitors from carbide derived carbon (CDC) films on silicon chips, J. Power Sources 225 (2013) 240–244.
[43] M.F. El-Kady, R.B. Kaner, Scalable fabrication of high-power graphene micro-supercapacitors for flexible and on-chip energy storage, Nat. Commun. 4 (1475) (2013) 1–9.
[44] M.F. El-Kady, V. Strong, S. Dubin, R.B. Kaner, Laser scribing of high-performance and flexible graphene-based electrochemical capacitors, Science 335 (2012) 1326–1330.
[45] W. Gao, N. Singh, L. Song, Z. Liu, A.L.M. Reddy, L. Ci, R. Vajtai, Q. Zhang, B. Wei, P.M. Ajayan, Direct laser writing of micro-supercapacitors on hydrated graphite oxide, Nat. Nanotechnol. 6 (2011) 496–500.
[46] Z.-S. Wu, K. Parvez, X. Feng, K. Müllen, Graphene-based in-plane micro-supercapacitors with high power and energy densities, Nat. Commun. 4 (2487) (2013) 1–8.
[47] Y. Chen, B. Xu, J. Xu, J. Wen, T. Hua, C.-W. Kan, Graphene-based in-planar supercapacitors by a novel laser-scribing, in-situ reduction and transfer-printed method on flexible substrates, J. Power Sources 420 (2019) 82–87.
[48] Z.-K. Wu, Z. Lin, L. Li, B. Song, K. Moon, S.-L. Bai, C.-P. Wong, Flexible micro-supercapacitor based on in-situ assembled graphene on metal template at room temperature, Nano Energy 10 (2014) 222–228.
[49] S. Gambhir, R. Jalili, D.J. Officer, G.G. Wallace, Chemically converted graphene: scalable chemistries to enable processing and fabrication, NPG Asia Mater. 7 (2015) 1–15.
[50] M. Horn, B. Gupta, J. MacLeod, J. Liu, N. Motta, Graphene-based supercapacitor electrodes: addressing challenges in mechanisms and materials, Curr. Opin. Green Sustain. Chem. 17 (2019) 42–48.
[51] X. Zhang, C. Yu, C. Wang, Z. Wang, J. Qiu, Chemically converting graphene oxide to graphene with organic base for Suzuki reaction, Mater. Res. Bull. 67 (2015) 77–82.
[52] S. Pei, H.-M. Cheng, The reduction of graphene, Carbon 50 (2012) 3210–3228.
[53] D. Li, M.B. Müller, S. Gilje, R.B. Kaner, G.G. Wallace, Processable aqueous dispersions of graphene nanosheets, Nat. Nanotechnol. 3 (2008) 101–105.
[54] J. Zhang, H. Yang, G. Shen, P. Cheng, J. Zhang, S. Guo, Reduction of graphene oxide via L-ascorbic acid, Chem. Commun. 46 (2010) 1112–1114.

[55] W. Chen, L. Yan, P.R. Bangal, Chemical reduction of graphene oxide to graphene by sulfur-containing compounds, J. Phys. Chem. C 114 (2010) 19885–19890.
[56] X. Fan, W. Peng, Y. Li, X. Li, S. Wang, G. Zhang, F. Zhang, Deoxygenation of exfoliated graphite oxide under alkaline conditions: a green route to graphene preparation, Adv. Mater. 20 (2008) 4490–4493.
[57] H.-J. Shin, K.K. Kim, A. Benayad, S.-M. Yoon, H.K. Park, I.-S. Jung, M.H. Jin, H.-K. Jeong, J.M. Kim, J.-Y. Choi, Y.H. Lee, Efficient reduction of graphite oxide by sodium borohydride and its effect on electrical conductance, Adv. Funct. Mater. 19 (2009) (1987-1992).
[58] X. Yang, C. Cheng, Y. Wang, L. Qiu, D. Li, Liquid-mediated dense integration of graphene materials for compact capacitive energy storage, Science 341 (2013) 534–537.
[59] E.G. Bushueva, P.S. Galkin, A.V. Okotrub, L.G. Bulusheva, N.N. Gavrilov, V.L. Kuznetsov, S.I. Moiseekov, Double layer supercapacitor properties of onion-like carbon materials, Phys. Status Solidi B 245 (2008) 2296–2299.
[60] M.E. Plonska-Brzezinska, A. Palkar, K. Winkler, L. Echegoyen, Electrochemical properties of small carbon nano-onion films, Electrochem. Solid St. 13 (2010) K35–K38.
[61] D. Pech, M. Brunet, H. Durou, P. Huang, V. Mochalin, Y. Gogotsi, P. Taberna, P. Simon, Ultra high power micrometre-sized supercapacitors based on onion-like carbon, Nat. Technol. 5 (2010) 645–651.
[62] P. Huang, D. Pech, R. Lin, J.K. McDonough, M. Brunet, P. Taberna, Y. Gogotsi, P. Simon, On-chip micro-supercapacitors for operation in a wide temperature range, Electrochem. Commun. 36 (2013) 53–56.
[63] C. Laurent, E. Flahaut, A. Peigney, The weight and density of carbon nanotubes versus the number of walls and diameter, Carbon 48 (2010) 2989–2999.
[64] S.H. Kim, G.W. Mulholland, M.R. Zachariah, Density measurement of size selected multiwalled carbon nanotubes by mobility-mass characterization, Carbon 47 (2009) 1297–1302.
[65] E. Frackowiak, F. Béguin, Carbon materials for the electrochemical storage of energy in capacitors, Carbon 39 (2001) 937–950.
[66] G. Lota, K. Lota, E. Frackowiak, Nanotubes based composites rich in nitrogen for supercapacitor application, Electrochem. Commun. 9 (2007) 1828–1832.
[67] Y. Wang, Y. Zhang, G. Wang, X. Shi, Y. Qiao, J. Liu, H. Liu, A. Ganesh, L. Li, Direct graphene-carbon nanotube composite ink writing all-solid-state flexible microsupercapacitors with high areal energy density, Adv. Funct. Mater. 30 (2020) 1907284.
[68] M. Beidaghi, C. Wang, Micro-supercapacitors based on interdigital electrodes of reduced graphene oxide and carbon nanotube composites with ultrahigh power handling peroformance, Adv. Funct. Mater. 22 (2012) 4501–4510.
[69] X.-L. Wu, A.-W. Xu, Carbonaceous hydrogels and aerogels for supercapacitors, J. Mater. Chem. A 2 (2014) 4852–4864.
[70] M.G. Saborio, P. Svelic, J. Casanovas, G. Ruano, M.M. Pérez-Madrigal, L. Franco, J. Torras, F. Estrany, C. Alemán, Hydrogels for flexible and compressible free standing cellulose supercapacitors, Eur. Polym. J. 118 (2019) 347–357.
[71] B.-S. Yin, S.-W. Zhang, K. Ke, Z.-B. Wang, Advanced deformable all-in-one hydrogel supercapacitor based on conducting polymer: toward integrated mechanical and capacitive performance, J. Alloys Compd. 805 (2019) 1044–1051.
[72] R.W. Pekala, J.C. Farmer, C.T. Alviso, T.D. Tran, S.T. Mayer, J.M. Miller, B. Dunn, Carbon aerogels for electrochemical applications, J. Non Cryst. Solids 225 (1998) 74–80.
[73] M.-F. Yan, L.-H. Zhang, R. He, Z.-F. Liu, Synthesis and characterization of carbon aerogel with different catalyst, J. Porous. Mater. 22 (2015) 699–703.

[74] Y.J. Lee, J.C. Jung, J. Yi, S.-H. Baeck, J.R. Yoon, I.K. Song, Preparation of carbon aerogel in ambient conditions for electrical double-layer capacitor, Curr. Appl. Phys. 10 (2010) 682–686.
[75] T. Bordjiba, M. Mohamedi, H. Dao Le, New class of carbon-nanotube aerogel electrodes for electrochemical power sources, Adv. Mater. 20 (2008) 815–819.
[76] C. Zhu, T. Liu, F. Qian, T. Yong-Jin Han, E.B. Duoss, J.D. Kuntz, C.M. Spadaccini, M.A. Worsley, Y. Li, Supercapacitors based on three-dimensional hierarchical graphene aerogels with periodic macropores, Nano Lett. 16 (2016) 3448–3456.
[77] Y. Zhang, X. Liu, S. Wang, L. Li, S. Dou, Bio-nanotechnology in high-performance supercapacitors, Adv. Energy Mater. 7 (2017) 1–23.
[78] M. Mladenov, P. Zlatilova, R. Raicheff, S. Vassilev, N. Petrov, K. Belov, V. Trenev, D. Kovacheva, Synthesis and characterization of novel nanostructured carbon for supercapacitors on the basis of biomaterials, Bulg. Chem. Commun. 40 (2008) 360–366.
[79] B.S. Yang, K.Y. Kang, Preparation of lignin-based carbon aerogels as biomaterials for nano-supercapacitor, J. Korean Phys. Soc. 71 (2017) 478–482.
[80] S. Herou, P. Schlee, A.B. Jorge, M. Titirici, Bio-mass-derived electrodes for flexible supercapacitors, Curr. Opin. Green Sustain. Chem. 9 (2018) 18–24.
[81] X. Wang, D. Kong, Y. Zhang, B. Wang, X. Li, T. Qiu, Q. Song, J. Ning, Y. Song, L. Zhi, All-biomaterial supercapacitor derived from bacterial cellulose, Nanoscale 8 (2016) 9146–9150.
[82] D. Pech, M. Brunet, P.L. Taberna, P. Simon, N. Fabre, F. Mesnilgrente, V.C. Conédéra, H. Durou, Elaboration of a microstructured inkjet-printed carbon electrochemical capacitor, J. Power Sources 195 (2010) 1266–1269.
[83] J. Wu, D. Yu, G. Wang, J. Yang, H. Wang, X. Liu, L. Guo, X. Han, Flexible microsupercapacitors based on naturally derived juglone, ChemPlusChem 83 (2018) 423–430.
[84] H. Pan, D. Wang, Q. Peng, J. Ma, X. Meng, Y. Zhang, Y. Ma, S. Zhu, D. Zhang, High-performance microsupercapacitors based on bio-inspired graphene microfibers, ACS Appl. Mater. Interfaces 10 (2018) 10157–10164.
[85] H.W. Shim, A.H. Lim, J.C. Kim, E. Jang, S.D. Seo, G.H. Lee, T.D. Kim, D.W. Kim, Scalable one-pot bacteria-templating synthesis route toward hierarchical, porous-Co3O4 superstructures for supercapacitor electrodes, Sci. Rep. 3 (2013) 1–9.
[86] Ł. Kolanowski, M. Graś, M. Bartkowiak, B. Doczekalska, G. Lota, Electrochemical capacitors based on electrodes made of lignocellulosic waste materials, Waste Biomass Valor. 11 (2020) 3863–3871.
[87] G.Z. Chen, Understanding supercapacitors based on nano-hybrid materials with interfacial conjugation, Prog. Nat. Sci. 23 (2013) 245–255.
[88] J. Guan, L. Yu, G.Z. Chen, Capacitive and non-capacitive faradaic charge storage, Electrochim. Acta 206 (2016) 464–478.
[89] A.C. Ferrari, J. Robertson, Interpretation of Raman spectra of disordered and amorphous carbon, Phys. Rev. B 61 (2000) 14095–14107.
[90] A.D. Smith, Q. Li, A. Vyas, M.M. Haque, K. Wang, A. Velasco, X. Zhang, S. Thurakkal, A. Quellmalz, F. Niklaus, K. Gylfason, P. Lundgren, P. Enoksson, Carbon-based electrode materials for microsupercapacitors in self-powering sensor networks: present and future development, Sensors 19 (2019) 1–17.
[91] C. Lethien, J.L. Bideau, T. Brousse, Challenges and prospects of 3D microsupercapacitors for powering the internet of things, Energy Environ. Sci. 12 (2019) 96–115.
[92] T. Huang, K. Jiang, D. Chen, G. Shen, Recent progress and perspectives of metal oxides based on-chip microsupercapacitors, Chin. Chem. Lett. 29 (2018) 553–563.

[93] N. Tang, X. Tian, C. Yang, Z. Pi, Facile synthesis of α-MnO_2 nanostructures for supercapacitors, Mater. Res. Bull. 44 (2009) 2062–2067.

[94] S. Peng, L. Li, Y. Hu, M. Srinivasan, F. Cheng, J. Chen, S. Ramakrishna, Fabrication of spinel one-dimensional architectures by single-spinneret electrospinning for energy storage applications, ACS Nano 9 (2015) 1945–1954.

[95] Q. Cheng, J. Tang, J. Ma, H. Zhang, N. Shinga, L.-C. Qin, Graphene and nanostructured MnO_2 composite electrodes for supercapacitors, Carbon 49 (2011) 2917–2925.

[96] Y. Lei, J. Li, Y. Wang, L. Gu, Y. Chong, H. Yuan, D. Xiao, Rapid microwave-assisted green synthesis of 3D hierarchical flower-shaped $NiCo_2O_4$ microsphere for high-performance supercapacitor, ACS Appl. Mater. Interfaces 6 (2014) 1773–1780.

[97] Y. Lan, X. Li, G. Li, Y. Luo, Sol-gel method to prepare graphene/Fe_2O_3 aerogel and its catalytic application for the thermal decomposition of ammonium perchlorate, J. Nanopart. Res. 17 (2015) 395.

[98] C.C. Liu, D.S. Tsai, D. Susanti, W.C. Yeh, Y.S. Huang, F.J. Liu, Planar ultracapacitors of miniature interdigital electrode loaded with hydrous RuO_2 and RuO_2 nanorods, Electrochim. Acta 55 (2010) 5768–5774.

[99] Z.-S. Wu, X. Feng, H.-M. Cheng, Recent advances in graphene-based planar micro-supercapacitors for on-chip energy storage, Natl. Sci. Rev. 1 (2014) 277–292.

[100] T.M. Dinh, A. Ahour, S. Vizireanu, G. Dinescu, L. Nistor, K. Armstrong, D. Guay, D. Pech, Hydrous RuO2/carbon nanowalls hierarchical structures for all-solid-state ultrahigh-energy density micro-supercapacitors, Nano Energy 10 (2014) 288–294.

[101] J. Wei, N. Nagarajan, I. Zhitomirsky, Manganese oxide films for electrochemical supercapacitors, J. Mater. Process. Technol. 186 (2007) 356–361.

[102] C. Arbizzani, M. Mastragostino, L. Meneghello, Polymer-based redox supercapacitors: a comparative study, Electrochim. Acta 41 (1996) 21–26.

[103] J.C. Carlberg, O. Inganäs, Poly(3,4-ethylenedioxythiophene) as electrode material in electrochemical capacitors, J. Electrochem. Soc. 144 (1997) L61–L64.

[104] W. Sun, R. Zheng, X. Chen, Symmetric redox supercapacitor based on microfabrication with three-dimensional polypyrrole electrodes, J. Power Sources 195 (2010) 7120–7125.

[105] N. Kurra, M.K. Hota, H.N. Alshareef, Conducting polymer micro-supercapacitors for flexible energy storage and Ac line-filtering, Nano Energy 13 (2015) 500–508.

[106] A.S. Sarac, H.D. Gilsing, A. Gencturk, B. Schultz, Electrochemically polimerized 2,2-dimethyl-3,4-propylenedioxythiophene on carbon fiber for microsupercapacitor, Prog. Org. Coat. 60 (2007) 281–286.

[107] Y. Mao, J. Xu, X. He, W. Yang, Y. Yang, L. Xu, Y. Zhao, Y. Zhou, All-solid-state flexible microsupercapacitors based on reduced graphene oxide/multi-walled carbon nanotube composite electrodes, Appl. Surf. Sci. 435 (2018) 1228–1236.

[108] A.T. Smith, A.M. LaChance, S. Zeng, B. Liu, L. Sun, Synthesis, properties and applications of graphene oxide/reduced graphene oxide and their nanocomposites, Nano Mater. Sci. 1 (2019) 31–47.

[109] Z.S. Wu, G. Shou, L.C. Yin, W. Ren, F. Li, H.M. Cheng, Graphene/metal oxide composite electrode materials for energy storage, Nano Energy 1 (2012) 107–131.

[110] J. Pu, X. Wang, R. Xu, S. Xu, K. Komvopoulos, Highly flexible, foldable, and rollable microsupercapacitors on an ultrathin polyimide substrate with high power density, Microsyst. Nanoeng. 4 (2018) 1–11.

[111] M. Kaempgen, C.K. Chan, J. Ma, Y. Cui, G. Gruner, Printable thin film supercapacitors using single-walled carbon nanotubes, Nano Lett. 9 (2009) 1872–1876.

[112] W. Yu, H. Zhou, B.Q. Li, S. Ding, 3D printing of carbon nanotubes-based micro-supercapacitors, ACS Appl. Mater. Interfaces 9 (2017) 4597–4604.

[113] J. Lin, C. Zhang, Z. Yan, Y. Zhu, Z. Peng, R.H. Hauge, D. Natelson, J.M. Tour, 3-Dimensional graphene carbon nanotube carpet-based microsupercapacitors with high electrochemical performance, Nano Lett. 13 (2013) 72–78.
[114] L. Huang, D. Chen, Y. Ding, S. Feng, Z.L. Wang, M. Liu, Nickel-cobalt hydroxide nanosheets on NiCo2O4 nanowires grown on carbon fiber paper for high-performance pseudocapacitors, Nano Lett. 13 (2013) 3135–3139.
[115] C. An, Y. Zhang, H. Guo, Y. Wang, Metal oxide-based supercapacitors: progress and prospectives, Nanoscale Adv. 1 (2019) 4644–4658.
[116] Y.G. Zhu, Y. Wang, Y. Shi, J.I. Wong, H.Y. Yang, CoO nanoflowers woven by CNT network for high energy density flexible micro-supercapacitor, Nano Energy 3 (2014) 46–54.
[117] R.B. Rakhi, W. Chen, D. Cha, H.N. Alshareef, Substrate dependent self-organization of mesopores cobalt oxide nanowires with remarkable pseudocapacitance, Nano Lett. 12 (2010) 2559–2567.
[118] J. Wang, F. Li, F. Zhu, O.G. Schmidt, Recent progress in micro-supercapacitor design, integration, and functionalization, Small Methods 3 (2018) 1–27.
[119] T.S. Tran, N.K. Dutta, N.R. Choudhury, Graphene-based inks for printing of planar micro-supercapacitors: a review, Materials 12 (2019) 1–21.
[120] K. Shen, J. Ding, S. Yang, 3D printing quasi-solid-state asymmetric micro-supercapacitors with ultrahigh areal energy density, Adv. Energy Mater. 8 (2018) 1–7.
[121] C. Zhang, L. McKeon, M.P. Kremer, S.H. Park, O. Ronan, A. Seral-Ascaso, S. Barwich, C. Coileain, N. McEvoy, H.C. Nerl, B. Anasori, J.N. Coleman, Y. Gogotsi, V. Nicolosi, Additive-free MXene inks and direct printing of micro-supercapacitors, Nat. Commun. 10 (2019) 1–9.
[122] Y. Liu, B. Zhang, Q. Xu, Y. Hou, S. Seyedin, S. Qin, G.G. Wallace, S. Beirne, J.M. Razal, J. Chen, Development of graphene oxide/polyaniline inks for high performance flexible microsupercapacitors via extrusion printing, Adv. Funct. Mater. 28 (2018) 1–12.
[123] L. Wei, N. Nitta, G. Yushin, Lithographically patterned thin activated carbon films as a new technology platform for on-chip devices, ACS Nano 7 (2013) 6498–6506.
[124] V. Presser, L. Zhang, J.J. Niu, J. McDonough, C. Perez, H. Fong, Y. Gogotsi, Flexible nano-felts of carbide-derived carbon with ultra-high power handling capability, Adv. Energy Mater. 1 (2011) 423–430.
[125] X. Feng, J. Ning, D. Wang, J. Zhang, J. Dong, C. Zhang, X. Shen, Y. Hao, All-solid-state planner micro-supercapacitor based on graphene/NiOOH/$Ni(OH)_2$ via mask-free patterning strategy, J. Power Sources 418 (2019) 130–137.

CHAPTER 2

Materials under research II: Silicon-based electrodes

Karolina Urszula Laszczyk[a] and Kazufumi Kobashi[b]
[a]Faculty of Microsystem Electronics and Photonics, Wroclaw University of Science and Technology, Wrocław, Poland
[b]CNT-Application Research Center, National Institute of Advanced Industrial Science and Technology, Ibaraki, Japan

2.1 Introduction

Microsupercapacitors (MSCs) are considered as an inherent part to complete the energy storage unit in self-powered micro- and nanosystems. The MSCs contribute to the development of sustainable portable electronic devices, including remote sensors and actuators, especially in the context of the Internet of Things. To make these systems achievable the MSCs must be yet integrated and packaged with the other electronic components on-chip. This becomes a big challenge considering technology compatibility. The majority of the sensing and actuating components are fabricated in crystal materials, e.g., silicon (Si). Meanwhile, most of the investigated materials for MSC electrodes are various carbon and nanocarbons. Though these materials are very essential in the performance, they are not well adapted to the semiconducting fabrication process. The reason is that their synthesis and fabrication are not compatible with the integrated circuit technology, because they require to be formed and sealed or packaged separately. The transfer of the nanocarbon material and next to its fabrication is rather tricky to be not destructive to the other electronics on-chip. On the other hand, the opposite manner rises the threat to destroy the fragile nanocarbon electrodes, which are not a solid structure, and due to defects or impurities, as well as intentionally added bond functional groups, are profound to reactivity with the processing chemistry. For these reasons, the combination of separately fabricated electronics and MSCs dramatically limits the miniaturization. In the search for the solution, scientists have started to investigate Si as a potential electrode material for batteries and supercapacitors [1]. Below, we would like to highlight the main aspects of Si as electrode materials. Especially, we would like to focus on porous silicon (PSi) and silicon nanowires (SiNWs) which offer

Microsupercapacitors
https://doi.org/10.1016/B978-0-08-102888-9.00008-2

structural control, which is very challenging for conventional nanocarbon materials, and provide the possible route to on-chip integration, where MSCs and silicon-based batteries, photovoltaics, sensors, and electronics are packed together lowering the necessary footprint area as it was proposed in [2, 3].

Additionally, PSi and SiNWs offer well-tailored hierarchical porosity that is controlled by growth process parameters bringing benefits from the high gravimetric capacities of the electrode structures. This constitutes a strong asset if compared to the other organic materials, which suffer from poor packing density and fragile networks.

2.2 Porous silicon

To make the crystalline Si porous, the simplest treatment employs the chemical or electrochemical etching, which is a common and simple method to be applied; the substrate is immersed in a mixture of hydrofluoric acid (HF) containing aqueous or non-aqueous electrolytes with the presence of a high positive (anodic) voltage or current. Hence, it is also called anodic etching. Next, the PSi is dried by simply evaporating solvent under atmospheric conditions. The drying process might be also performed under supercritical conditions or washing with a liquid with low surface tension, such as pentane [4]. The latest enables avoiding the microscale cracking of the surface [1]. Other methods successfully reported are chemical stain etching [5], chemical vapor etching [6], laser-induced etching [7], metal-assisted etching [8]. For non-chemical etching: spark processing [9] and reactive ion etching [10]. Using one of the mentioned methods we can obtain the formation of the anisotropic structures, which indeed form a layer on silicon, along the normal to its surface. In most cases, the pores propagate in one of the directions (⟨100⟩), irrespective of crystal orientation. The properties of the PSi layer, i.e., the shape and the depth of the pores, porosity, specific surface area (SSA) by the control of process parameters, such as the anodization current density, the concentration of the etch solution, the etch time, the dopant type in the silicon and its amount. For example, the dopant modification results in various morphology of the porous structure: in a moderately and highly doped p-type Si, the pores form branches, while in p-type Si with low doping levels, we get the interconnected network of nanometer-sized Si ligaments (Fig. 2.1) [11, 12]. For the morphology, pore sizes and porosity—defined as the percentage coverage with the pores inside the etched layer—are also classified. Interestingly, the PSi layer could be a highly

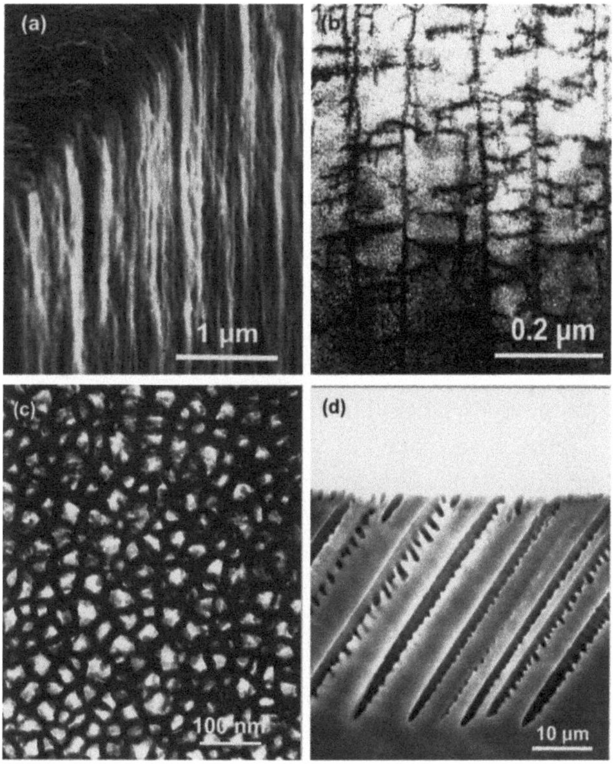

Fig. 2.1 The SEM images of the different PSi layers made of n- or p-silicon presenting various morphology of the pores. *(Reprinted with permission from V. Parkhutik, Porous silicon Ð mechanisms of growth and applications, Solid State Electron. 43 (1999) 1121–1141).*

ordered three-dimensional matrix if electrochemical etching is used [13]. SSA of PSi was reported to be 910 $m^2\ cm^{-3}$ or 812 $m^2\ g^{-1}$ [14] and depends on porosity (meanwhile SSA of porous activated carbon reaches up to 2760 $m^2\ g^{-1}$ [15] and of activated porous graphene—3100 $m^2\ g^{-1}$ [16]), but paradoxically, as the porosity becomes higher the SSA decreases [17]. Additionally, silicon-based materials suffer from reactivity [18, 19]; the so-called "aging effect" of PSi manifests in reaction with the ambient air [20] and in general with other chemical substances, liquid, or gases [21–23]. The reactivity of surface-bound silicon atoms causes changes in PSi composition and its properties, such as electrical resistivity [4], but it is the most cumbersome in terms of electrochemical stability [1]. Despite these issues, the fact Si electrodes and electronics have a common material substrate and compatible technology makes Si attractive to use for energy storage electrodes [24]. PSi has found its application in anodes of metal-ion batteries [25], mainly because of its attractively high theoretical capacity (3579 mA h g^{-1} for $Li_{15}Si_4$) [26]. Meanwhile, for the supercapacitors, the new trend provides towards the prevention of surface oxidation and methods of stabilizing the porous layer as well as the idea to use PSi as a substantial builder of pseudocapacitive electrodes for SCs.

2.3 PSi for supercapacitor electrodes

Besides PSi possesses low chemical and mechanical stability, it has also poor conductivity. Hence, to make PSi attractive its conductive passivation is needed to increase the performance of the electrodes and to make them electrochemically stable. Although with the coating we gain the chemical stability, we get poorer SSA; in the case of TiN, SSA was noticed with a double decline (from 241 to 113 m^2 cm^{-3}) [2]. It is a specially crucial issue because the passivating film plays a role in electric double layer formation. The first attempt was to cover the PSi surface with a sputtered gold [1, 27], an atomic deposited layer of TiN [28], or diamond [29]. Further, PSi was to employ the film inhibiting faradaic reaction using ruthenium oxide [30] as well as metal oxides [31, 32].

The difference between pure PSi and PSi coated with the few-nanometers graphene led to their electrochemical test results. Oakes et al. [33] demonstrated graphene-coated PSi that exhibited about a twice greater electrochemical window, about 50 times lower equivalent series resistance, and 10–40 times the higher volumetric and specific energy density. Wu et al. [34] performed a coating of graphene on the surface of a PSi matrix: PSi was formed by electrochemical etching and further, a graphene film was directly grown in a Ni vapor-assisted chemical vapor deposition (CVD) process, at various temperatures around 1000 °C (close to the melting temperature of bulk silicon, i.e., 1400 °C.). To reduce the surface energy the different reorganization of silicon pores took place, hence the morphology of the graphene-coated PSi layer changed: from a meso-, through a hybrid- to a macroporous structure. Such prepared electrodes immersed in 0.5 M aqueous Na_2SO_4 solution resulted mostly in a rectangular shape of cyclic voltammograms, with various performances depending on the type of porous structure. For example the areal capacitance for the mesoporous structure (8.48 mF cm^{-2} at 5 mV s^{-1}) was 21 times higher compared to hybrid-porous structure (0.396 mF cm^{-2} at 5 mV s^{-1}), but with the low retention (0.1% vs 87.5%). Though the mesoporous layer had higher SSA than the hybrid-porous layer, it was more pronounced to oxidation, due to thinner graphene coating.

It should be noticed here, that the nanoscale Si was found to act as a catalyst for the formation of the graphene coating at temperatures 900 °C and below, while the formation of the graphene in the same condition failed when proceeded on a bulk silicon [35]. Moreover, as was described by Rowland [1], non-conformal coating of the graphene forms so-called oxide islands, especially pronounced to faradaic reactions.

2.4 PSi for microsupercapacitors electrodes

PSi materials for MSC electrodes include mainly plain and doped silicon nanowires (SiNWs) [36, 37], silicon carbide nanowires [38], and SiNWs with a conducting polymer coating [39]. There are few reports of electrodes for MSCs made of plasma-etched PSi [40] or silicon microparticle [41] SiNWs are high aspect ratio structures forming either a loosely arranged mesh of wires (Fig. 2.2A) or an oriented forest [42] (Fig. 2.2B), which diameters are in the range of tens of nanometers and density of 10^7–10^8 NWs cm^{-2} [37]. The densely agglomerated and curled SiNWs with additionally grown branches at the nanowires sidewalls are known as silicon nanotrees (SiNTr) (Fig. 2.2C) [43, 44]. The formation of the SiNWs-based electrodes included a growth process in a CVD reactor. Nonetheless, SiNWs, in general, could be formed by evaporation [45], laser ablation [46], molecular beam epitaxy [47], and highly pressurized supercritical organic fluids enriched with a liquid silicon precursor [48]. The coating is a separate process that usually bases on a dedicated CVD method and requires chemical washing of SiNWs before the deposition to remove the impurities.

Aradilla et al. [49] examined the MSC electrodes made of the bare SiNWs grown in a CVD reactor by using the vapor-liquid-solid (VLS) method via gold catalysis on a highly doped n-Si (111) substrate. The electrodes (every 1 cm^2) were tested in a sandwich configuration. The areal capacitance reached ~23 μF cm^{-2}, while the energy density was 0.053 μW h cm^{-2}. The authors obtained 80% enhancement of the energy density and stability; it reaches 8 million charge-discharge cycles with quite a flat retention of about 25% of the initial capacitance. The tests were performed in ionic liquid $PYR_{13}TFSI$. For comparison, the SiNWs electrodes tested by Thissandier et al. [24]

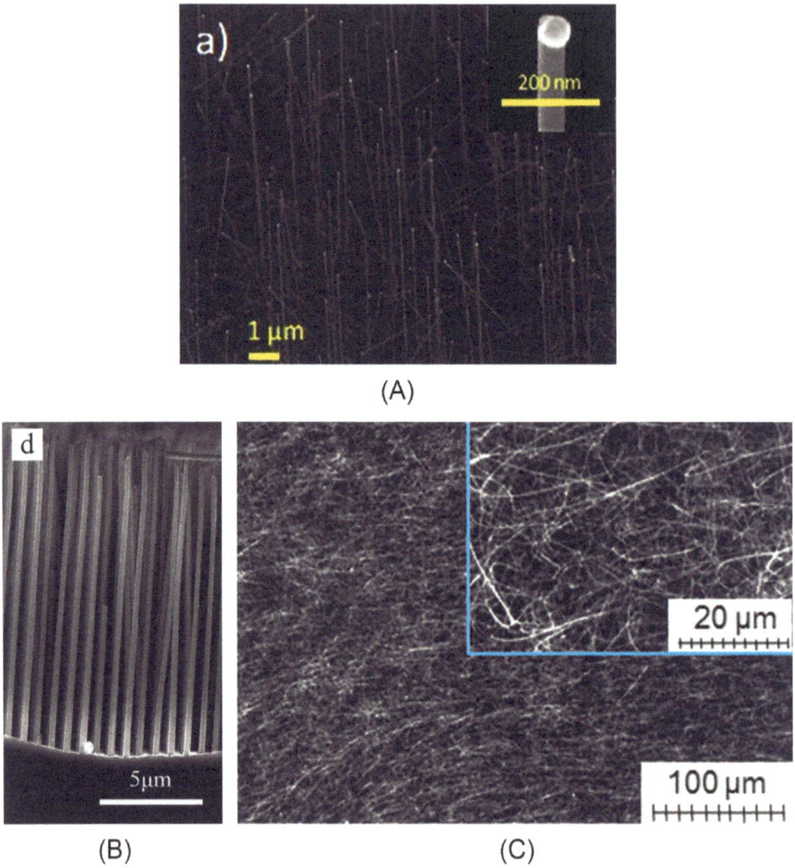

Fig. 2.2 The SEM image of the SiNW mesh (A), SINWs forest (B), and SiNWTr (C). *(Reprinted with permission from F. Thissandier, N. Pauc, T. Brousse, P. Gentile, S. Sadki, Micro-ultracapacitors with highly doped silicon nanowires electrodes, Nanoscale Res. Lett. 8 (2013) 23; B. Li, G. Niu, L. Sun, Lu Yaoa, C. Wang, Y. Zhanga, Design optimization and antireflection of silicon nanowire arrays fabricated by Au-assisted chemical etching, Mater. Sci. Semicond. Process. 82 (2018) 1–8; A. Platek, J. Piwek, K. Fic, T. Schubert, P. Gentile, G. Bidan, E. Frackowiak, Electrochemical performance of silicon nanostructures in low-temperature ionic liquids for microelectronic applications, J. Mater. Chem. A 5 (2017) 22708).*

resulted in the comparable areal capacitance (26 µF cm^{-2}) using 1 M NEt_4BF_4 in PC. Moreover, the mentioned group found the capacitance to become higher (up to 440 µF cm^{-2}) with the increase of the length of the SiNWs (from 5 to 50 µm) as the long SiNWs contributed to the enlargement of the surface to be accessed by ions. On the other hand, following Gardner et al. [50], the capacitance of SiNW electrodes scaled approximately linearly with the pore depth: from 3 to 6 mF cm^{-2}, for 2 μm deep pores, up to 28 mF cm^{-2} measured, for 12 μm deep pores. A significant increase in capacitance was reported for the pseudocapacitive SiNWs electrodes using poly (3,4-ethylene dioxythiophene) (PEDOT) [39] or ruthenium (Ru) nanoparticles [51]. In the latest, the SiNWs were decorated with Ru nanoparticles at room temperature utilizing electroless deposition technique by simply immersing the substrate with SiNWs in liquid medium (an aqueous solution of $RuCl_3 \times H_2O/HF$). Fig. 2.3 presents an example of the decorated SiNWs with the Ru nanoparticles having a diameter of 3–6 nm. The electrodes exhibited a specific capacitance of 36.25 mF cm^{-2} and retention higher than 87% upon 25,000 charge-discharge cycles in aqueous Na_2SO_4 electrolyte. In the case of PEDOT [39], the 300 nm thick film was electrochemically deposited on SiNWs from the same ionic liquid ($PYR_{13}TFSI$) to be used

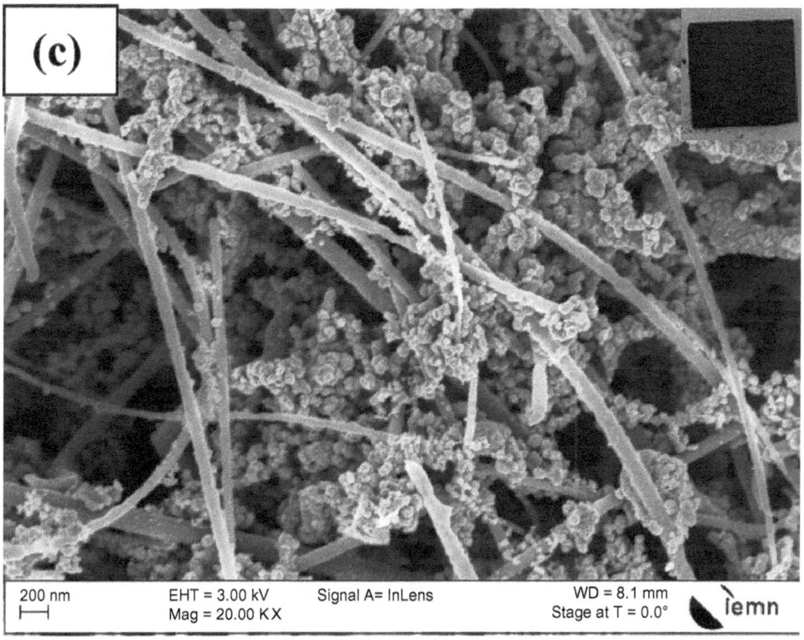

Fig. 2.3 The SEM image of SiNWs decorated with Ru nanoparticles. *(Reprinted with permission from Y. Bencheikh, M. Harnois, R. Jijie, A. Addad, P. Roussel, S. Szunerits, T. Hadjersi, S. El Hak Abaidia, R. Boukherrou, High performance silicon nanowires/ ruthenium nanoparticles micro-supercapacitors, Electrochim. Acta 311 (2019) 150–159).*

as an electrolyte further for testing achieving almost three-order higher areal capacitance (9 mF cm^{-2}) compared to pure SiNWs reported earlier, and with the ~20% retention of the capacitance, but with shorter stable working time, i.e., up to 3500 charge-discharge cycles. Yet, higher areal capacitance (~325 mF cm^{-2}) and ~83% retention for 5000 cycles were obtained by Alper et al. [38] by coating SiNWs with the carbon sheath and immersing the electrodes in ionic liquid (1-ethyl-3-methylimidazolium bis-(trifluoromethylsulfonyl) imide) (EMIM-TFSI). The electrodes tested in aqueous electrolytes exhibited slowly increasing capacitance during cycling due to the slow infiltration of the water into the small hydrophobic pores. Similarly, testing of MnO_2-coated SiNWs [52] in various electrolytes showed the importance of the right choice of the electrolyte. Therefore, the excellent wettability of the electrode surface, as well as electrochemical compatibility between the electrolyte and electrodes, are of high importance.

Another carbon coating, that was successfully applied for SiNWs electrodes of MSCs, is graphene, which demonstrates high power and energy densities if applied for MSC electrodes. Aradilla et al. [53] demonstrated the vertically aligned graphene nanosheets (VAGNs) on Si and employed them in a symmetric MSC. The VAGNs were grown on SINWs using a catalyst-free method based on electron cyclotron resonance-plasma enhanced chemical vapor deposition (ECR-CVD) technique and the sheet thickness was 12 μm. The MSC composed of SiNWs-VAGNS in the ionic liquid electrolyte ($PYR_{13}TFSI$) resulted in high areal capacitance, energy, and power: 2 mF cm^{-2}, 4 mW h cm^{-2}, and 4 mW cm^{-2}, respectively. Meanwhile, the capacitance retention was 80% of initial capacitance with cycling up to 150,000 times. However, still, the unsolved problem is how to make a selective deposition of material on a particular place on a chip, and so far presented CVD-based methods are very difficult to be the solution. Soam et al. [54] addressed this problem by application of electrophoretic deposition (EPD) process to coat selectively SiNWs with graphene oxide (GO) nanosheets (Fig. 2.4), which was converted to conductive-reduced graphene oxide (rGO) in the next step, i.e., UV light exposure (photo-reduction). To avoid the excess deposition of rGO, EPD was carried out at 10 V for a very short time (5 s). As in the case composite matrix of graphene and carbon nanotubes, SiNWs prevented the restacking and aggregation of the rGO nano-sheets. The areal capacitance of the MSC was reported to be 240 mF cm^{-2} and was four times improved compared to the bare SiNWs. Additionally, it was demonstrated a negligible drop of capacitance up to 1000 cycles. In both recalled examples, the addition of

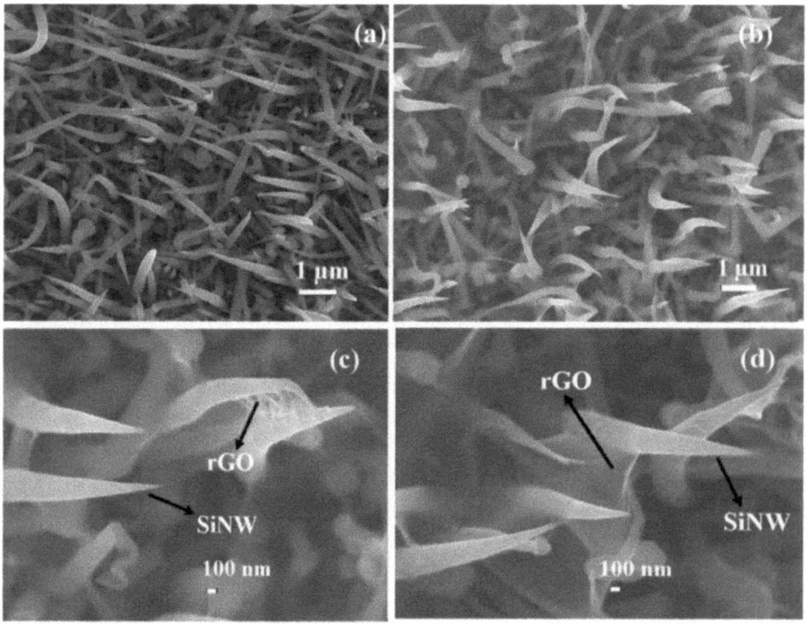

Fig. 2.4 SEM images of as grown SiNWs (A) and after the EPD of GO and its photoreduction to rGO (B–D). *(Reprinted with permission from A. Soam, P. Kavle, A. Kumbhar, R.O. Dusane, Performance enhancement of micro-supercapacitor by coating of graphene on silicon nanowires at room temperature, Curr. Appl. Phys. 17 (2017) 314–320).*

graphene indeed increases the capability of Si-based electrodes to be useful for MSCs, taking benefit of extremely high SSA, hence theoretically extremely high specific capacitance ($550\ F\ g^{-1}$ [55]). Combining the three-dimensional structure in Si substrate with novel coatings might expand the nominal capacitance of the MSC electrodes which, due to the two-dimensional nature of nanoscale materials, is limited. Table 2.1 compares the performance of silicon-based electrodes tested for the supercapacitor and MSC.

2.5 Summary

We presented briefly the promising material for MSC electrodes that consist of porous silicon and silicon nanowires and that could be coated with different active materials such as graphene or pseudocapacitive film or nanoparticles. Though carbon materials are the most widely used for MSC electrodes, their application for on-chip electronics is limited due to the

Table 2.1 Properties of selected MSC electrodes in literature.

Electrode material	Electrolyte	Capacitance	Cycling stability	References
SiNWs/PEDOT	$PYR_{13}TFSI$	~8 mF cm^{-2} (1 mA cm^{-2})	20% after 35,000 cycles (1 mA cm^{-2})	[39]
SiNWTr	EMIM TSFI + AMIM TFSI	800 μF cm^{-2} (600 mV s^{-1})	44%	[44]
Graphene SiNWs	EMIM TFSI	240 μF cm^{-2} (40 mV s^{-1})	Stable at 10,000 cycles	[54]
Graphene SiNWs	0.5 M Na_2SO_4	8.19 mF cm^{-2} (5 mV s^{-1})	Stable at 10,000 cycles	[55]
SiNWs/Ru nanoparticles	1 M Na_2SO_4	36.25 mF cm^{-2} (1 mA cm^{-2})	80% after 25,000 cycles (at 1 mA cm^{-2})	[51]
SiNWs/MnO_2	1 M Na_2SO_4	21.296 mF cm^{-2} (10 mV s^{-1})	Significant drop (>50%) after 50 cycles	[32]
SiNWs/MnO_x	0.1 M EMIM-TFSI	~400 mF cm^{-2}	82% after 10,000 cycles	[31]
Silicon microparticles	1 M Na_2SO_4	0.58 mF cm^{-2} (10 mV s^{-1})	80% after 1000 cycles	[41]

difficulty in assembly with the integrated circuits. Since then, Si is a substrate material in the semiconducting industry with well-adapted large-scale technology. It seems the Si-based energy storage unit including MSC should contribute to the progress in the on-chip integration. Bare SiNWs or PSi have indeed lower performance compared to the nanocarbon or pseudocapacitive electrodes, however, the fact, Si is amenable for coating with various materials gives the perspective of strengthening the MSC performance and application in large-scale industrial fabrication.

References

[1] S.E. Rowlands, R.J. Latham, W.S. Schlindwein, Supercapacitor devices using porous silicon electrodes, Ionics 5 (1999) 144–149.

[2] D.S. Gardner, C.L. Pint, C.W. Holzwarth, W. Jin, Z. Chen, J.L. Gustafson, Integration of energy storage devices onto substrates for microelectronics and mobile devices, U.S. Patent #9,245,695, filed December 21, 2011, issued January 26, 2016.

[3] H. Durou, D. Pech, D. Colin, P. Simon, P.-L. Taberna, M. Brunet, Wafer-level fabrication process for fully encapsulated micro-supercapacitors with high specific energy, Microsyst. Technol. 18 (2012) 467–473.

[4] G. Korotcenkov, B.K. Cho, Silicon porosification: state of the art, Crit. Rev. Solid State Mater. Sci. 35 (2010) 153–260.

[5] D. Dimova-Malinovska, M. Sendova-Vassileva, N. Tzenov, M. Kamenova, Preparation of thin porous silicon layers by stain etching, Thin Solid Films 297 (1997) 9–12.

[6] A. Ben Jaballah, M. Hassen, M. Hajji, M. Saadoun, B. Bessais, H. Ezzaouia, Chemical vapour etching of silicon and porous silicon: silicon solar cells and micromachining applications, Phys. Status Solidi A 202 (2005) 1606–1610.

[7] N. Noguchi, I. Suemune, Luminescent porous silicon synthesized by visible light irradiation, Appl. Phys. Lett. 62 (1993) 1429–1431.

[8] X. Li, P.W. Bohn, Metal-assisted chemical etching in HF H2O2 produces porous silicon, Appl. Phys. Lett. 77 (2000) 2572–2574.

[9] J.V.S. John, J.L. Coffer, Y.G. Rho, R.F. Pinizzotto, Formation of rare-earth oxide doped silicon by spark processing, Appl. Phys. Lett. 68 (1996) 3416–3418.

[10] S. Shingubara, S. Maruo, T. Yamashita, M. Nakao, T. Shimizu, Reduction of pitch of nanohole array by self-organizing anodic oxidation after nanoimprinting, Microelectron. Eng. 87 (2010) 1451–1454.

[11] V. Parkhutik, Porous silicon Ð mechanisms of growth and applications, Solid State Electron. 43 (1999) 1121–1141.

[12] P.C. Searsont, J.M. Macaulay, The fabrication of porous silicon structures, Nanotechnology 3 (1992) 1558–1565.

[13] S. Matthias, F. Müller, U. Gösele, Simple cubic three-dimensional photonic crystals based on macroporous silicon and anisotropic posttreatment, J. Appl. Phys. 98 (2005). 023524-1-023524-4.

[14] L.T. Canham, A.1. Groszek, Characterization of microporous Si by flow calorimetry: comparison with hydrophobic SiO2 molecular sieve, J. Appl. Phys. 72 (1992) 1558–1565.

[15] M. Xu, D. Li, Y. Yan, T. Guo, H. Pang, H. Xue, Porous high specific surface area-activated carbon with co-doping N, S and P for high-performance supercapacitors, RSC Adv. 7 (2017) 43780–43788.

[16] Y.W. Zhu, S. Murali, M.D. Stoller, K.J. Ganesh, W.W. Cai, P.J. Ferreira, A. Pirkle, R.-M. Wallace, K.A. Cychosz, M. Thommes, D. Su, E.A. Stach, R.S. Ruoff, Carbon-based supercapacitors produced by activation of graphene, Science 332 (2011) 1537–1541.
[17] G. Korotcenkov, Porous Silicon: From Formation to Application: Formation and Properties, CRC Press, vol. 1, 2016.
[18] M.J. Sailor, Porous Silicon in Practice: Preparation, John Wiley & Sons, Characterization and Applications, 2012.
[19] J.C. Flake, M.M. Rieger, G.M. Schmid, P.A. Kohl, Electrochemical etching of silicon in nonaqueous electrolytes containing hydrogen fluoride or fluoroborate, J. Electrochem. Soc. 146 (1999) 1960–1965.
[20] J. Salonen, V.P. Lehto, E. Laine, The room temperature oxidation of porous silicon, Appl. Surf. Sci. 120 (1997) 191–198.
[21] Y.H. Ogata, T. Tsuboi, T. Sakka, S. Naito, Oxidation of porous silicon in dry and wet environments under mild temperature conditions, J. Porous. Mater. 7 (2000) 63–66.
[22] U. Frotscher, U. Rossow, C. Ebert Mpietryga, W. Richter, M.G. Berger, R. Arens-Fischer, H. Munder, Investigation of different oxidation processes for porous silicon studied by spectroscopic ellipsometry, Thin Solid Films 276 (1996) 36–39.
[23] M. Yamana, N. Kashiwazaki, A. Kinoshita, T. Nakano, M. Yamamoto, C.W. Walton, Porous silicon oxide layer formation by the electrochemical treatment of a porous silicon layer, J. Electrochem. Soc. 137 (1990) 2925.
[24] F. Thissandier, A. Le Comte, O. Crosnier, P. Gentile, G. Bidan, E. Hadji, T. Brousse, S. Sadki, Highly doped silicon nanowires based electrodes for microelectrochemical capacitor applications, Electrochem. Commun. 25 (2012) 109–111.
[25] M. Ge, X. Fang, J. Rong, C. Zhou, Review of porous silicon preparation and its application for lithium-ion battery anodes, Nanotechnology 24 (2013) 422001.
[26] W. An, B. Gao, S. Mei, B. Xiang, J. Fu, L. Wang, Q. Zhang, P.K. Chu, K. Huo, Scalable synthesis of ant-nest-like bulk porous silicon for high-performance lithium-ion battery anodes, Nat. Commun. 10 (2019) 1447.
[27] S. Desplobain, G. Gautier, J. Semai, L. Ventura, M. Roy, Investigation of porous silicon electrode material in electrochemical capacitors, Phys. Status Solidi 4 (2007) 2180–2184.
[28] K. Grigoras, J. Keskinen, L. Grönberg, E. Yli-Rantala, S. Laakso, H. Välimäki, P. Kauranen, J. Ahopelto, M. Prunnila, Conformal titanium nitride in a porous silicon matrix: a nanomaterial for in-chip supercapacitors, Nano Energy 26 (2016) 340–345.
[29] F. Gao, G. Lewes-Malandrakis, M.T. Wolfer, W. Müller-Sebert, P. Gentile, D. Aradilla, T. Schubert, C.E. Nebel, Diamond-coated silicon wires for supercapacitor applications in ionic liquids, Diamond Relat. Mater. 51 (2015) 1–6.
[30] R. Warren, F. Sammoura, F. Tounsi, M. Sanghadasac, L. Lin, Highly active ruthenium oxide coating via ALD and electrochemical activation in supercapacitor applications, J. Mater. Chem. A 3 (2015) 15568–15575.
[31] S. Ortaboy, J.P. Alper, F. Rossi, G. Bertoni, G. Salviati, C. Carraro, R. Maboudian, MnOx-decorated carbonized porous silicon nanowire electrodes for high performance supercapacitors, Energ. Environ. Sci. 10 (2017) 1505–1516.
[32] F. Moulai, T. Hadjersi, M. Ifires, A. Khen, N. Rachedi, Enhancement of electrochemical capacitance of silicon nanowires arrays (SiNWs) by modification with manganese dioxide MnO2, Silicon 11 (2019) 2799–2810.
[33] L. Oakes, A. Westover, J.W. Mares, S. Chatterjee, W.R. Erwin, R. Bardhan, S.M. Weiss, C.L. Pint, Surface engineered porous silicon for stable, high performance electrochemical supercapacitors, Sci. Rep. 3 (2013) 3020.
[34] T. Wu, C. Chang, C. Wang, S. Parwaiz, C.-C. Lai, Y.-Z. Chen, S.-Y. Lu, Y.-L. Chueh, Few-layer graphene sheet-passivated porous silicon toward excellent electrochemical double-layer supercapacitor electrode, Nanoscale Res. Lett. 13 (2018) 242.

[35] S. Chatterjee, R. Carter, L. Oakes, W.R. Erwin, R. Bardhan, C.L. Pint, Electrochemical and corrosion stability of nanostructured silicon by graphene coatings: toward high power porous silicon supercapacitors, J. Phys. Chem. C 118 (2014) 10893–10902.
[36] F. Thissandier, N. Pauc, T. Brousse, P. Gentile, S. Sadki, Micro-ultracapacitors with highly doped silicon nanowires electrodes, Nanoscale Res. Lett. 8 (2013) 23.
[37] N. Berton, M. Brachet, F. Thissandier, J. Le Bideau, P. Gentile, G. Bidan, T. Brousse, S. Sadki, Wide-voltage-window silicon nanowire electrodes for micro-supercapacitors via electrochemical surface oxidation in ionic liquid electrolyte, Electrochem. Commun. 41 (2014) 31–34.
[38] J.P. Alper, M. Vincent, C. Carraro, R. Maboudian, Silicon carbide coated silicon nanowires as robust electrode material for aqueous micro-supercapacitor, Appl. Phys. Lett. 100 (2012) 163901.
[39] D. Aradilla, G. Bidan, P. Gentile, P. Weathers, F. Thissandier, V. Ruiz, P. Gomez-Romero, T.J.S. Schubert, H. Sahine, S. Sadki, Novel hybrid micro-supercapacitor based on conducting polymer coated silicon nanowires for electrochemical energy storage, RSC Adv. 4 (2014) 26462–26467.
[40] P. Lu, P. Ohlckers, L. Müller, S. Leopold, M. Hoffmann, K. Grigoras, J. Ahopelto, M. Prunnila, X. Chen, Nano fabricated silicon nanorod array with titanium nitride coating for on-chip supercapacitors, Electrochem. Commun. 70 (2016) 51–55.
[41] R. Kumar, M. Singh, A. Soam, Study on electrochemical properties of silicon micro particles as electrode for supercapacitor application, Surf. Interface 19 (2020) 100524.
[42] B. Li, G. Niu, L. Sun, L. Yaoa, C. Wang, Y. Zhanga, Design optimization and anti-reflection of silicon nanowire arrays fabricated by Au-assisted chemical etching, Mater. Sci. Semicond. Process. 82 (2018) 1–8.
[43] A. Platek, J. Piwek, K. Fic, T. Schubert, P. Gentile, G. Bidan, E. Frackowiak, Electrochemical performance of silicon nanostructures in low-temperature ionic liquids for microelectronic applications, J. Mater. Chem. A 5 (2017) 22708.
[44] F. Thissandier, P. Gentile, T. Brousse, G. Bidan, S. Sadki, Are tomorrow's micro-supercapacitors hidden in a forest of silicon nanotrees? J. Power Sources 269 (2014) 740–746.
[45] Q. Gu, H. Dang, J. Cao, J. Zhao, S. Fan, Silicon nanowires grown on iron-patterned silicon substrates, Appl. Phys. Lett. 76 (2000) 3020–3021.
[46] Y.F. Zhang, Y.H. Tang, N. Wang, D.P. Yu, C.S. Lee, I. Bello, S.T. Lee, Silicon nanowires prepared by laser ablation at high temperature, Appl. Phys. Lett. 72 (1998) 1835–1837.
[47] P. Das Kanungo, N. Zakharov, J. Bauer, O. Breitenstein, P. Werner, U. Goesele, Controlled in situ boron doping of short silicon nanowires grown by molecular beam epitaxy, Appl. Phys. Lett. 92 (2008) 263107.
[48] J.D. Holmes, K.P. Johnston, R.C. Doty, B.A. Korgel, Control of thickness and orientation of solution-grown silicon nanowires, Science 287 (2000) 1471–1473.
[49] D. Aradillaa, P. Gentile, G. Bidan, V. Ruiz, P. Gómez-Romero, T.J.S. Schubert, H. Sahine, E. Frackowiak, S. Sadki, High performance of symmetric micro-supercapacitors based on silicon nanowires using N-methyl-Npropylpyrrolidinium bis (trifluoromethylsulfonyl)imide as electrolyte, Nano Energy 9 (2014) 273–281.
[50] D.S. Gardner, C.W. Holzwarth, Y. Liu, S.B. Clendenning, W. Jin, B.-K. Moon, C. Pint, Z. Chen, E.C. Hannah, C. Chen, C. Wang, E. Mäkilä, R. Chen, T. Aldridge, J.L. Gustafson, Integrated on-chip energy storage using passivated nanoporous-silicon electrochemical capacitors, Nano Energy 25 (2016) 68–79.
[51] Y. Bencheikh, M. Harnois, R. Jijie, A. Addad, P. Roussel, S. Szunerits, T. Hadjersi, S. El Hak Abaidia, R. Boukherrou, High performance silicon nanowires/ruthenium nanoparticles micro-supercapacitors, Electrochim. Acta 311 (2019) 150–159.

[52] D.P. Dubal, D. Aradilla, G. Bidan, P. Gentile, T.J.S. Schubert, J. Wimberg, S. Sadki, P. Gomez-Romero, 3D hierarchical assembly of ultrathin MnO2 nanoflakes on silicon nanowires for high performance micro-supercapacitors in Li-doped ionic liquid, Sci. Rep. 5 (2015), 09771.
[53] D. Aradilla, M. Delaunay, S. Sadki, J.-M. Gerard, G. Bidan, Vertically aligned graphene nanosheets on silicon using an ionic liquid electrolyte: towards high performance on-chip micro-supercapacitors, J. Mater. Chem. A 3 (2015) 19254.
[54] A. Soam, P. Kavle, A. Kumbhar, R.O. Dusane, Performance enhancement of micro-supercapacitor by coating of graphene on silicon nanowires at room temperature, Curr. Appl. Phys. 17 (2017) 314–320.
[55] Z.-S. Wu, K. Parvez, X. Feng, K. Mullen, Graphene-based in-plane micro-supercapacitors with high power and energy densities, Nat. Commun. 4 (2013) 2487.

CHAPTER 3

The role and the necessary features of electrolytes for microsupercapacitors

Vidyanand Vijayakumar[a,b,c], Meena Ghosh[b,c], Roby Soni[d], Bihag Anothumakkool[e], Sreekumar Kurungot[c], Martin Winter[a,f,g], and Jijeesh Ravi Nair[a]

[a]Helmholtz Institute Münster, IEK-12, Forschungszentrum Jülich GmbH, Münster, Germany
[b]Academy of Scientific and Innovative Research (AcSIR), Ghaziabad, Uttar Pradesh, India
[c]Physical and Materials Chemistry Division, CSIR-National Chemical Laboratory, Pune, India
[d]Electrochemical Innovation Lab, Department of Chemical Engineering, University College London, London, United Kingdom
[e]TNO Holst Centre, Dutch National Institute for Applied Scientific Research, Eindhoven, The Netherlands
[f]Institute of Physical Chemistry, University of Münster, Münster, Germany
[g]MEET Battery Research Center, Münster, Germany

Abbreviations

([AMIM][TFSI])	1-allyl-3-methylimidazolium bis(trifluoromethylsulfonyl)imide
([NEt_3H][TFSI])	triethylammonium bis(trifluoromethylsulfonyl) imide
([PIP_{13}][FSI])	*N*-methyl-*N*-propylpiperidinium bis(fluorosulfonyl)imide
([PYR_{14}][FSI])	*N*-butyl-*N*-methylpyrrolidinium bis(fluorosulfonyl)imide
([PYR_{14}][TFSI])	*N*-methyl-*N*-butylpyrrolidinium bis(trifluoromethane)sulfonylimide
[BMIM][BF_4]	1-butyl-3-methylimidazolium tetrafluoroborate
[BMIM][PF_6]	1-butyl-3-methylimidazolium hexafluorophosphate
[EMIM][BF_4]	1-ethyl-3-methylimidazolium tetrafluoroborate
[EMIM][TFSI]	1-ethyl-3-methylimidazolium bis(trifluoromethylsulfonyl)imide
2D	two dimensional
3D	three dimensional
AC	activated carbon
ACN	acetonitrile
AG	activated graphene
ALD	atomic layer deposition
CMC	carboxymethyl cellulose
CMES	carbon microelectrochemical system
CNF	carbon nanofiber
CNT	carbon nanotube
CPE	composite polymer electrolyte
CV	cyclic voltammetry
CVD	chemical vapor deposition
DEC	diethyl carbonate
DMC	dimethyl carbonate
DMDMS	dimethyldimethoxysilane

Microsupercapacitors
https://doi.org/10.1016/B978-0-08-102888-9.00003-3

DVD	digital versatile disc
EC	ethylene carbonate
EDOT	ethylenedioxythiophene
EES	electrochemical energy storage
ESR	equivalent series resistance
Fr	flash reduction
GO	graphene oxide
GPE	gel polymer electrolytes
GQD	graphene quantum dot
HMPP	2-hydroxy-2-methylpropiophenone
HPE	hybrid polymer electrolyte
HQ	hydroquinone
IDA	interdigitated array
IoTs	internet of things
ISE	inorganic solid electrolyte
LbL	layer-by-layer
LiPON	lithium phosphorous oxy-nitride
LISICON	lithium super ionic conductor
LiTf	lithium triflate
LSG	laser-scribed graphene
LTO	lithium titanium oxide ($Li_4Ti_5O_{12}$)
MP	MnO_2/PEDOT
MSC	microsupercapacitor
MWCNT	multiwalled carbon nanotube
OLC	onion-like carbon
PAN	polyacrylonitrile
PANI	polyaniline
PC	propylene carbonate
PDMS	polydimethylsiloxane
PE	polymer electrolytes
PEDOT	poly(3,4-ethylenedioxythiophene)
PEGDA	poly(ethylene glycol) diacrylate
PEO	polyethylene oxide
PET	polyethylene terephthalate
PMMA	polymethylmethacrylate
ppy	polypyrrole
PS	polystyrene
PVA	polyvinyl alcohol
PvDF -HFP	poly(vinylidene fluoride)-*co*-hexafluoropropylene
PvDF	poly(vinylidene fluoride)
RF	radio frequency
rGO	reduced graphene oxide
RT	room temperature
RTIL	room temperature ionic liquids
SC	supercapacitor
SEM	scanning electron microscopy
SiCNW	SiC nanowire

SPEs	solid polymer electrolytes
SSE	solid-state electrolyte
SWCNT	single walled carbon nanotube
$TEABF_4$	tetraethylammonium tetrafluoroborate
T_g	glass transition temperature
TMOS	tetramethyl orthosilicate
UV	ultraviolet
YSZ	yttria-stabilized zirconia

3.1 Introduction

Despite the advances pertaining to the electrode materials and device fabrication technologies, one of the bottlenecks pushing back the full-fledged application and commercialization of microsupercapacitors (MSCs) is the lack of suitable electrolytes [1, 2]. Generally, an electrolyte is expected to be safe, economic as well as easily integrable with the electrochemical device without compromising the performance so that it can be imparted to commercial applications. In the context of an electrolyte, safety refers to more than one aspect such as nonflammability, environmental friendliness, chemically inert to the other cell components such as separator, electrode substrate and packaging materials, nontoxicity, free from leakage, evaporation, and gas evolution during the operation of the electrochemcial cell, and so on [2–4]. In majority of the cases, conventional electrolytes are electrolytic solution-type, which are liquids in the service-temperature range consisting of mobile ions generated by means of the dissociation of salts/acids/alkalis (electrolyte solutes) in solvents such as water (aqueous) or organic molecules (nonaqueous, also called aprotic solvents) [5]. These mobile ions can direct the movement of charges and the property is known as ionic conduction. In other words, all the electrolytes are ionic conductors.

In the case of liquid electrolytes, they demand sophisticated and expensive packaging norms (in the context of this chapter, packaging refers to all the processes involved in the encapsulation of the fabricated electrochemical device followed by sealing) to be integrated with microdevices due to their free-flowing nature despite high ionic conductivity ($\geq 10^{-3}$ S cm^{-1} at room temperature (RT)) [5–7]. Still, the risks of leakage, corrosive and toxic nature, and even flammability associated with several aqueous and nonaqueous electrolytes remain a grave concern [8, 9]. Moreover, liquid electrolytes limit the prospect of flexibility (due to the lack of solid-like operability and possible exudation of liquid electrolyte from the device during mechanical deformations), which is very much important in the industry of smart and flexible electronics [10].

Considering the constraints imparted by liquid electrolytes, well known from battery and supercapacitor (SC) R&D [11–14], a transition to solid-state electrolytes (SSEs) is highly desirable for fueling the progress of MSCs [15]. SSEs possess advantages over liquid electrolytes in terms of safety, mechanical, and thermal properties. Since they are in most cases inherently less reactive compared to liquid components, the side reactions with the cell components will be minimized leading to improved shelf-life of the MSC device. They enhance the high temperature usability of the MSC devices, at which conventional aqueous and organic solvent-based liquid electrolytes decompose, ignite, and/or evaporate [16]. Moreover, SSEs can be easily integrated with flexible architectures avoiding the constraints imposed by free-flowing nature of liquid electrolytes as already mentioned.

SSEs are broadly classified into solid polymer electrolytes (SPEs) and inorganic solid electrolytes (ISEs). An SPE can be defined as a polymer system, which can facilitate the conduction of ions within a polymeric matrix. In the case of SPEs, the polymer matrix helps in the dissociation of the salt and the ion conduction generally occurs through hopping mechanism, which is coupled or decoupled with the segmental motion of polymer chains (or polymer chain dynamics) [17, 18]. The first of its kind was SPEs proposed by P.V. Wright et al. in 1973 with alkali-metal salts incorporated into a semi-crystalline polyethylene oxide (PEO) [19, 20]. However, high crystallinity of the polymer matrix in SPEs arrests the polymer chain dynamics and leads to very low ionic conductivity (PEO-based Li^+ ion conducting SPEs display low ionic conductivity in the order of 10^{-8}–10^{-4} S cm^{-1} at room temperature (RT)) [21]. In order to overcome the limitation of ionic conductivity in SPEs, the concept of gel polymer electrolytes (GPEs) is introduced [22]. In a GPE, the polymer matrix encompasses a liquid solvent/plasticizer along with the conducting species, which facilitates ion mobility and increases the amorphous regions (or in other words, segmental mobility of polymer chains is improved due to reduced crystallinity) of the macromolecular system [22–25]. Aqueous and organic solvents or even room temperature ionic liquids (RTILs, they are room temperature molten salts, which can be also used directly as liquid electrolytes in electrochemical devices) can be used for gelating the polymer matrix (SPEs and GPEs are together called polymer electrolytes (PEs)) [26]. In the context of MSCs, when RTIL is used as the gelating agent, the formed gel electrolytes are called ionogels [27]. Due to advantages pertaining to ionic conductivity, most of the available reports on MSCs are dominated by GPEs, ionogels, and liquid electrolytes.

On the other hand, inorganic crystalline, polycrystalline, and amorphous solids, ceramics, glasses, glass-ceramics, etc., possessing mobile ions are known as inorganic solid-state electrolytes (ISEs) [16, 28]. In the case of crystalline ISEs, the movement of mobile ions takes place through hopping mechanism where the ions migrate between the rigid crystal lattices. However, the ion conduction mechanism in amorphous ISEs is not well established [16]. Several ISEs are reported to possess RT ionic conductivities comparable to that of liquid electrolytes. For example, several sulfide glasses, thio-LISICON, and glass-ceramics can exhibit RT ionic conductivities in the order of 10^{-2}–10^{-3} S cm^{-1} [28, 29]. However, high-cost issues pertaining to tedious processing, and most importantly, the high charge-transfer resistance due to inferior electrode/electrolyte interface (explained in Section 3.3) still remain the key challenges to overcome for ISEs to be extensively employed in electrochemical devices [16, 30–32]. Both SPEs and ISEs can be together called all-SSEs since they are free from any solvent or other components in liquid phase. The classification of electrolytes in the context of MSCs is summarized in Fig. 3.1. The general properties of these various types of electrolytes under discussion are tabulated in Table 3.1.

The primary role of an electrolyte in any electrochemical device is to act as a mediator of charge transfer in the form of ions. How fast the charge can be transferred between a pair of electrodes decides how fast the energy can be released or stored [5]. This is also called power capability or more specifically power density of the electrochemical energy device. This is in fact dependent upon the ionic conductivity of the electrolyte. Most of the all-SSEs exhibit low ionic conductivity compared to the liquid electrolyte

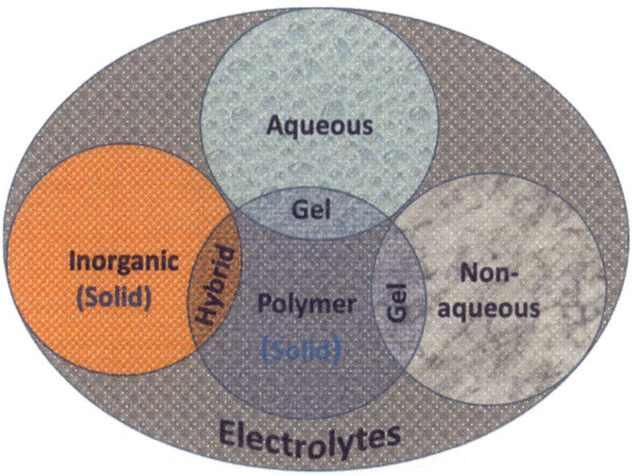

Fig. 3.1 A general classification of electrolytes used in MSCs.

Table 3.1 General properties of various types of electrolytes used in electrochemical energy storage devices.

Electrolyte	Ionic conductivity	Mechanical strength	Flammability	Toxicity
Aqueous	High	Low	None	Low
Nonaqueous	High	Low	High	Medium
Gel polymer	High	Low	Medium	Medium
Solid polymer	Low	High	Low	Medium
Composite polymer	Medium	High	Low	Medium

counterpart hampering the power output of MSC devices. In the case of ISEs, they have been employed for the fabrication of thin-film MSCs [33]. However, SPEs are yet to be used for MSC fabrication owing to the surplus of available GPEs. However, recent advancements in SPEs with high ionic conductivity matching with GPEs, yet solvent-free processing along with device fabrication techniques such as in situ polymerization, could be useful for future MSCs [34–37]. Toward designing high-energy MSC devices, redox-active liquid electrolytes and GPEs can also be employed [38–41].

3.2 Types of MSCs and diffusion of electrolyte ions

MSCs can be broadly classified into 2D MSCs and 3D MSCs depending on the microelectrode architecture [33, 42–44]. In the context of MSCs, a 3D microelectrode is nothing but a collection of nanoscale or microscale building blocks with a surface exposed in 3D, whereas a 2D microelectrode is usually a thin film with a planar surface. Although 2D microelectrodes are typically thin films, recently, thick films are also employed for increasing the loading per unit footprint so that higher energy density than with 2D-thin-film MSCs can be achieved. In a recent review by Brousse et al., thick-film electrodes were classified as those with a thickness $\geq 5 \mu m$ [1]. The two main topologies adopted for the fabrication of MSCs devices are parallel plate and interdigitated configuration [1]. MSCs adopting parallel plate topology are sandwiched MSCs (also called stacked- or parallel plate-MSCs), which can be considered as conventional parallel plate macro-supercapacitors (SCs) reduced into microdimensions.

In sandwiched MSCs, the microelectrodes are stacked over each other where the contact between the electrodes is avoided by means of a separator (usually the separator is soaked in a liquid electrolyte prior to cell fabrication) to prevent the short circuit. If SSEs are used, the requirement of a separator can be avoided since the SSE plays the dual role of an electrolyte as well as separator. Usually, a conformal coating of the SSE is good enough to avoid the device short circuit, which can help in further reducing the thickness of MSC devices [45]. The schematic representation of 2D- and 3D-sandwiched MSC devices are presented in Fig. 3.2A and D, respectively. The sandwiched configuration is considered as the most effective topology when the available footprint area is limited. The critical drawback of the sandwiched MSC configuration is its large volume owing to the stacking of multiple electrodes. In the case of in-plane topology (MSCs possessing this topology can be called in-plane MSCs), unlike a sandwiched MSC, both microelectrodes are patterned over the same plane of a single substrate in an interdigitated fashion (Fig. 3.2B and C). Compared to the in-plane geometry of the aforementioned MSCs, the sandwiched MSCs can be considered to possess an out-of-plane geometry. In the case of in-plane interdigitated MSCs, the active electrode area is reduced at least by a factor of two compared to the sandwiched MSC owing to the spacing between two neighboring microelectrodes [1, 33].

The nature of ion diffusion in the case of MSCs can depend upon whether it possesses an in-plane or stacked topology (Fig. 3.2E–G). As it can be seen, in the case of sandwiched MSCs, the ion diffusion occurs vertically to the substrate/current collector, i.e., across the microdevice (Fig. 3.2E and F), whereas in the case of in-plane MSCs, the ion diffusion takes place parallel to the planar substrate (Fig. 3.2G) [15, 46–48]. In the context of Fig. 3.2E and F, the ion diffusion in 2D-nanosheet-like microelectrodes (e.g., graphene) is represented. However, similar type of directionality in ion diffusion is applicable to any types of electrode materials when they are fabricated in stacked or in-plane topology. However, it should be noted that this directionality is in fact a bulk property when whole MSC device is considered but the nature of ion diffusion within the interior of the active material may vary. For example, Fig. 3.2E represents the microelectrodes of a sandwiched 2D MSC with 2D nanosheets facing to each other in a stacked fashion. Here, the ion diffusion across the nanosheets is found to be hindered. At the same time, when similar device is fabricated with the same 2D-nanosheets aligned in a vertical fashion, the ion diffusion is found to be facilitated between the intersheet spaces (Fig. 3.2F). In the case of in-plane MSCs with 2D-nanosheets arranged in stacked manner (Fig. 3.2G), the ion diffusion is facilitated throughout the electrode material unlike in the case of sandwiched MSC represented in Fig. 3.2E. This

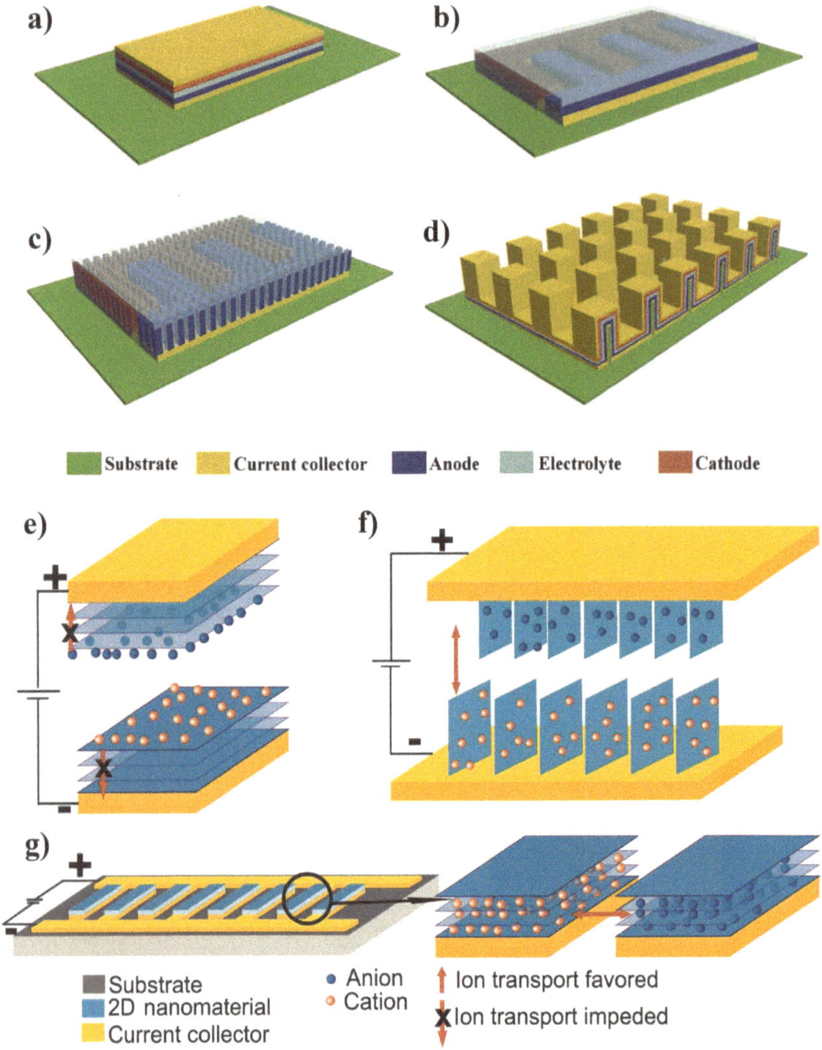

Fig. 3.2 Four types of MSC configurations: (A) 2D-sandwich (parallel plate), (B) 2D-in-plane (interdigitated), (C) 3D-in-plane (interdigitated), and (D) 3D-sandwich [33], (E) and (F) representing the nature of ion transport in sandwiched MSCs, where ion diffusion occurs vertically to the substrate, or in other words across the electrodes. (G) The ion diffusion in the case of in-plane MSCs occurs parallel to the planar substrates [46]. *(A)–(D) Reproduced with permission from S. Zheng, X. Shi, D. Das, Z.-S. Wu, X. Bao, The road towards planar microbatteries and micro-supercapacitors: from 2D to 3D device geometries, Adv. Mater. 31 (50) (2019) 1900583, https://doi.org/10.1002/adma.201900583, Copyright (2019), Wiley-VCH; (E)–(G) Reproduced with permission from reference B. Mendoza-Sánchez, Y. Gogotsi, Synthesis of two-dimensional materials for capacitive energy storage, Adv. Mater. 28 (29) (2016) 6104–6135, https://doi.org/10.1002/adma.201506133, Copyright (2016), Wiley-VCH.*

difference in the nature of ion diffusion in the nanoscale dimension in turn results in different electrochemical performance exhibited by each type of the aforementioned devices under consideration. Recently, the aspect ratio of microelectrodes affecting the rate performance of MSC in liquid electrolytes is also studied by using carbon nanotube (CNT)-based microelectrodes as a model system [49]. The results suggest that the increase in aspect ratio of the microelectrodes can help in achieving MSCs with high rate capability. At the same time, there are several other studies related to the confinement of ions in the nanopores of electrode materials and ion dynamics in carbon-based materials using sophisticated in situ techniques such as nuclear magnetic resonance (NMR) spectroscopy, small-angle X-ray scattering experiments (SAXS), and electrochemical quartz crystal microbalance (EQCM) [50–53]. All these results shine light on the importance of various electrode geometries, patterns, designing, and fabrication methods in the context of SCs and MSCs with which maximum utilization of the active material by the electrolyte (solid, liquid, and gel) and an optimum device performance can be achieved. An apt example is the evolution of 3D microelectrode designs by which the active material utilization can be improved so that the energy density of an in-plane MSCs can match with 2D MSCs in sandwiched configuration. Such basic studies are further fueling the development of MSCs with improved performance so that potential industrial scope can be envisaged in the future.

3.3 Challenges during the transition from liquid to SSEs

3.3.1 Liquid electrolytes and electrode/electrolyte interfaces

In any electrochemical device, electrode/electrolyte interfaces are the most important factor influencing the performance metrics [30, 54, 55]. In the case of MSCs as well, the requirement is the same irrespective of the device configuration. The main intention behind the transition from 2D to 3D MSCs is to take advantage of the "height" or "thickness," which is the third dimension, that a high mass loading of the active material can be achieved in a minimum area footprint of the device [42, 52]. However, achieving high mass loading at the electrode level alone cannot ensure a high-energy density; rather how much of the active material is utilized during the charge-discharge process is more important. When a liquid electrolyte (aqueous, organic, or ionic liquid) is employed for the fabrication of MSCs, it can effectively contribute to wetting of the electrode surface and bulk (the latter, through sufficient electrode porosity) [34, 35]. This property of the liquid electrolyte is presented in Fig. 3.3A. The liquid electrolyte owing to its

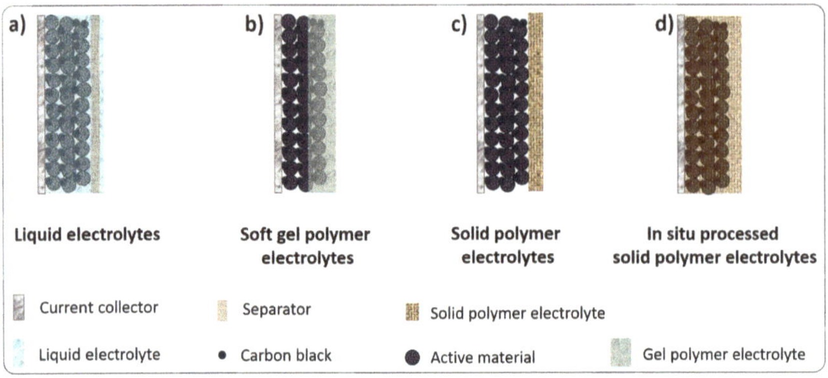

Fig. 3.3 Electrode/Electrolyte interfaces in (A) liquid electrolyte, (B) soft gel polymer electrolytes, (C) solid polymer electrolytes, and (D) in situ processed solid polymer electrolytes.

inherent free-flowing nature as well as its low viscosity can easily penetrate deep into the electrode thickness (or bulk regions of the electrode material). This enhances the active material utilization resulting in best electrochemical characteristics. The enhanced electrochemical performance indirectly indicates that electrode/electrolyte interfaces in liquid electrolytes are the most conformal and continuously intimate throughout operation that can be achieved in an electrochemical cell. Especially, in the case of a 3D or thick-film MSCs, such an intimate and effective electrode/electrolyte interface is inevitable for achieving increased energy density and power density by maximizing the utilization of the active material.

3.3.2 Solid-state electrolytes and electrode/electrolyte interfaces

The fabrication of liquid electrolyte-based MSC devices is very trivial in approach, where a porous separator (e.g., polypropylene, glass-fiber, cellulose-based filter paper, etc.) soaked with a suitable electrolyte is used for sandwiched MSCs. In the case of interdigitated MSCs, the electrodes are simply submerged in the liquid electrolyte followed by packaging. Although these methods are suitable in the lab scale, the inferior safety associated with liquid electrolytes casts a shadow over the reliability of such MSC designs. As already mentioned above, the ultimate target toward safe and upscalable MSCs is to go for all-solid-state designs. But along with the low ionic conductivity of all-SSEs, the critical factor impeding such designs to enter the mainstream is their inferior electrode/electrolyte interface as compared to liquid electrolytes.

Since all-SSEs are generally used as thin films separating the two electrodes in MSCs, the interaction between electrode and electrolyte is confined to the active materials at the surface of the electrode and not to the interior, which leads to low active material utilization and reduced energy density. Poor current rate performance in the case of all-SSE-based MSCs or any other electrochemical devices in general, when thick- or 3D electrodes are employed, is due to the poor interaction between the electrolyte components and active materials present at the bulk of the electrode [56]. Mostly ISEs are employed for the fabrication of all-solid-state MSCs. As of now, SPEs have not been employed for MSC applications. However, the applications of all-SSEs are mostly confined to sandwiched thin-film MSCs due to the sleek nature of thin films, which provides a short-diffusion path, hence can offset the limitations arising from the low ionic conductivity of all-SSEs [57]. However, their limitations often prevent them from being used for in-plane 2D- and 3D-MSCs.

In this aspect, moving from all-SSEs to quasi-solid-state GPEs is attempted and it is generally considered that they are highly suitable for MSCs. Although the ultimate aim of achieving all-solid-state MSCs is compromised, a variety of GPEs has been used in MSC fabrication and found to perform excellently. The ionic conductivity of many of the GPEs is far higher than known for all-SSEs, and sometimes also comparable to liquid electrolytes [13,58,59]. Owing to the presence of liquid phase, they can wet the electrode surface more effective than all-SSEs, at the same time, providing solid-like operability and electrode/electrolyte interfaces close to liquid electrolytes. The electrode/electrolyte interface, which can be achieved with GPEs and all-SSEs when used as thin-film or viscous solution (in the case of GPEs) is schematically shown in Fig. 3.3B and C. Compared to the electrode/electrolyte interface achieved with liquid electrolytes, all-SSEs can access only the active material on the surface of the electrode material, whereas the electrode/electrolyte interface of the GPEs lies intermediate between liquid and all-SSEs.

Recent reports on the in situ generation of SPEs and GPEs by ultraviolet (UV) light or thermal exposure can help in further improving the electrode/electrolyte interface [37,60,61]. In this method, unlike the conventional strategy of using a SPE or GPE as a film, a reactive solution containing polymerizable monomers, conducting salt, solvent (if any), and polymerization initiator (photo, thermal, or ionic initiators) is uniformly spread over the microelectrode surface. Due to the liquid nature of the reactive solution, it behaves as if a conventional liquid electrolyte easily impregnates into the bulk and interiors of the electrode material. This is followed by triggering the polymerization by external energy source so that the generation of PE is directly taken place on the microelectrode with the electrolyte having accessibility to the bulk and interiors of the electrode material contributing to maximum active material utilization. By means of adopting this method, even the electrode/electrolyte interface in an SPE can be made close to that of liquid electrolytes in MSC device. In Fig. 3.3D, the in situ generated electrode/electrolyte interface in a SPE integrated electrode is presented, which possess an ideal electrode/electrolyte interface as in the case of liquid electrolytes. Fig. 3.4 summarizes the major breakthroughs in the field of MSCs in the context of electrolyte research since the inception of ISE-based thin-film SCs (thin-film SCs can be considered as the predecessor of presently using MSCs) in early 2000s to the latest models of lithium and zinc-ion MSCs of the late 2019.

3.4 Electrolytes for MSCs

3.4.1 ISEs

Inorganic electrolytes are historically important since earlier models of all-solid-state-sandwiched thin-film MSC appeared in the reports of Kim et al. in 2003 using amorphous LiPON (chemical composition is $Li_xPO_yN_z$)-based ISE (Fig. 3.5) [63]. In fact, this report is considered as the first work coining the term MSC followed by which other types of electrolytes (liquid electrolytes, PEs, etc.) were used for the fabrication of the same. In this work, tungsten-cosputtered RuO_2 (W-RuO_2) on Pt current collector is used as the electrode material. The fabricated MSC device displayed an areal capacitance of 54 mF cm^{-2}. However, the very low ionic conductivity of LiPON (~10^{-6} S cm^{-1} at room temperature (RT)) leads to poor rate capability of the device. The low RT ionic conductivity of the ISE electrolyte is generally considered as an intricacy toward its transformation from lab to pilot production. Prior to this report, LiPON has been used for the fabrication of thin-film microbatteries [71, 72]. As already mentioned, although this report coined the term MSC for the first time, there had been already a few other reports from the same group on thin-film supercapacitor employing sputtered cobalt oxide (Co_3O_4) and RuO_2 and similar LiPON ISEs [62,73,74].

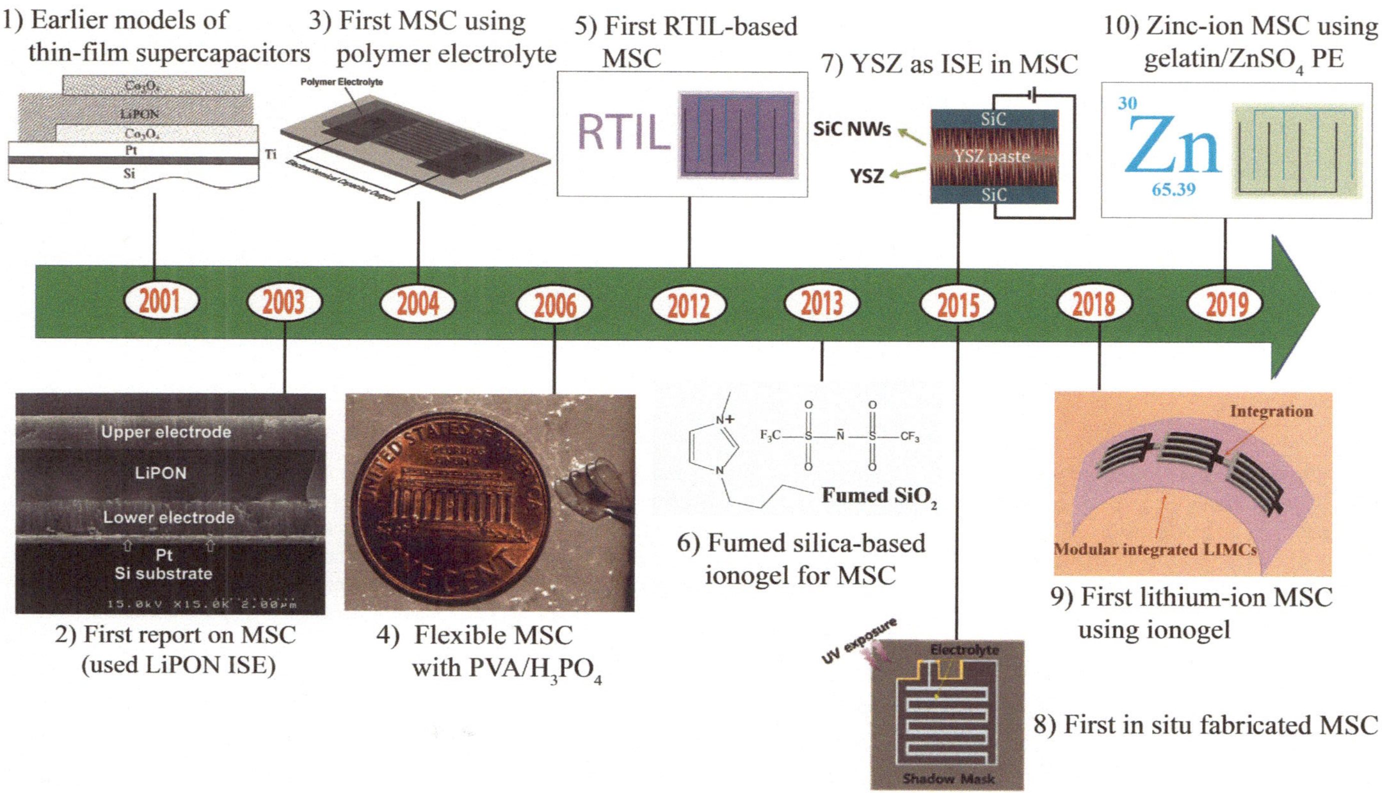

Fig. 3.4 Major breakthroughs in the electrolytes for MSC technology since its inception in the early 2000s–2019. (1) Early models of thin-film SCs [62], (2) first report on MSC [63], (3) first MSC using PE [64], (4) flexible MSC based on PVA-H_3PO_4 [65], (5) RTIL-based MSC [66], (6) fumed silica-based ionogel [67], (7) YSZ-based MSC [68], (8) in situ fabricated MSC [37], (9) lithium-ion MSC [69], and (10) zinc-ion MSC [70].

(Continued)

Fig. 3.4, cont'd *((1) Reprinted from H.-K. Kim, T.-Y. Seong, J.-H. Lim, W. Ii Cho, Y. Soo Yoon, Electrochemical and structural properties of radio frequency sputtered cobalt oxide electrodes for thin-film supercapacitors, J. Power Sources 102 (1) (2001) 167–171, https://doi.org/10.1016/S0378-7753(01)00864-3, Copyright (2001), with permission from Elsevier; (2) Reprinted from H.-K. Kim, S.-H. Cho, Y.-W. Ok, T.-Y. Seong, Y.S. Yoon, All solid-state rechargeable thin-film microsupercapacitor fabricated with tungsten cosputtered ruthenium oxide electrodes, J. Vacuum Sci. Technol. B 21 (3) (2003) 949–952, Copyright 2003, American Vacuum Society; (3) Reprinted from J.-H. Sung, S.-J. Kim, K.-H. Lee, Fabrication of all-solid-state electrochemical microcapacitors, J. Power Sources 133 (2) (2004) 312–319, https://doi.org/10.1016/j.jpowsour.2004.02.003, Copyright (2004), with permission from Elsevier; (4) Reprinted from J.-H. Sung, S.-J. Kim, S.-H. Jeong, E.-H. Kim, K.-H. Lee, Flexible micro-supercapacitors, J. Power Sources 162 (2) (2006) 1467–1470, Copyright (2006), with permission from Elsevier; (7) Reprinted with permission from C.-H. Chang, B. Hsia, J.P. Alper, S. Wang, L.E. Luna, C. Carraro, S.-Y. Lu, R. Maboudian, High-temperature all solid-state microsupercapacitors based on SiC nanowire electrode and YSZ electrolyte, ACS Appl. Mater. Interfaces 7 (48) (2015) 26658–26665, https://doi.org/10.1021/acsami.5b08423, Copyright (2015). American Chemical Society; (8) Reprinted with permission from D. Kim, G. Lee, D. Kim, J.S. Ha, Air-stable, high-performance, flexible microsupercapacitor with patterned ionogel electrolyte, ACS Appl. Mater. Interfaces 7 (8) (2015) 4608–4615, https://doi.org/10.1021/am5077843, Copyright (2015). American Chemical Society; (9) Republished with permission of Royal Society of Chemistry, from S. Zheng, J. Ma, Z.-S. Wu, F. Zhou, Y.-B. He, F. Kang, H.-M. Cheng, X. Bao, All-solid-state flexible planar lithium ion micro-capacitors, Energ. Environ. Sci. 11 (8) (2018) 2001–2009, https://doi.org/10.1039/C8EE00855H, Copyright (2018); permission conveyed through Copyright Clearance Center, Inc.)*

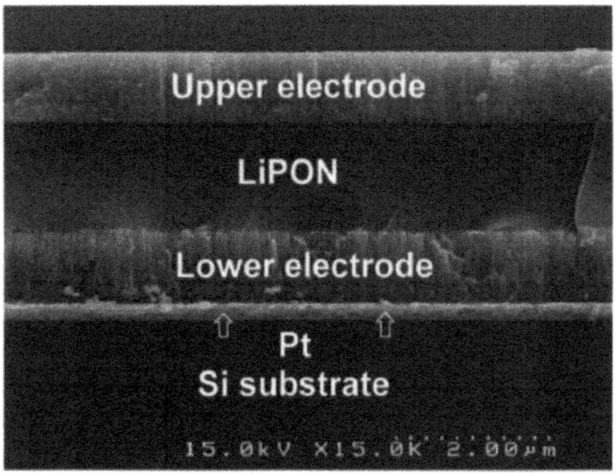

Fig. 3.5 Cross-sectional scanning electron microscopy (SEM) image representing the electrode/electrolyte interfaces in a LIPON-based thin-film MSC [63]. *Reprinted from H.-K. Kim, S.-H. Cho, Y.-W. Ok, T.-Y. Seong, Y.S. Yoon, All solid-state rechargeable thin-film microsupercapacitor fabricated with tungsten cosputtered ruthenium oxide electrodes, J. Vacuum Sci. Technol. B 21 (3) (2003) 949–952, Copyright 2003, American Vacuum Society.)*

Ta_2O_{5+x} solid-oxide thin-film possessing proton conductivity is used as an ISE for thin-film-sandwiched MSC fabrication [75]. Here, RuO_2 is deposited onto a Pt current collector by means of radiofrequency (RF) sputtering. Later, the same sputtering method is used for depositing a thin film of Ta_2O_5 over RuO_2. The sputtering is carried out in a hydrogen atmosphere to induce proton conductivity. However, the specific capacitance value achieved for the MSC device is very low, which makes the electrolyte not feasible for practical application. Apart from the aforementioned reports in the early 21st century, the research in the field of MSC based on ISEs has been dormant till Chang et al. reported a yttria-stabilized zirconia (YSZ)-based electrolyte for MSC in 2015 [68]. YSZ, an oxide ion conducting SSE, is used for high temperature sandwiched MSC based on SiC nanowire (SiCNW) electrodes grown on Si wafer. Considering the nanowire architecture of SiC, it can be considered as a 3D-sandwiched MSC. In other words, this is the only 3D MSC based on an all-solid-state electrolyte available in the literature so far. Here, the coating of YSZ electrolyte over the 3D microelectrode is achieved by means of a sol-gel method (Fig. 3.6). The 3D MSC displayed an areal capacitance of 92 $\mu F\ cm^{-2}$ when operated at 450 °C.

LiPON-based sandwiched thin-film MSC made reentry into the MSC arena in a report by Schmidt et al. in 2017 [76], which is based on Co_3O_4 as

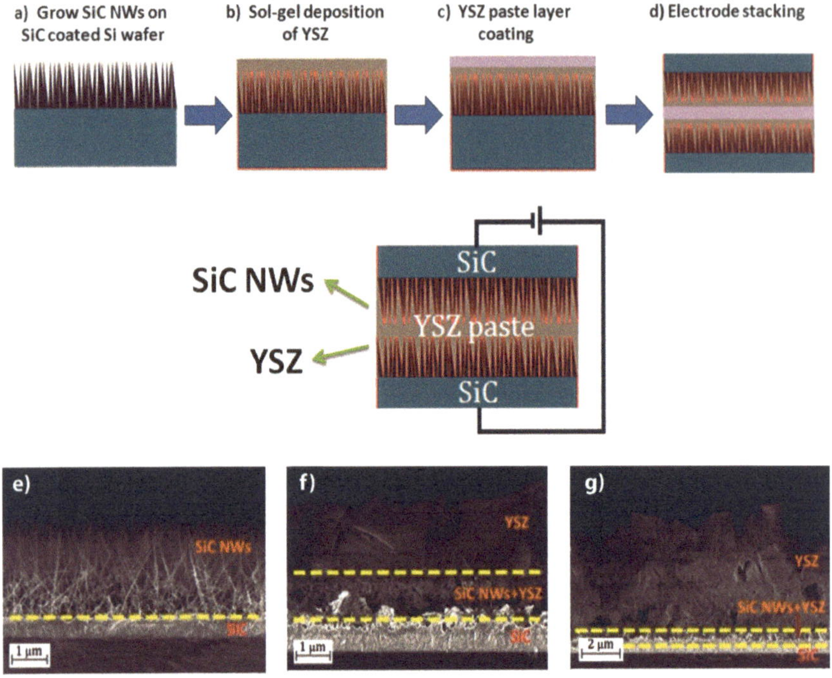

Fig. 3.6 (A–D) Scheme representing the stepwise fabrication of SiCNW|YSZ|SiCNW sandwiched MSC device; cross-sectional SEM images of (E) SiCNW microelectrode, (F) YSZ over SiCNW (one cycle of deposition), and (G) YSZ over SiCNW after 20 cycles of deposition [68]. *(Reprinted with permission from C.-H. Chang, B. Hsia, J.P. Alper, S. Wang, L.E. Luna, C. Carraro, S.-Y. Lu, R. Maboudian, High-temperature all solid-state microsupercapacitors based on SiC nanowire electrode and YSZ electrolyte, ACS Appl. Mater. Interfaces 7 (48) (2015) 26658–26665, https://doi.org/10.1021/acsami.5b08423, Copyright (2015). American Chemical Society.)*

the electrode material. The device fabrication is carried out on a Si/SiO_2 substrate over which Cr (current collector) is deposited by electron beam evaporation. The cross-section of the sandwiched device components is shown in Fig. 3.7. With low discharge current densities at RT, the MSC device exhibits a volumetric capacitance of about 14 F cm^{-3}. The work claims to possess the highest energy (8 mW h cm^{-3}) and power (15 W cm^{-3}) densities higher than many of the MSCs already published [77]. Recently, it is seen that techniques such as atomic layer deposition (ALD) are being used for tuning the thickness and properties of LiPON, which has been culminated in areal capacitance value of 100 μF cm^{-2} [78].

The aforementioned reports sum up the research works on ISE-based all-solid-state MSCs (summarized in Table 3.2). Ever since the introduction in

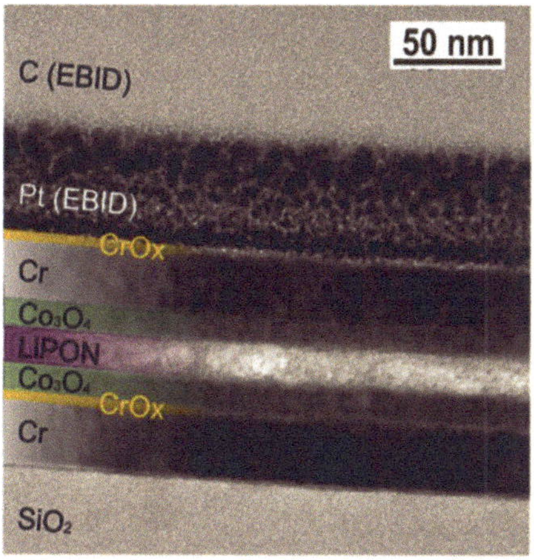

Fig. 3.7 Cross-sectional transmission electron microscopy (TEM) image of the sandwiched MSC device. The interface between different components of the MSC device is also shown in the image [76]. *(Reprinted from T. Göhlert, P.F. Siles, T. Päßler, R. Sommer, S. Baunack, S. Oswald, O.G. Schmidt, Ultra-thin all-solid-state micro-supercapacitors with exceptional performance and device flexibility, Nano Energy 33 (2017) 387–392, https://doi.org/10.1016/j.nanoen.2017.01.054, Copyright (2017), with permission from Elsevier.)*

the early 2000s, there has not been any significant progress in this research area. The low ionic conductivity, as well as tedious synthesis methods, pulls back the practicality of ISEs in the context of MSCs. With the surge in interest in all-solid-state alkali-ion batteries (e.g., lithium-ion and Na-ion batteries), there have been many new Li^+-ion conducting solid-state ISEs introduced in the past several years [79–81]. However, they are yet to experiment with MSC fabrication. Despite the availability of several Li^+ or Na^+-ion conducting ISEs, H^+, and OH^-, ion conducting ISEs are very rare. The most explored H^+-conducting ISEs are Keggin-type heteropolyacids (e.g., Vanadium heteropoly acids and tungsten heteropoly acids), which need high humidification conditions to exhibit proton conductivity, hence not suitable for MSCs [82, 83]. The requirement of high humidification in the case of Keggin-type heterpoly acids is due to the vehicle-type proton conducting mechanism, which follows through the interaction between proton and water molecules [84]. This interaction results in the generation of species such as H_3O^+, H_3O^+, $H_5O_2{}^+$, and $H_9O_4{}^+$ by means of which proton conduction occurs and the amount of which in fact depend upon the water content within the crystal structure (proton conduction mechanism in GPEs is explained in Section 3.4.3.1).

Table 3.2 Components, fabrication, characteristics, and performance summary of ISE-based MSCs.

Electrolyte	Electrolyte fabrication method	Ion	σ (S cm^{-1})	Electrode	Substrate	Microelectrode preparation method	Capacitance (areal/ volumetric)	Voltage (V)	Refs.
LiPON	RF sputtering	Li^+	$\sim 10^{-6}$ (RT)	W-RuO_2	Pt	Reactive sputtering	54 mF cm^{-2}	0–2	[63]
Yttria-stabilized zirconia (YSZ)	Sol-gel deposition method	Oxides	–	SiCNW	Si	CVD	92 μF cm^{-2} (450 °C)	−1–1	[68]
LiPON	Li_3PO_4 target sputtered in N_2 atm	Li^+	6.8×10^{-8} (RT)	Co_3O_4	Si/SiO_2/Cr	Reactive sputtering	14 F cm^{-3}	0–2	[76]
LiPON	ALD	Li^+	–	TiN and Ti	Si	–	100 μF cm^{-2}	0–3	[78]

Although the research in MSCs started with ISEs, its unreliability toward practical applications forced the researchers to go for alternative electrolytes. Hence, conventional liquid electrolytes started to be adopted for MSC fabrication. Since the introduction of the concept of MSCs, there has been a surge in reports based on aqueous, organic, or ionic liquid-based liquid electrolytes for MSC fabrication. The following section attempts to give a detailed idea of various liquid electrolytes that have been used for the fabrication of MSCs with several examples quoted from literature.

3.4.2 Liquid electrolyte for MSCs

Inorganic acids, base, or salts dissolved in an excess amount of solvents (protic or aprotic) are the commonly used liquid electrolytes for MSCs. For the fabrication of the MSC device, the liquid electrolytes are dropped onto the electrodes followed by encapsulation of the device. Although liquid electrolytes are not commercially viable in terms of MSCs, they have helped in providing insights into the potential of several types of electrode materials as well as device fabrication strategies that can be used in MSC technology. Hence, liquid-electrolyte-based MSC devices have played a critical role in the evolution of current MSC-based technology mainly depended upon quasi-solid-state GPEs. This section delivers brief overview of the advantages and drawbacks of various types of liquid electrolytes (i.e., aqueous, organic, and ionic liquids) explored to date.

3.4.2.1 Aqueous electrolytes

Aqueous electrolytes are benefitted with high ionic conductivity (0.5–1 S cm^{-1} at 25°C), environmental benignity and easy handling, which simplify the device assembling process [85]. The low viscosity (1 centipoise (cP) at for 1 M H_2SO_4 and >2.5 cP for 1 M $LiClO_4$/ethylene carbonate (EC) + dimethyl carbonate (DMC) and 1 M $TEABF_4$/propylene carbonate (PC) at RT) and faster ion mobility result in low equivalent series resistance (ESR) in MSC devices, contributing to high specific capacitance, energy, and power density [86–88]. However, aqueous electrolytes do not meet the commercial requirements due to their low electrochemical stability limiting the operating voltage window of the device in between 0 and 1.23 V [89]. In the case of a SC device, the energy density (E) is usually represented by means of the following equation:

$$E = \left(\frac{1}{2}\right) CV^2 \tag{3.1}$$

where C represents the specific capacitance and V represents the operating voltage window of the SC device. According to this equation, it can be seen that doubling the voltage window can result in four times increment in the energy density of the device provided that "C" is kept constant. Therefore, nonaqueous electrolytes exhibiting high electrochemical stability are used with which SC devices operable beyond 2.5 V can be realized and are preferred over aqueous electrolytes [34]. However, nonaqueous electrolytes require complex purification treatments as well as a controlled environment to keep away from impurities and moisture making them more expensive. In this aspect, aqueous electrolytes are still important when moving toward cheaper devices in the context of MSCs. Normally, a diluted solution of H_2SO_4, KOH, or inorganic salts in water is employed as electrolytes together with various electrode materials, both pseudocapacitive as well as carbon-based electrode materials. This class of electrolytes is further grouped into three categories: acidic, alkaline, and neutral. As the name implies, acidic electrolytes are usually made by dissolving inorganic acids (H_2SO_4, H_3PO_4, etc.) in water. They are mainly employed for conducting protons (H^+ ions). Similarly, aqueous solution of molecules such as NaOH, KOH, and LiOH is known as alkaline electrolytes, which can conduct hydroxide (OH^-) ions. Due to the corrosive nature of acidic and alkaline electrolytes, they can result in reduced life-span of the electrochemical device, which can be overcome by employing neutral electrolytes (aqueous solutions of salts (NaCl, KCl, Na_2SO_4, etc.)).

Acidic liquid electrolytes

Acidic liquid electrolytes, in particular, 1 M H_2SO_4, are frequently used in MSCs because of their excellent ionic conductivity (0.8 S cm^{-1} for 1 M H_2SO_4 at 25°C) [90]; they enable high capacitance and low ESR values depending on the electrode materials employed. Due to their low hydrated radius (hydrated radius represents the size of the ion including the closely bounded water molecules surrounding them), protons can easily diffuse into the carbon micropores, which enhance the utilization of electrode surface for ion adsorption. During the past several years, different types of carbons, such as carbide-derived carbon [91], activated carbon (AC) [92], carbon nanotubes (CNT) [93], and so on, are tested as electrode materials in H_2SO_4-based liquid electrolyte for SC applications [94]. In a recent report, carbide-derived carbon film (8 μm thickness) is prepared through a chlorination process using magnetron sputtered TiC film as the starting material. It is observed that for a microporous carbon electrode, a higher specific capacitance than in neutral electrolytes can be achieved with 1 M H_2SO_4. The free-standing elastic carbon film exhibits high areal capacitance of 240 mF cm^{-2} in 1 M H_2SO_4 electrolyte [95]. To improve the capacitance of carbon-based MSCs, the electrolyte can be further modified with redox-active species (e.g., KI, hydroquinone, etc.), which can eventually take part in the Faradaic reactions at the electrode/electrolyte interface and contribute to the enhanced capacitance value [96–98]. The addition of a redox-active KI additive in 1 M H_2SO_4 or to the active electrode material is found to enhance the capacitance (ca. four times improvement) and operating voltage window (0–0.8 V in the absence of KI to 0–1.2 V in the presence of KI) of the printed graphene electrode [41].

Acidic electrolytes are also the most suitable choice for certain pseudo-capacitive materials such as RuO_2 and conducting polymers (e.g., polyaniline-PANI and polypyrrole-ppy, etc.) where H^+-ion plays an important role in their capacitive properties [99–101]. Mesoporous RuO_x electrodes fabricated through electrodeposition process is assembled into an interdigitated MSC and characterized in 0.5 M H_2SO_4. The MSC can deliver 12.6 mF cm^{-2} areal capacitance in full cell configuration [102]. A high power planar MSC (maximum power density 600 W cm^{-3}) is reported in 1 M H_2SO_4 where exfoliated graphene/PANI composite is employed as electrode material [103]. In the first report on liquid electrolyte-based MSC (Fig. 3.8), 0.1 M H_3PO_4 is used to carry out the electrochemical analysis of ppy-based microelectrode [104]. In the same

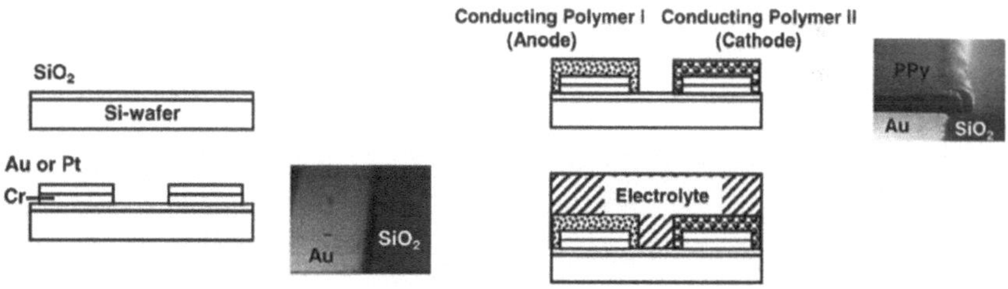

Fig. 3.8 The procedure adopted in 2003 for the fabrication of first liquid electrolyte-based in-plane MSC by Sung et al., using electrodeposited ppy-based microelectrodes. Both acidic and organic electrolytes are used for MSC fabrication [104]. This work was published followed by the first ever report on MSC as already explained in Section 3.4.1 and Fig. 3.5 where LiPON was used as the electrolyte. *(Reprinted from J.-H. Sung, S.-J. Kim, K.-H. Lee, Fabrication of microcapacitors using conducting polymer microelectrodes, J. Power Sources 124 (1) (2003) 343–350, https://doi.org/10.1016/S0378-7753(03)00669-4, Copyright (2003), with permission from Elsevier.)*

work, the MSC is also fabricated by means of an organic liquid electrolyte, which is explained later in Section 3.4.2.2.

Depending on the properties of electrode materials and the respective charge storage mechanism, acidic electrolytes could offer additional benefits over other electrolytes. For example, specific surface functionalities (such as quinone) on the carbon material show prominent redox contribution in 1 M H_2SO_4 [105]. Apart from H_2SO_4, solution of few other inorganic acids such as phosphoric acid, perchloric acid, and tetrafluoroboric acid are also occasionally used as the electrolyte [13,64]. However, unlike the carbon electrode, the choice of pseudocapacitive materials for acidic electrolytes is limited since many of them (e.g., oxides based on Ni, Co, V, etc.) are relatively unstable at low pH conditions [106]. Table 3.3 summarizes few reports on MSC using acidic liquid electrolytes.

Alkaline electrolyte

Alkaline electrolytes have notable importance in MSC research. Among alkaline electrolytes, aqueous solution of KOH (ranging from 1 to 6 M) is of particular interest. Similar to acidic electrolytes, the high ionic conductivity of alkaline electrolytes (0.6 S cm^{-1} for 6 M KOH at 25°C) improves the ESR (<1 Ω) and rate performance of MSCs [107]. In addition, carbon materials can undergo Faradaic reactions in KOH electrolyte involving the functional groups attached to electrode material (such as –COOH group in graphene, nitrogen-doped carbon) and electrolyte ions, resulting in improved capacitance [108].

Recently, very high capacitance of 639.56 mF cm^{-2} is achieved with a 3D-printed graphene aerogel/multiwalled carbon nanotube (MWCNT) composite-based MSC in 6 M KOH [109–111]. Besides carbon materials, alkaline electrolytes are extensively applied with transition metal compounds such as NiO_x, Fe_2O_3, $Ni(OH)_2$, $Co(OH)_2$, and VN [112]. The specific capacitance obtained for magnetron sputtered VN thin-film electrode (thickness 594 nm) in 1 M KOH is 16 mF cm^{-2} [113]. Later, the performance is improved to 40 mF cm^{-2} by increasing the thickness to 2 μm and tuning the porosity of thin-film VN electrode in an interdigitated planar MSC [114]. As a comparison, the same material is tested in 1 M tetraethylammonium tetrafluoroborate ($TEABF_4$)/acetonitrile (ACN) electrolyte, but no Faradaic property is observed followed by drastic drop in capacitance. This confirms the importance of OH^- ions in the electrolyte to facilitate the redox reactions of VN. Composite electrodes, viz., reduced graphene oxide (rGO)/MnO_2 [115], laser-scribed graphene (LSG)/Co_3O_4 [116], and

Table 3.3 Components, fabrication, characteristics, and performance summary of acidic liquid electrolyte-based MSC devices.

Electrolyte	Electrode	Substrate	Microelectrode fabrication method	Capacitance (areal/ volumetric)	Voltage (V)	Refs.
1 M H_2SO_4	KI absorbed on graphene	Nitrocellulose filter membrane	Transfer technique using a wax printer and stamping mechanism	130 mF cm^{-3}	0–1.2	[41]
1 M H_2SO_4	AC film	Si/SiO_2	Lithography patterning	250 F cm^{-3}	0–0.6	[92]
1 M H_2SO_4	Carbide-derived carbon films	Si/SiO_2	DC magnetron sputtering of TiC followed by chlorination	350 F cm^{-3}	0–0.8	[95]
0.5 M H_2SO_4	RuO_x	Commercialized interdigitated array (IDA)	Electrodeposition and electro-oxidation	12.6 mF cm^{-2}	0–1	[102]
1 M H_2SO_4	Graphene/ PANI composite	Si/SiO_2	Pressure-assisted dry transfer method	75 F cm^{-3}	0–0.8	[103]

3D-Co_3O_4/sputtered Pt nanoparticle [117], are few other reports where KOH electrolytes have been used. Recently, a screen-printable asymmetric MSC using KOH electrolyte and a composite of $Ti_3C_2T_x$ MXene (negative electrode) and Co-Al-layered hydroxide (positive electrode) is also reported [118]. The MSC device in 6 M KOH liquid electrolyte exhibited a specific capacitance of ~40.0 mF cm^{-2} at a current density of 0.75 mA cm^{-2}. In the same report, the performance of the MSC in KOH-based GPE is also presented.

In contrast to acidic electrolytes, alkaline electrolytes allow the use of cost-effective support materials, such as Ni foam, as the current collector for MSC. For example, $NiCo_2S_4$ is directly grown on interdigitated 3D-Ni foam by a simple hydrothermal process and later used as the positive electrode for fabricating hybrid MSC (carbon nanofiber (CNF) is used as a negative electrode) with KOH electrolyte [119]. Similarly, $Ni(OH)_2$ nanowires are deposited over Ni fibers to obtain $Ni(OH)_2$ nanowire/Ni fiber coaxial cable-like electrode for wire-shaped MSC, which can be directly taken for electrochemical analysis in 6 M KOH (Table 3.4) [120].

Neutral electrolyte

Neutral electrolytes (conductivity of ≈0.10 S cm^{-1} at 25°C) are of significant interest in MSC research owing to their noncorrosive nature compared to the above-discussed strong acidic or alkaline electrolytes [121]. Aqueous salt solutions of LiCl, $LiNO_3$, $LiClO_4$, Na_2SO_4, Li_2SO_4, etc., are the most common examples of neutral electrolytes. Owing to the low concentration of H^+ and OH^-, these electrolytes can shift the oxygen and hydrogen evolution reaction toward higher overpotential, suppressing the water decomposition [13, 122–124]. Besides, the electrochemical performance of these electrolytes depends on the properties of the corresponding electrolyte ions. For examples, in case of sulfate electrolyte, the rate performance of carbon electrodes follows the order $Li_2SO_4 < Na_2SO_4 < K_2SO_4$, which is correlated to the mobility of respective solvated cations, i.e., $Li^+ < Na^+ < K^+$ [125]. Regarding the size of ions, the small-sized Cl^- ion shows higher specific capacitance compared to larger SO_4^{2-} ions, revealing the impact of anion size on device performance [126]. Instead of using conventional monovalent cations, bivalent cations (Mg^{2+}, Ca^{2+}, Ba^{2+})-based neutral electrolytes are reported to boost the capacitance of MnO_2 by two times [127].

Similar to other aqueous electrolytes, the reports on neutral electrolyte-based MSCs have utilized both carbon and pseudocapacitive electrode materials. Laser-reduced rGO-based rGO|graphene oxide (GO)|rGO MSCs

Table 3.4 Components, fabrication, characteristics, and performance summary of alkaline liquid electrolytes-based MSC devices.

Electrolyte	Electrode	Substrate	Electrode fabrication method	Capacitance (areal/volumetric)	Voltage (V)	Refs.
6 M KOH	Graphene aerogel/MWCNT	PET	3D printing	639 mF cm^{-2}	0–1.2	[111]
1 M KOH	VN	Glass	DC reactive sputtering	250 F cm^{-3}	0–1.2	[113]
1 M KOH	Co_3O_4/Pt nanonetworks	Cr/Au over Si/SiO_2	Rapid temperature annealing	23 F cm^{-3}	0–0.8	[117]
6 M KOH	$Ti_3C_2T_x$MXene and Co-Al layered hydroxide (positive electrode)	Au over PET or commercial printing paper	Screen printing	40.0 mF cm^{-2}	0.4–1.4	[118]

with in-plane and sandwiched configurations are reported with 1.0 M Na_2SO_4 electrolyte. The comparison data given in this study further reveal a superior power density (~9.4 W cm^{-3}) of the resulting device compared to that of 1 M $TEABF_4$/ACN organic electrolyte [128]. MSC with even higher power density (~77 W cm^{-3}) is reported with 3 M KCl electrolyte and interdigitated electrode of rGO and carbon nanotube (CNT) fabricated by electrostatic spray-deposition technique [129]. 3D MSC based on self-supported AC electrode in 1 M $NaNO_3$ electrolyte have achieved 90.7 mF cm^{-2} areal capacitance with 51.5 mW cm^{-2} power density [130]. Regarding pseudocapacitive materials, the performance of a 3D-LSG/MnO_2 composite electrode is analyzed in 1 M Na_2SO_4 [131]. The as-fabricated asymmetric MSC device (LSG||LSG-MnO_2) (operating voltage window of 0–2 V) exhibits ultrahigh stack capacitance of ~5 F cm^{-3}. As another example, 1 M KCl is used to characterize interdigitated MSC comprising electrodeposited ppy over 3D-CMES (CMES: carbon microelectrochemical system) within the voltage window of 0–0.8 V [132]. Table 3.5 summarizes a few of the neutral liquid electrolyte-based MSC reports.

In conclusion, despite having some breakthrough improvements in terms of operating voltage window, capacitance, and energy density, aqueous electrolytes are yet to be employed for the fabrication of commercial SC/MSC devices. Further research in this field is necessary to realize aqueous electrolytes as a potential competitor to traditional SCs with organic electrolytes and invoking the same for realization of commercial MSCs.

3.4.2.2 Organic electrolytes

Typical organic electrolytes are a solution of conductive salt(s) dissolved in aprotic organic liquids [133–135]. Therefore, similar to aqueous electrolytes, the electrochemical performance of the organic electrolytes is dependent on the size of ions, ion-solvent interaction, diffusion rate of ions, and conductivity [2, 136]. In the context of MSCs, propylene carbonate (PC) and ACN are the common solvents used for electrolyte preparation. They can be operated at wider operating voltages than aqueous electrolytes leading to the realization of high-voltage MSC devices (with approximate device operating voltage window within 0–2.5 V). Among them, ACN-based electrolytes with low viscosity (1 M $TEABF_4$/ACN has a viscosity of 0.62 cP) and high ionic conductivity (0.06 S cm^{-1} for 1 M $TEABF_4$/ACN and 0.01 S cm^{-1} for 1 M $TEABF_4$/PC at 25°C) are the preferred choice for commercial devices, despite the drawbacks of low flash point

Table 3.5 Components, fabrication, characteristics, and performance summary of neutral liquid electrolyte-based MSC devices.

Electrolyte	Electrode	Substrate	Microelectrode fabrication method	Capacitance (areal/volumetric)	Voltage (V)	Refs.
1 M Na_2SO_4	rGO	Free-standing GO film	Direct laser writing	~1 mF cm^{-2}	0–1	[128]
3 M KCl	rGO/CNT composite	Ti/Au current collector over Si/SiO_2	Electrostatic spray deposition	6.1 mF cm^{-2}	0–1.2	[129]
1 M $NaNO_3$	AC	Ti/Au current collector over Si	Inject-dry process	90.7 mF cm^{-2}	0–1	[130]
1 M Na_2SO_4	LSG (negative) and LSG-MnO_2 (positive) electrodes	Au over PI or PET substrate	LSG by LightScribe DVD burner and MnO_2 by electro-deposition	[a]~5 F cm^{-3}	0–2	[131]
1 M KCl	ppy	Carbon-microelectrochemical system (C-MEMS) current collector over Si/SiO_2	Electropolymerization	~78 mF cm^{-2}	0–0.8	[132]

[a]Stack capacitance.

(5.5°C) and high toxicity [87, 88, 137–139]. However, compared to a single solvent system, binary or ternary mixture of solvents is more promising as an organic electrolyte to achieve improved ionic conductivity and suitable viscosity. With respect to this, the use of carbonate-based cosolvents such as ethylene carbonate (EC), dimethyl carbonate (DMC), diethyl carbonate (DEC) is also reported [140]. The advantage of organic electrolytes lies in the ability to sustain high voltage, a critical parameter for the realization of high-energy MSCs. To further improve the electrochemical stability of organic electrolytes, RTILs are often used as the ion source in organic solvents over the conventional inorganic salts (few examples are discussed in Section 3.4.2.3).

The pioneering work of Sung et al. utilizes 0.5 M $TEABF_4$ in ACN for interdigitated MSC fabrication in addition to 1 M H_3PO_4 (Fig. 3.8). The operating voltage window of the interdigitated symmetric MSC (ppy||ppy) is 0–0.8 V and a further improvement in operating voltage window (0–1.4 V) is accomplished by taking the advantage of an asymmetric cell configuration (ppy||Pt) [104]. For MSCs or conventional SCs, Li-based compounds (mainly $LiClO_4$) and tetraalkylammonium cation-based salts (such as $TEABF_4$) are mostly used in organic electrolytes. The salt concentration, degree of dissociation, and ion pair interaction have a profound influence on the conductivity of the electrolyte. Normally, the large size of the solvated ions and high viscosity of organic electrolyte induces lower ionic conductivity than in aqueous electrolytes. In fact, the RT ionic conductivity of $TEABF_4$/ACN is 0.06 S cm^{-1}, a value much lower than that of 1 M H_2SO_4 solution [138].

Among organic carbonate solvents, PC is one of the most used candidates for SC application. MSC device using PC with 1 M $TEABF_4$ operates within 0–2 V with TiC-derived carbon film [141] and at 0–2.5 V with an inkjet-printed AC electrode (Fig. 3.9A) [142]. Further enhancement of the operating voltage window of the device (0–3 V) is observed with 1 M $TEABF_4$/PC electrolyte and nanodiamond-derived onion-like carbon (OLC) electrode (Fig. 3.9B) [143]. This electrolyte is also used to demonstrate the advantage of a wafer-level encapsulation strategy of liquid organic electrolytes for fully encapsulated MSC devices operating within 0–2.5 V. However, on repeated use, such high operating voltage window often causes undesired self-discharge and aging of the device, creating serious concern regarding the reliability in practical applications [144]. To overcome this, significant developments have been made in both the electrode and electrolyte materials. For instance, graphene-based MSCs fabricated by

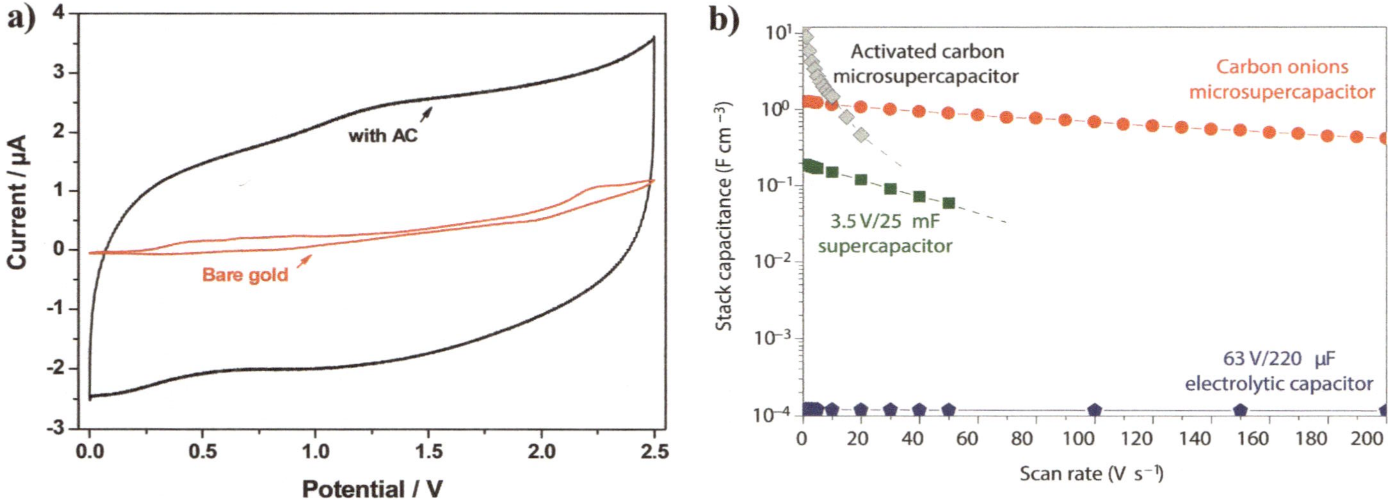

Fig. 3.9 Cyclic voltammetry (CV) profiles of MSCs with organic electrolytes displaying high operating voltage window: (A) an inkjet-printed AC-based MSC displaying a voltage window of 0–2.5 V with 1 M $TEABF_4$-PC liquid electrolyte [142] and (B) comparison of carbon onion microsupercapacitors characteristics as stack capacitance versus scan rate against other energy storage devices, *inset*: nanodiamond derived onion-like carbon (OLC)-based MSC displaying an improved voltage window (0–3 V) with the same electrolyte [143] when performed CV at a scan rate of 1 V s^{-1}. *((A) Reprinted from D. Pech, M. Brunet, P.-L. Taberna, P. Simon, N. Fabre, F. Mesnilgrente, V. Conédéra, H. Durou, Elaboration of a microstructured inkjet-printed carbon electrochemical capacitor, J. Power Sources 195 (4) (2010) 1266–1269, Copyright (2010), with permission from Elsevier; (B) Reprinted by permission from Springer Nature: Nature: Nature Nanotechnolgy, D. Pech, M. Brunet, H. Durou, P. Huang, V. Mochalin, Y. Gogotsi, P.-L. Taberna, P. Simon, Ultrahigh-power micrometre-sized supercapacitors based on onion-like carbon, Nat. Nanotechnol. 5 (9) (2010) 651–654, https://doi.org/10.1038/nnano.2010.162, Copyright (2010).)*

layer-by-layer (LbL) printed technique [145] and laser writing technique show reduced self-discharge [146] when used with advanced GPEs.

Table 3.6 summarizes few reports on the organic liquid electrolyte-based MSC devices in the literature.

3.4.2.3 Room temperature ionic liquids (RTILs)

RTILs (used as solute as well as "solvent") are room temperature molten salts featured by outstanding electrochemical stability widening the operating voltage window (voltage window within 0–4 V can be achieved) of the SC device surpassing that of organic electrolytes [147]. Additionally, RTILs possess negligible volatility, wide operational temperature ranges, and hardly flammable nature. Both liquid state at RT and ionic nature of RTILs have brought a unique opportunity to use them as electrolytes without the requirement of any additional solvents and salts [148]. Aprotic RTILs are of particular interest for high-energy MSC compared to other types (e.g., protic and zwitterionic). Imidazolium, tetraalkylammonium, and pyrrolidinium cations and BF_4^-, $TFSI^-$ and PF_6^- anion-based RTILs are used in literature with carbon-based as well as pseudocapacitive electrode materials [66]. The unique combination of large charge-delocalized anions and cations offers weak ion-pair interaction, ensuring independent mobility of the ions when they are used as electrolyte [147, 149, 150].

Although RTILs represent many advantages over aqueous and organic electrolytes, their viscous nature often imposes low conductivity ($\sim$14 mS cm^{-1} at 25°C) [151]. Mixing RTILs with organic solvents (e.g., ACN and PC) is a common practice to improve the ionic conductivity. A carbide-derived free-standing carbon film delivers 160 F cm^{-3} volumetric capacitance in 2 M 1-ethyl-3-methylimidazolium tetrafluoroborate in ACN ([EMIM][BF_4]-ACN) electrolyte and exhibits good capacitive behavior within 0–3 V window [95]. In another report, the voltage window of the device is further extended to 0–3.2 V by introducing *N*-methyl-*N*-butyl pyrrolidinium bis(trifluoromethane) sulfonylimide ([PYR_{14}][TFSI]) [152]. However, considering the safety concerns associated with additional flammable organic solvents, designing of solvent-free RTIL electrolytes is more desirable. RTIL carrying a cation with ether bond in the alkyl side chain is found to have higher ionic conductivity as compared to that of a similar cation of the pure alkyl side chain [153]. As a result, low resistance and improved device capacitance are achieved with the ether-bond-containing RTILs.

Table 3.6 Components, fabrication, characteristics, and performance summary of organic liquid electrolyte-based MSC devices.

Electrolyte	Electrode	Substrate	Electrode fabrication method	Capacitance (areal/volumetric)	Voltage (V)	Refs.
1 M $TEABF_4$ in PC	Carbide-derived carbon films	Si/SiO_2	DC magnetron sputtering of TiC followed by chlorination at 450°C	37 F cm^{-3}	0–2	[141]
1 M $TEABF_4$–PC	AC	Ti/Au over Si/SiO_2	Inkjet printing	2.1 mF cm^{-2}	0–2.5	[142]
1 M $TEABF_4$ in PC	OLC	Ti/Au over Si/SiO_2	Electrophoretic deposition	1.3 F cm^{-3}	0–3	[143]

3D MSC based on interdigitated AC electrode and 1-butyl-3-methylimidazolium tetrafluoroborate [BMIM][BF_4] electrolyte operates within 2 V along with 330 mJ cm^{-2} energy density [66]. In another work, the same electrolyte is used with a carbon-based thick electrode (thickness 200 μm) to fabricate a 3 V device delivering 1400 mJ cm^{-2} (0.38 mW h cm^{-2}) energy density [154]. Another RTIL, [BMIM][PF_6], also provides a 0–3 V window for interdigitated hybrid MSC based on phosphorene and graphene as electrode materials (Fig. 3.10A) [155]. The flexibility of the as-fabricated MSC device as well as the representative CV within an operating voltage window of 0–3 V is also displayed (Fig. 3.10B). It should be noted that the leakage

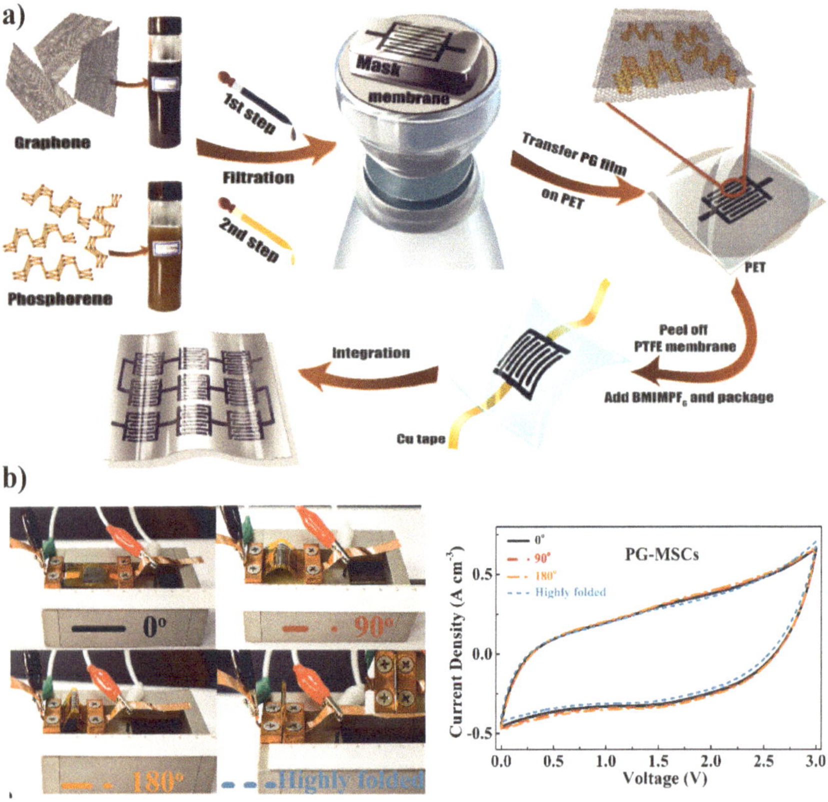

Fig. 3.10 (A) The fabrication of interdigitated MSC based on graphene and phosphorene using [BMIM][PF_6] ionic liquid and (B) flexibility of the device during different mechanical deformation and the CV representing the 0–3 V operating voltage window of MSC device [155]. *(Reprinted with permission from H. Xiao, Z.-S. Wu, L. Chen, F. Zhou, S. Zheng, W. Ren, H.-M. Cheng, X. Bao, One-step device fabrication of phosphorene and graphene interdigital micro-supercapacitors with high energy density, ACS Nano 11 (7) (2017) 7284–7292, https://doi.org/10.1021/acsnano.7b03288, Copyright (2017). American Chemical Society.)*

of liquid electrolyte (RTIL) is prevented here during the mechanical deformations of the device, which is leveraged from the packaging-norms adopted. The cell voltage of 3D-intergited MSC using pseudocapacitive MnO_2 can be extended to 1.5 V using [EMIM][TFSI] RTIL, which is otherwise difficult to achieve in traditional aqueous electrolyte [156]. Other RTILs with pyrrolidinium, sulfonium, and ammonium cations, which can potentially enhance the voltage window of MSCs is also known. Recently, protic RTIL, triethylammonium bis(trifluoromethylsulfonyl) imide ([NEt_3H][TFSI]), is used with silicon nanowire-based symmetric MSC delivering a 4 V working voltage [157]. Interestingly, the device shows extraordinary cycling stability with only a 27% loss of original capacitance over 5×10^6 charge/discharge cycles.

As mentioned earlier, RTILs can withstand a wide range of temperature, which is essential for the safe operation of electronic devices at both low and high temperatures. This property is found to be improved when two miscible RTILs are mixed together at an optimum ratio. For example, a binary composition of ([EMIM][TFSI]) and 1-allyl-3-methylimidazolium bis(trifluoromethylsulfonyl) imide ([AMIM][TFSI]) [158] operates within the temperature range from −40°C to 20°C, which is better than that of its individual components. A further improvement in temperature range is achieved with a eutectic mixture of *N*-methyl-*N*-propylpiperidinium bis (fluorosulfonyl)imide ([PIP_{13}][FSI]) and *N*-butyl-*N*-methylpyrrolidinium bis(fluorosulfonyl) imide ([PYR_{14}][FSI]) that can work from −50° to 100°C between 0 and 3.5 V window [159]. Table 3.7 summarizes several reports on MSCs based on RTIL.

3.4.3 Polymer electrolytes for MSCs

The challenges in ISEs and liquid electrolytes have catalyzed the research interest toward the development of PEs for energy storage devices. As already mentioned in the previous sections, the field of PEs started with the discovery of solid-state alkali metal-ion conduction in PEO. However, such all-solid-state PEs possess several limitations similar to that of the ISEs, such as low ionic conductivity and issues associated with the electrode / electrolyte interface. Therefore, similar to ISEs, all-solid-state MSCs based on SPEs are rarely reported. However, PEs provide the opportunity to design GPEs, which can maintain high ionic conductivity and better electrode/ electrolyte interfaces compared to ISEs and SPEs. Therefore, it is generally considered that GPEs are very much suitable for the fabrication of all kinds of MSCs.

Table 3.7 Components, fabrication, characteristics, and performance summary of RTIL-based MSC devices.

Electrolyte	Electrode	Substrate	Electrode fabrication method	Capacitance (areal/volumetric)	Voltage (V)	Refs.
$[EMIM][BF_4]$-ACN	Carbide-derived carbon films	Si/SiO_2	DC magnetron sputtering of TiC followed by chlorination at 50/700/800°C	160 F cm^{-3}	0–3	[95]
$[BMIM][BF_4]$	AC	Ti/Au current collector over glass	Slurry coating	91 mF cm^{-2}	−1–1	[66]
$(BMIM][PF_6]$	Phosphorene and graphene	PET	Mask-assisted vacuum filtration	9.8 mF cm^{-2}	0–3	[155]
[EMIM][TFSI]	MnO_2	Pt current collector over Si	Electrodeposition	19.5 mF cm^{-2}	0–1.5	[156]
$([NEt_3H][TFSI])$	Si nanowire	n-Type Si	CVD	17 μF cm^2	0–4	[157]
([EMIM][TFSI]) + ([AMIM][TFSI])	Si nanotrees	Si	CVD	[a]377 μF cm^{-2}	0–3	[158]

[a]At −40°C.

3.4.3.1 SPEs

The ionic conduction mechanism in the case of PEO-based PEs in the context of Li^+-ion conduction is provided in Fig. 3.11A (similar mechanism can be also applied to explain the conduction of other types of ions (Na^+, Mg^{2+}, etc.) in SPEs). It is generally agreed that the ionic conduction in PEO and PEO-like SPEs occurs through the hopping mechanism taking advantage of the segmental motion of polymer chains [160]. Other than PEO, there are several other polymer matrixes that have been used for the preparation of SPEs. However, the ionic conductivity of classical SPEs based on the

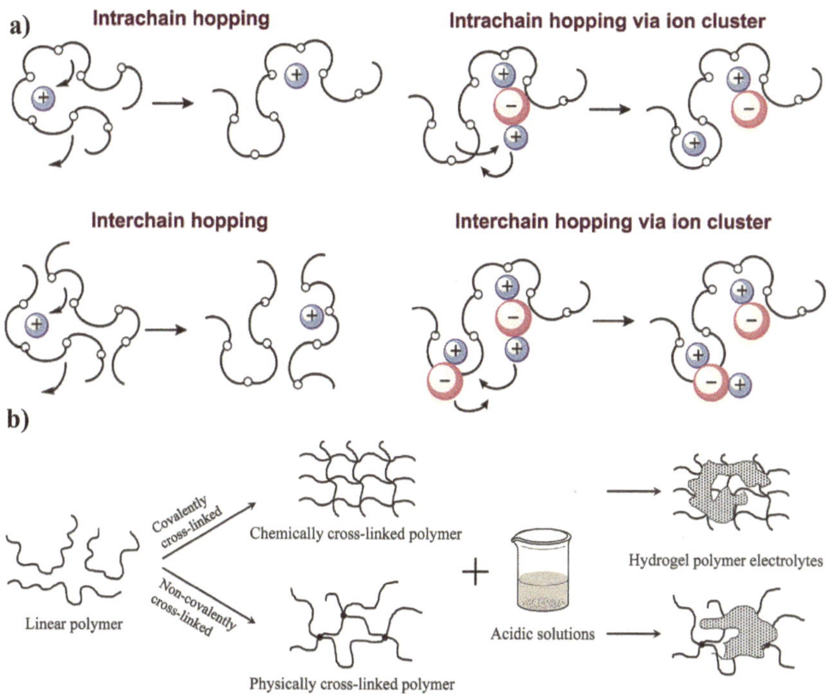

Fig. 3.11 (A) Generally accepted Li^+-ion conducting mechanisms in PEO-based SPEs [21] and (B) schematic illustration of the preparation of a GPE using an acidic electrolyte solution [166]. *((A) Republished with permission of Royal Society of Chemistry, from Z. Xue, D. He, X. Xie, Poly(ethylene oxide)-based electrolytes for lithium-ion batteries, J. Mater. Chem. A 3 (38) (2015) 19218–19253, https://doi.org/10.1039/C5TA03471J, Copyright (2015); permission conveyed through Copyright Clearance Center, Inc. (B) Republished with permission of Royal Society of Chemistry, from H. Gao, K. Lian, Proton-conducting polymer electrolytes and their applications in solid supercapacitors: a review, RSC Adv. 4 (62) (2014) 33091–33113, https://doi.org/10.1039/C4RA05151C, Copyright (2014); permission conveyed through Copyright Clearance Center, Inc.)*

PEO and other polymer matrixes is as low as $\sim 10^{-6}$ S cm^{-1} at RT [21,161,162]. The major reason for the low ionic conductivity exhibited by SPEs is due to the highly crystalline nature of the polymer matrix. By improving the amorphous character of the polymer matrix, the ionic conductivity can be improved. Amorphous SPEs can benefit from the improved segmental motion of the polymer chains resulting in enhanced ion conduction. The amorphous character of SPE can be tracked from its glass transition temperature (T_g) value. For instance, SPE possessing low T_g value indicates its improved amorphous character and in turn can help in achieving high ionic conductivity. There have been many attempts in this direction dedicated to improving the ionic conductivity of SPEs that it can match with liquid electrolytes. Although the ionic conductivity of SPEs is still lagging behind state-of-the-art liquid electrolytes, many recent reports suggest that SPEs with room temperature Li^+-ion conductivity in the range of 1–0.1 mS cm^{-1} have been achieved [163–165].

The attempts on improving the ionic conductivity of SPEs are mainly pertaining to tuning the amorphous character. Cross-linking of the polymer matrix is reported to help in reducing the crystalline nature of SPEs. Preparation of block copolymers, fully or semiinterpenetrated polymers, blending of two different polymers, organic-inorganic hybrid polymers, etc., are also attempted by various research groups to control the polymer morphology, hence enhance the ionic conductivity [167]. Several reports are available on the synthesis of composite SPEs (CPEs) where inorganic fillers such as Al_2O_3, SiO_2, and CeO_2 are mixed with a solid polymer matrix [168–170]. CPEs by combining a solid polymer matrix with currently available ISEs have also been attempted; they are also called hybrid SPEs (HPEs) [171, 172]. Considering the current situation of SPEs, there have been enormous developments in the past two decades resulting in systems that surpass ISEs. These SPEs are heavily employed for lithium-ion and lithium-metal battery fabrication [173] as well as other alternative battery systems [174, 175]. Unfortunately, these are yet to be investigated in details for MSC fabrication, whereas for batteries as well as all-solid-state MSCs, there should be serious attempts to develop an industrially competitive process based on SPEs.

3.4.3.2 GPEs

Compared to SPEs, GPEs are extensively used for the fabrication of all types of MSCs. GPEs are intermediate between all-SSEs and liquid electrolytes. They possess the benefits of mechanical stability of SSEs and the ionic conductivity of liquid electrolytes. Moreover, their flexible nature helps in easy integration with MSCs devices. In fact, all the polymer matrixes, which have been employed for the preparation of SPEs, can be used for GPEs as well. Another advantage is the possibility of designing GPEs with aqueous and organic solvents or RTILs depending on the nature of the electrode material under consideration. Considering the available literature reports, the most used polymer matrix for the preparation of aqueous GPEs is polyvinyl alcohol (PVA). Using PVA, acidic, alkaline, and neutral GPEs have been realized. In the case of ionogels or other organic electrolyte-based GPEs, poly(vinylidene fluoride (PvDF), poly(vinylidene fluoride-*co*-hexafluoropropylene (PvDF-HFP), polymethylmethacrylate (PMMA), etc., are used. Considering the extensive use of GPEs for the fabrication of MSCs, the following section is focused on providing a detailed impression regarding GPE-based MSCs to the readers, thus aqueous-, organic-, and ionic liquid-based GPEs (also called ionogels) are summarized below. The preparation of a GPE using an acidic liquid electrolyte is schematically illustrated in Fig. 3.11B. Although the illustration in Fig. 3.11B depicts the use of an acidic solution for the preparation of proton-conducting GPE, alkaline, neutral, and organic GPEs can also be prepared in a similar way.

3.4.3.3 Aqueous GPEs

Aqueous GPEs are mainly of three types: acidic, alkaline, and neutral GPEs. They are prepared by encapsulating the liquid electrolytes in a suitable polymer matrix. For example, the most used polymer matrix for the preparation of aqueous GPE is PVA. In a typical procedure for the preparation of PVA-based acidic GPE, PVA is first dissolved in water (1 g in 10 mL water) by heating at a temperature of 60–80°C with continuous stirring [176]. Once PVA is dissolved, the solution turns to clear and transparent. Later, the PVA solution is allowed to cool down to RT followed by the addition of the desired amount of concentrated inorganic acids such as H_2SO_4 and H_3PO_4 under stirring. The viscous acidic solution of PVA can be applied over the electrode surface prior to device fabrication. This type of proton-conducting GPEs is every important for MSCs since most of the pseudocapacitive electrodes require protons to facilitate/enable their Faradaic reactions [166,177]. Similarly, KOH-based alkali and neutral GPEs can be also prepared.

In the case of neutral GPEs based on PVA, salts such as LiCl, NaCl, and Na_2SO_4 are extensively been used. By using polymer matrices other than PVA, other GPEs can be also realized [166].

The ion transport mechanism in proton conducting electrolytes is extensively studied and it is generally agreed to be occurring through three well-known mechanisms, viz., Grotthuss mechanism, diffusion or vehicle mechanism, and direct transport/hopping assisted by the segmental motion of polymer chains (Fig. 3.12A–D) [166]. In the case of Grotthuss-type mechanism for proton transport, the proton moves from one site to another through a hydrogen-bonded network (Fig. 3.12A) [178]. This occurs through braking and making of covalent bonds. This mechanism was established by Theodore von Grotthuss in 1804 to explain the proton transport involved between water molecules. Later it was extended to liquid electrolytes followed by PEs (Grotthuss-type mechanism is also used to explain the hydroxide (OH^-) ion conduction in alkaline liquid electrolytes and PEs (Fig. 3.12D)) [179,180]. In the case of proton conducting GPE, the hydrogen bonding often originates from the liquid-phase (usually water) entrapped in the polymer matrix or in other words a hydrated environment is necessary for facile proton conduction (e.g., Nafion) [181]. Such PEs are widely used for devices in bigger dimension such as fuel cells where the degree of hydration can be managed during the course of operation (with a water management system). However, if the polymer matrix itself possesses inter/intramolecular hydrogen bonding, the proton conduction can be further enhanced since the polymer matrix can also contribute to the hydrogen bonding network so that improved proton conduction in an environment with minimal hydration/liquid phase can be achieved. In the context of MSCs, such GPEs with minimal liquid content are inevitable considering their microscale dimension without any scope for hydration management during normal operation conditions.

On the contrary, if the hydrogen bonding is weak, the proton conduction predominantly follows another mechanism known as vehicle type (Fig. 3.12B) [182]. In this case, the proton conduction occurs through a "vehicle" (water molecules/cluster) in the form of H_3O^+, $H_5O_2^+$, $H_9O_4^+$, etc., rather than a hydrogen bonded network (vehicle-type proton conduction in the case of Keggin-type heteropoly acids is already explained in Section 3.4.1). In this mechanism, the proton conduction is largely dependent upon the diffusion rate of the vehicle. Along with the Grotthuss and vehicle-type mechanisms, the polymer chain dynamics can also assist in the proton transport in GPEs as displayed in Fig. 3.12C (similar mechanism in the context of alkali-metal ion and other ion conducting SPEs is already explained with the help of Fig. 3.11A). This is known as direct transport or hopping assisted by segmental motion of polymer chains.

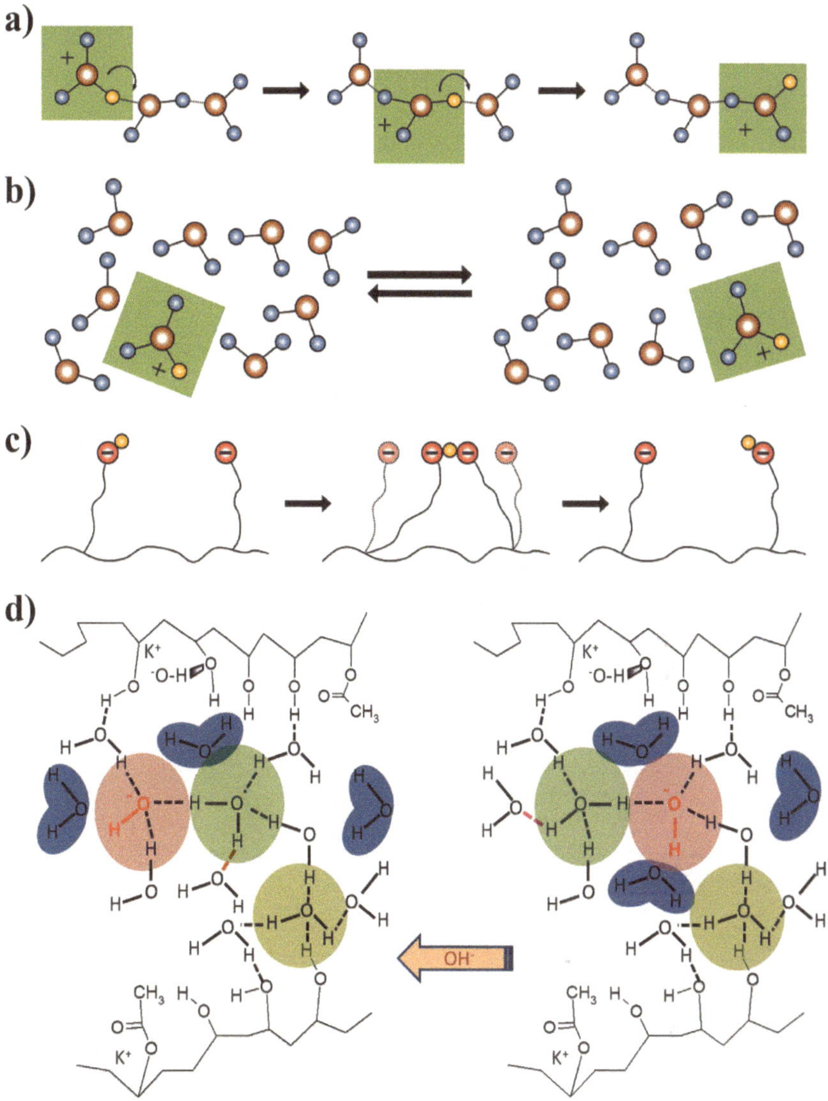

Fig. 3.12 Various types of proton conduction mechanisms involved in acidic GPEs. (A) Grotthuss-type mechanism in the presence of a strong hydrogen bonding, (B) vehicle-type mechanism in presence of a weak hydrogen bonding, (C) proton conduction assisted by segmental motion of polymer chains [166], and (D) Grotthuss-type mechanism involved in the OH^- ion transport of polyvinyl alcohol (PVA)-KOH GPEs. The OH^- transport direction is indicated by *an arrow*. The OH^- conduction involves the breaking of a hydrogen bond (*pink*) followed by the formation of a new hydrogen bond (*orange*) [179]. *((A–C) Republished with permission of Royal Society of Chemistry, from H. Gao, K. Lian, Proton-conducting polymer electrolytes and their applications in solid supercapacitors: a review, RSC Adv. 4 (62) (2014) 33091–33113, https://doi.org/10.1039/C4RA05151C, Copyright (2014); permission conveyed through Copyright Clearance Center, Inc; (D) Reprinted from F. Santos, J.P. Tafur, J. Abad, A.J. Fernández Romero, Structural modifications and ionic transport of PVA-KOH hydrogels applied in Zn/Air batteries, J. Electroanal. Chem. 850 (2019) 113380, https://doi.org/10.1016/j.jelechem.2019.113380, Copyright (2019), with permission from Elsevier.)*

3.4.3.4 Acidic GPEs for MSCs

Acidic GPEs are historically important since the first report on flexible MSCs uses a PVA-H_3PO_4-based GPE in 2006 (by Sung et al.) [65]. The interdigitated flexible MSC is fabricated by a combination of photolithography and electropolymerization techniques. Au is used as the current collector and ppy as the active electrode material over the Si/SiO_2 substrate. Prior to this work, the same group demonstrated the use of PVA-H_3PO_4 in MSC for the first time in 2004, although the flexible aspect was not mentioned [64]. Whatsoever, the aforementioned report in 2004 can be considered as the first report on a PE-based MSC. The method used for the flexible-MSC device fabrication as it appears in their report is presented in Fig. 3.13A.

In the report by El-Kady et al., graphene-based MSC is fabricated using PVA-H_2SO_4 electrolyte. The interdigitated in-plane MSC is fabricated by using a consumer-grade LightScribe digital versatile disc (DVD) burner [67]. The device fabrication is shown in Fig. 3.13B where the GPE is used as a film over the microelectrode assembly. In another work from the same group, PVA-H_3PO_4-based GPE is also used for the fabrication of LSG-based MSC device [146].

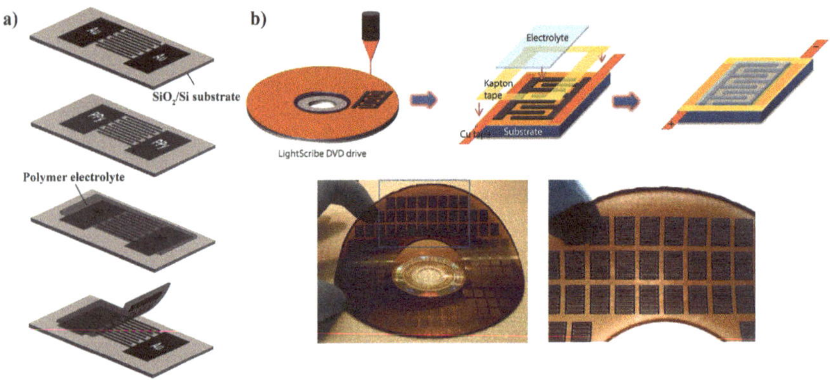

Fig. 3.13 (A) The steps involved in the fabrication of first flexible MSC in 2003 reported by Sung et al. using PVA-H_3PO_4 GPE [65] and (B) graphene-based MSC devices fabricated over a DVD by laser scribing method [67]. *((A) Reprinted from J.-H. Sung, S.-J. Kim, S.-H. Jeong, E.-H. Kim, K.-H. Lee, Flexible micro-supercapacitors, J. Power Sources 162 (2) (2006) 1467–1470, Copyright (2006), with permission from Elsevier; (B) Reprinted by permission from Springer Nature: Nature: Nature Communications, M.F. El-Kady, R.B. Kaner, Scalable fabrication of high-power graphene micro-supercapacitors for flexible and on-chip energy storage, Nat. Commun. 4 (1) (2013) 1475, https://doi.org/10.1038/ncomms2446, Copyright (2013).)*

Kang et al. also used PVA-H_2SO_4 for the preparation of a graphene-based in-plane MSC. Here, the GO is converted into rGO by a flash-reduction (Fr) method [183]. Hence, the synthesized rGO is termed as FrGO. The device exhibited a specific capacitance of 36.90 mF cm^{-2} (at 0.2 mA cm^{-2}) and good cycling stability over 8000 cycles. Another work on graphene-based MSC using PVA-based acidic GPEs includes separate reports from Wu [184] and Yoo et al. [48] using PVA-H_2SO_4 and PVA-H_3PO_4, respectively. The former used spin coating of GO on Si substrate. The in-plane-interdigitated MSC delivered high rate capability providing operability at high scan rates of 1000 V s^{-1}. In the latter report, pristine graphene films are achieved by chemical vapor deposition (CVD) method, whereas multilayer graphene films obtained by a combination of LbL assembly and chemical reduction [185, 186].

Lochmann et al. demonstrated an MSC made from nanoporous carbon using nanoimprint lithography with a line width down to 500 nm [187]. The device with nitrogen-doped carbon exhibits 1.38 F cm^{-3} volumetric capacitance with PVA-H_2SO_4 GPE. Harsem et al. prepared graphene ink to print MSC using an inkjet printer [188]. The CV features of the MSC fabricated with PVA-H_2SO_4 GPE found to be close to ideal rectangular shape; rate capability is also found to be very high with 54% capacitance retention when the current was increased to 200 times. The MSC also had high bending and impact tolerance showing little to no change in performance when impacted with a hammer.

Apart from in-plane MSC designs, PVA-based acidic GPEs are used for 3D MSC fabrication as well. In a report by Pu et al., MSC is fabricated by 3D microelectrodes consisting of carbon-based active materials (graphene or activated carbon). In this work, it is shown that the use of GPE allows the infusion of electrolyte ions into the 3D-electrode surface and enhances active material utilization [189]. This is not possible in the case of ISE or SPE film, especially in 3D configuration. These reports underline the utility of PVA-based acidic GPEs in the fabrication of various types of MSCs designs with carbon-based electrode materials. Apart from the carbon-based MSCs, PVA-based acidic GPEs are also used for the fabrication of MSCs with pseudocapacitive materials such as conducting polymers and metal oxides. However, acidic GPEs are not suitable for some metal oxides like vanadium oxide and manganese oxides when long-term durability is desirable as they dissolve and change into sulfates.

PEDOT-based in-plane-interdigitated flexible MSCs are fabricated by a combination of photolithography for patterning and electrodeposition of poly(3,4-ethylenedioxythiophene) (PEDOT) by Kurra et al. [190]. In this work, PVA-H_2SO_4 GPE is used. This device showed excellent frequency response with 500 V s^{-1} ultrahigh scan rate achieved at a crossover frequency of 400 Hz at a phase angle of −45°. An interesting work on MSCs fabricated using a cost-effective and simple pen lithography was demonstrated by Lee et al. [191]. The microelectrode is composed of graphene/PEDOT composite, and for the device fabrication, PVA-H_2SO_4 GPE is used. For device fabrication, PVA-H_2SO_4 GPE is used. The flexible device showed a volumetric capacitance of 7.7 F cm^{-3} and a maximum energy and power density of 1.5 mW h cm^{-3} and 141 W cm^{-3}, respectively. This technique, although simple and cost-effective, formed nonuniform films and uneven films of the active material, unlike other lithography methods.

Shujiang Ding adopted 3D printing for the fabrication of CNT-based MSC with PVA-H_3PO_4 GPE [109,192]. Similarly, Yun et al. also reported direct 3D printing of a graphene oxide hydrogel for the fabrication of in-plane-interdigitated MSC using PVA-H_3PO_4 [193]. The MSC device exhibited high areal capacitance of 101 mF cm^{-2} at a current density of 0.5 mA cm^{-2}. Liu et al. developed symmetric and asymmetric in-plane MSCs with GO-based negative electrode and PANI/GO composite-based positive electrode [194]. Here, the electrolyte used is PVA-H_3PO_4. In the symmetric device, the maximum voltage window achieved is 0–0.8 V, whereas the asymmetric MSC helped to improve the voltage up to 0–1.2 V. However, the asymmetric device exhibited 100% capacitance retention over 5000 cycles. Yang et al. fabricated MnO_2/OLC composite-based MSC using PVA-H_3PO_4 as the GPE [195]. Over the interdigitated Ag current collector, the MnO_2/OLC composite electrode is patterned by screen-printing process. Later, the GPE is applied over the in-plane MSC electrode assembly for device fabrication and testing. Apart from MnO_2, other acid-compatible electrodes such as RuO_2 are also used for acidic GPE-based MSCs [196]. Table 3.8 summarizes several aqueous acidic GPE-based MSC devices in the literature.

3.4.3.5 Neutral GPEs for MSCs

Neutral GPEs are significant where electrode materials are prone to degradation and dissolution in acidic and alkaline GPEs as they are potentially less corrosive. For asymmetric MSC design, neutral GPEs are suitable choice for their wide operating voltage window. Elshof et al. used a PVA-LiCl GPE for the fabrication of MSC using inkjet-printed MnO_2 microelectrodes. The MSC exhibits a volumetric capacitance of 2.4 F cm^{-3} along with the power

Table 3.8 Components, fabrication, characteristics, and performance summary of aqueous acidic GPE-based MSC devices.

Salt	Polymer matrix	Electrode	Substrate	Microelectrode fabrication method	Capacitance (areal/volumetric)	Voltage (V)	Refs.
H_3PO_4	PVA	Graphene and rGO	Au current collector	CVD for graphene films, LbL and chemical reduction for rGO	[a]80 μF cm^{-2} and [b]394 μF cm^{-2}	0–1	[48]
H_3PO_4	PVA	LSG	PET	Laser writing by LightScribe DVD burner	1.5–3 mF cm^{-2}	0–1	[146]
H_2SO_4	PVA	rGO	Au current collector over PET	Flash reduction	37 mF cm^{-2}	0–1	[183]
H_2SO_4	PVA	rGO	Si	Spin coating, plasma treatment	80.7 μF cm^{-2}	0–1	[184]
H_2SO_4	PVA	PEDOT	Glass, polyehtylenenaphthalate	Electrodeposition	33 F cm^{-3}	0–0.8	[190]
H_3PO_4	PVA	GO/PANI and rGO	PET, glass, PDMS	Extrusion printing	[c]38 F cm^{-3} and [d]25 F cm^{-3}	0–0.8 and 0–1.2	[194]
H_3PO_4	PVA	MnO_2/OLC	Si, glass, printing paper	Screen printing	7.04 mF cm^{-2}	0–0.8	[195]

[a] 1–2 Graphene layers.
[b] Multilayer rGO.
[c] Symmetric MSC.
[d] Asymmetric MSC.

and energy density of 0.018 W cm^{-3} and 0.18 mW h cm^{-3}, respectively [197]. Carboxymethyl cellulose (CMC)-Na_2SO_4 neutral GPE is used by Morag et al. for the fabrication of flexible asymmetric MSCs. The microelectrodes are fabricated on a freestanding hollow nickel microfiber coated by MnO_2 (positive electrode) and ppy/rGO composite (negative electrode). The MSC device exhibited an energy density of 0.11 mW h cm^{-2} and power density 37.5 mW cm^{-2} with excellent cycling stability [198].

Giannakou et al. reported inkjet-printed NiO-based MSC using PVA-$MgClO_4$ GPE [199]. The $Mg(ClO_4)_2$ electrolyte is chosen after testing the performance of MSC in several electrolytes based on LiOH, KOH, NaOH, KCl, $NaNO_3$, $NaClO_4$, $Mg(ClO_4)_2$, and poly(4-styrenesulfonic acid). The device based on PVA-$MgClO_4$ is operated within a voltage window of 0–1.5 V. The areal and volumetric capacitances are 155 mF cm^{-2} and 705 F cm^{-3}, respectively. Zheng et al. reported a high-energy asymmetric MSC in a sandwiched configuration using PVA-LiCl GPE [200]. In this work, a composite electrode containing MnO_2/PEDOT (MP) nanosheets is employed as a positive electrode, whereas exfoliated graphene is used as the negative electrode. In the case of conventional macro-SCs or MSCs, porous separators soaked in liquid electrolytes/viscous GPEs are employed for the device fabrication. However, in this work, the potential of employing boron nitride (BN) as the separator is demonstrated. BN is casted as thin layer between the electrode materials so that monolithic MSCs could be prepared. The as-fabricated MSC is operated within a voltage window of 0–1.8 V. The exfoliated graphene||MP asymmetric MSC delivers a high-energy density of 8.6 mW h cm^{-3} in addition to possessing excellent flexibility. Similar to BN, recently other materials such as GO are also used as separator for the fabrication of MSCs. However, the ion conducting mechanism in such 2D materials is still unclear [33].

An in-plane-interdigitated asymmetric MSC based on MnO_2-ppy and V_2O_5-PANI composites as positive and negative electrodes, respectively, is reported by Yue et al. [201]. Here also PVA-LiCl is employed as the electrolyte. It is worth mentioning that the integration of a photo-detector and a wireless charging coil with the MSC device is also demonstrated in this work. Interdigitated MSC based on ppy is reported by Sun et al. using PVA/$LiClO_4$ electrolyte [202]. The device displayed a specific capacitance of 0.03 F cm^{-2} with 2 mW cm^{-2} power density.

Very recently, a new type of SC device known as zinc-ion hybrid supercapacitors is receiving immense attention, which is operated by means of the conduction of Zn^{2+} ions between respective electrodes. In this scenario, a recent report suggests the development of a Zn^{2+}-ion conducting GPE for the fabrication of a flexible zinc-ion MSC for the first time [70]. Here, a natural polymer, viz., gelatin, is used for realizing the gel electrolyte with 1 M $ZnSO_4$ solution as the liquid electrolyte. The MSC possessed an interdigitated design in asymmetric configuration where electroplated zinc and laser-processed CNT is used as the negative and positive microelectrodes, respectively. Recently, other polymers such as Nafion, PVA, and polyacrylonitrile are also known to facilitate Zn^{2+}-ion conduction [203–205]. However, they are yet to be explored for the fabrication of zinc-ion MSCs. The device displayed an areal capacitance of 83.2 mF cm^{-2} at 1 mA cm^{-2} current density. In Table 3.9, a summary of the aqueous neutral GPE-based MSCs is provided.

3.4.3.6 Alkaline GPEs for MSC

In alkaline GPEs, KOH is the most commonly used hydroxide due to its high conductivity and inexpensiveness. Other hydroxides that can be utilized in alkaline GPEs for MSCs include LiOH and NaOH. Alkaline GPEs are particularly suitable for heteroatom-doped carbon and functionalized carbon materials as alkaline electrolytes have been shown to participate in pseudocapacitive behavior. Alkaline GPEs are compatible with some metal oxide and hydroxide such as Co_3O_4, NiO, $NiCo_2O_4$, $Ni(OH)_2$, etc. The OH^- ion plays a significant role in their psuedocapacitive behavior through oxide hydroxide transition; therefore, these materials can achieve high capacitance in alkaline GPEs. In 2015, Shi fabricated MoS_2@Sulfonated rGO-based flexible MSC using PVA-KOH GPE. The optimized device showed capacitance of 6.65 mF cm^{-2} with energy and power density of 0.58 mW h cm^{-2} and 13.4 mW cm^{-3}, respectively [206]. A facile screen-printing technology is used by Zhu et al. for the fabrication of MSC based on cobalt oxide (CoO) nanoflowers woven in a CNT network [207]. Here also, PVA-KOH is used as the GPE. The device is operated at a voltage of 1.2 V with an energy density of 3.48 mW h cm^{-3}. PVA-KOH is also used for the fabrication of all-quasi-solid-state wire-shaped MSCs [208]. In this work, the microelectrode consists of graphene quantum dot (GQD)/Co $(OH)_2$ electrodeposited on nanoporous Au–Ag alloy wire. The MSC is fabricated in an asymmetric configuration (activated carbon (AC)|GQD/ Co $(OH)_2$). The device exhibited excellent capacitance retention of 99.5% for over 10,000 cycles at a current density of 2 A cm^{-3}.

In the report of Feng et al., a graphene/NiOOH/$Ni(OH)_2$ composite electrode is prepared by a combination of chemical vapor deposition,

Table 3.9 Components, fabrication, characteristics, and performance summary of aqueous neutral GPE-based MSCs.

Salt	Polymer matrix	Electrode	Substrate	Microelectrode fabrication method	Capacitance (areal/volumetric)	Voltage (V)	Refs.
LiCl	PVA	PEDOT: PSS/δ-MnO_2	PI	Inkjet printing	2.4 F cm^{-3}	0–0.8	[197]
Na_2SO_4	CMC	MnO_2 (positive) and ppy/rGO (negative)	Ni	Electrodeposition	370 mF cm^{-2}	0–1.5	[198]
$MgClO_4$	PVA	NiO	Ag current collector over PET	Inkjet printing	155 mF cm^{-2}	0–1	[199]
LiCl	PVA	MnO_2/PEDOT (positive) and exfoliated graphene (negative)	PET, paper, nylon membrane	Mask-assisted printing	9.6 mF cm^{-2}	0–1.8	[200]
LiCl	PVA	MnO_2-ppy (positive) &V_2O_5_PANI (negative)	Au current collector over PI	Electrodeposition	[a]7.5 mF cm^{-2} and [b]6.5 mF cm^{-2}	[b]0–1.6	[201]
$ZnSO_4$	Gelatin	Zn (negative) and CNT (positive)	PI	Electrodeposition (Zn) and laser processing of CNT	83.2 mF cm^{-2}	0.2–1.8	[70]

[a]Symmetric MSC.
[b]Asymmetric MSC.

mask-free patterning, and spray-coating methods [209]. The MSC device fabrication is completed by means of PVA-KOH electrolyte. The MSC device exhibited specific capacitance of 0.75 mF cm^{-2} at 5 mV s^{-1} (7.54 F cm^{-3}) with a maximum energy density of 1.04 mW h cm^{-3}. Gu et al. used microelectronic photo-lithography technology combined with plasma etching technique for the fabrication of rGO/Fe_2O_3-based MSC in PVA-KOH GPE [210]. The device exhibits 11.57 F cm^{-3}-specific capacitance at a scan rate of 200 mV s^{-1}. A photo-detector based on CdS nanowires is integrated into the MSC device and an effective photo-detecting system is fabricated. Similarly, in the work of Xu et al., an rGO-based MSC is fabricated with PVA-KOH and later a CdS nanowire-based photo-detecting system is integrated [211]. Few reports on aqueous alkaline GPE-based MSC devices are summarized in Table 3.10.

3.4.3.7 Organic GPEs

This section covers selected reports, where organic solvent-based GPEs are used for the fabrication of MSCs. The main advantage of using organic GPEs arise from their high electrochemical stability compared to aqueous GPEs. Most of the reports on organic GPEs are based on PMMA and carbonate-based organic solvents with lithium-conducting salts. Compared to the surfeit of reports in aqueous GPEs, the reports on organic GPEs are less. Most of the researchers prefer ionogels over conventional GPEs. Despite the fact that ionogels are also a type of organic GPE, in the context of this chapter, they are explained in a separate category due to the availability of many numbers of reports on the same.

As already mentioned in the acidic GPE section, the works of Sung et al. involve the use of an organic GPE for the first time for MSC fabrication. They used a GPE system based on polyacrylonitrile (PAN) and carbonate-based liquid electrolyte (lithium triflate (LiTf) in a mixture of EC and PC) for MSC fabrication (ppy is used as electrode material) in interdigitated fashion [64]. A PMMA-PC-$LiClO_4$ GPE is used for the fabrication of stretchable and patchable array of high-performance MSC by Lee et al. [212]. For GPE synthesis, 2.13 g of $LiClO_4$ is first dissolved in PC. Later, 2 g of PMMA is added to it followed by stirring and heating at 100°C. The interdigitated Mn_3O_4/MWCNT microelectrode is fabricated by a combination of photolithography and LbL assembly (Fig. 3.14A–D). The most important aspect of this organic GPE-based MSC device is the nonrequirement of encapsulation. Since a high boiling PC solvent is used, the evaporation of solvent is minimized here compared to aqueous GPEs.

Table 3.10 Components, fabrication, characteristics, and performance summary of aqueous alkaline GPE-based MSC devices.

Salt	Polymer matrix	Electrode	Substrate	Microelectrode fabrication method	Capacitance (areal/ volumetric)	Voltage (V)	Refs.
KOH	PVA	MoS_2/rGO	PI	Gravure printing	6.56 mF cm^{-2}	0–0.8	[206]
KOH	PVA	CoO/CNT	Ag current collector over PET	Screen printing	17.40 F cm^{-3}	0–1.2	[207]
KOH	PVA	GQD/Co$(OH)_2$	Au-Ag nanowire	Electrodeposition	10 F cm^{-3}	0–1.4	[208]
KOH	PVA	Graphene/ NiOOH/Ni$(OH)_2$	Ti/Au current collector over Si/SiO_2	Mask-patterning and electro spinning	7.54 F cm^{-3}	0–1	[209]
KOH	PVA	rGO/Fe_2O_3	PET	Spin coating	11.57 F cm^{-3}	0–1	[210]

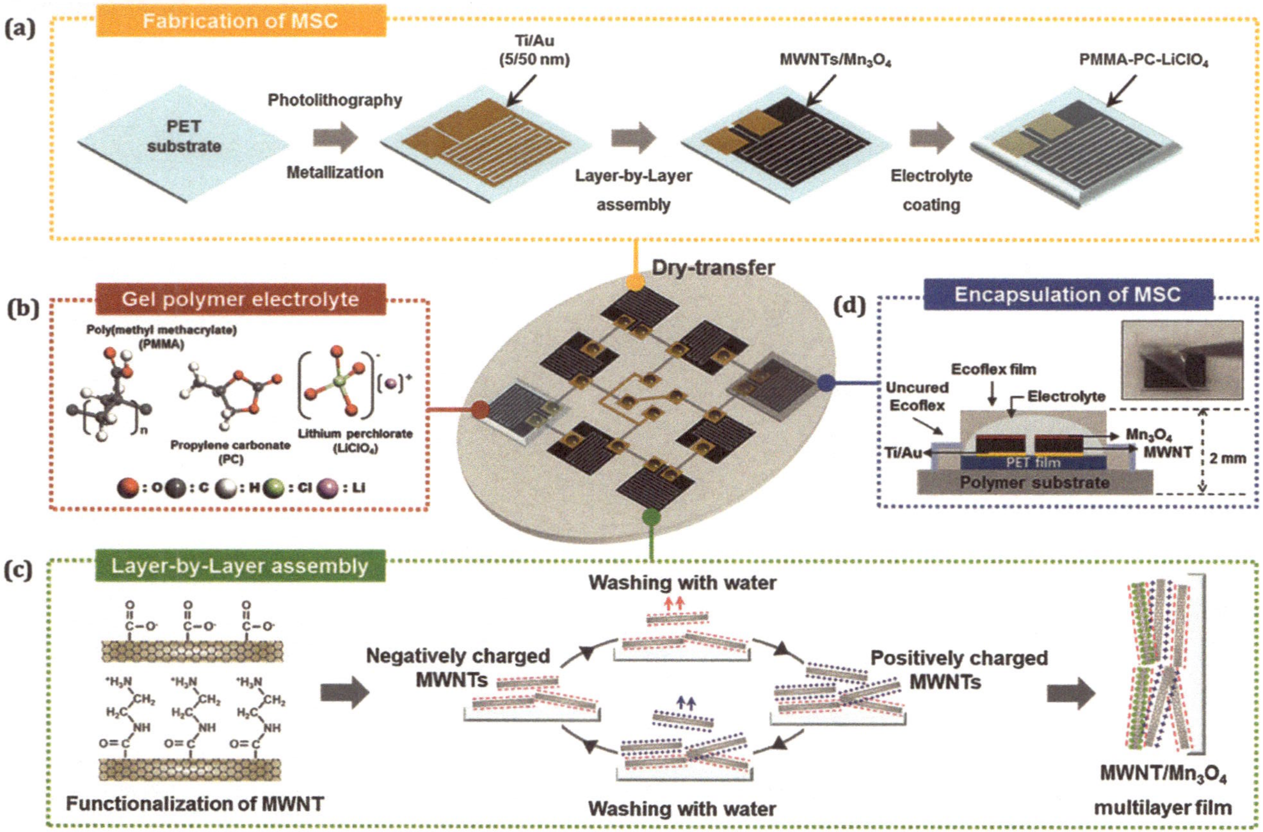

Fig. 3.14 (A) LbL-assisted fabrication of $MWCNT/Mn_3O_4$ microelectrode-based MSC device using PMMA-PC-$LiClO_4$ organic GPE, (B) the chemical structure of the components in the GPE, (C) the schematic illustration of LbL assembly method leading to $MWCNT/Mn_3O_4$ film, and (D) the cross-sectional scheme representing the encapsulation of the MSC device along with the optical image of the assembled image (*inset*) [212]. *(Republished with permission of Royal Society of Chemistry, from G. Lee, D. Kim, D. Kim, S. Oh, J. Yun, J. Kim, S.-S. Lee, J.S. Ha, Fabrication of a stretchable and patchable array of high performance micro-supercapacitors using a non-aqueous solvent based gel electrolyte, Energ. Environ. Sci. 8 (6) (2015) 1764–1774, https://doi.org/10.1039/C5EE00670H, Copyright (2015); permission conveyed through Copyright Clearance Center, Inc.)*

This, in turn, helps to achieve 85% of initial capacitance even after 2 weeks in air atmosphere without packaging. Additionally, 8.9 F cm^{-3} of specific capacitance is obtained at a current density of 0.1 A cm^{-3}.

In the report of Yun et al., asymmetric MSC based on MnO_2 nanoball deposited MWCNT as positive and V_2O_5 wrapped MWCNT as negative is reported [213]. They also used PMMA-PC-$LiClO_4$ as the GPE. The MSC is fabricated on a PET substrate deposited with Au current collector in interdigitated fashion by photolithography and electron-beam evaporation techniques. MWCNT is spray-coated over the current collectors. The MnO_2 and V_2O_5 electrode materials are later electrodeposited, and the device fabrication is completed by means of applying the GPE followed by encapsulation. The MSC device is used for powering a patterned Graphene/ZnO UV sensor, which is integrated into the final device. To improve the capacitance of carbon-based solid-state MSC, Kim et al. have prepared a redox-active gel-polymer electrolyte by introducing hydroquinone (HQ) into PMMA-PC-$LiClO_4$ matrix [40]. With a spray-coated MWCNT-COOH electrode, this electrolyte can enhance the specific capacitance of the MSC device by eleven times as compared to the HQ-free counterpart. Several organic GPE-based MSC devices are summarized in Table 3.11.

3.4.3.8 Ionogels

Confining RTILs into a solid matrix (or ionogel) is considered as a promising alternative to aqueous and organic GPEs in the MSC field. Like other GPEs, ionogels preserve the properties of RTILs in terms of ionic conductivity (in the range of 1–10 mS cm^{-1} at RT) and operating voltage beyond 3 V, while allowing desired shape compatible to the flexible microdevice [214]. Sol-gel process is widely used for ionogel preparation where porous-silica is used as the confinement matrix (Fig. 3.14A). Unlike classical sol-gel process in which the low-volatile solvent is evaporated to get porous oxide architectures, RTILs as solvents permit the retention of the gel form itself and hence called ionogel. In this case, the stability of ionogels depends on whether the silica matrix is organically modified or not. At the same time, the relationship between the ionic liquid content and the texture of silica matrix is also known. The physical properties of the ionogels such as thermal stability and ionic conductivity also depend on the confinement matrix. Recently, other approaches such as swelling a polymer matrix with RTIL or preparation of ionogels by in situ polymerization (explained in Section 3.3.2) are also popular [215].

Table 3.11 Components, fabrication, characteristics, and performance summary of organic GPE-based MSC devices.

Salt	Solvent	Polymer matrix	Electrode	Substrate	Microelectrode fabrication method	Capacitance (areal/volumetric)	Voltage (V)	Refs.
$LiClO_4$ (HQ as additive)	PC	PMMA	MQCNT-COOH	Ti/Au current collector over PET	Spray coating	[a]25 mF cm^{-2} and [b]5 mF cm^{-2}	0–1.2	[40]
$LiClO_4$	PC	PMMA	Mn_3O_4/MWCNT	Ti/Au current collector over PET	LbL assembly	8.9 F cm^{-3}	− 2.4–2.4	[212]
$LiClO_4$	PC	PMMA	MWCNT/MnO_2 (positive) and MWCNT/V_2O_5 (negative)	Au current collector over PET	Spray coating and electro-deposition	2.67 mF cm^{-2}	0–1.6	[213]

[a]Double-sided MSC.
[b]Single-sided MSC.

The concept of ionogels for fabricating MSC is first demonstrated with light-scribed interdigitated graphene-based in-plane MSC (Figs. 3.13B and 3.15A). This ionogel consists of fumed silica nanopowder as encapsulating matrix in combination with 1-ethyl-3-methylimidazolium bis(trifluoromethylsulfonyl)imide [EMIM][TFSI] RTIL [67]. Later, the same is used in on-chip-integrated MSC designed by lithographic patterning of photo-resist derived porous carbon. The device delivers a maximum energy density of 3 mW h cm^{-3} and a power density of 26 W cm^{-3}, the former is a value higher than commercial Li-ion thin-film battery [216]. Since SCs and MSCs are generally high-power devices, raising their energy density par with high-energy Li-ion batteries is an important but challenging task. In this aspect, the aforementioned work stands out since it could surpass the energy density of thin-film Li-ion batteries by combining a high-voltage ionogel in combination by invoking sophisticated MSC device fabrication strategy. In a recent report, a fumed silica-[EMIM][BF_4] electrolyte is integrated with pseudocapacitive micro-electrode based on $Cu(OH)_2$@FeOOH. The fabricated 1.5 V planar MSC device utilizes cost-effective copper as the current collector, which is prepared by electroless copper deposition technique [217].

Use of different polymer matrices, viz., PVA and PVdF-HFP, is also another common approach to obtain physical gel with RTILs, which can be directly drop-casted onto the microelectrode. A solid-state planar MSC based on VS_2 ultrathin nanosheets is designed with PVA-[BMIM][BF_4] gel electrolyte. However, despite the use of RTIL-based electrolyte, the selected working voltage in this report is only 0.6 V [218]. In contrast, a 3 V 2D-planar MSC is demonstrated in polystyrene (PS)-PMMA-[EMIM][TFSI] ionogel coupled with a spray-coated SWCNT-COOH electrode. Here, the ionogel is prepared by dissolving a PS-PMMA polymer and [EMIM][TFSI] RTIL in the ACN solvent followed by evaporating off the solvent from the electrolyte mixture [219]. Recently, PVDF-HFP-based ionogel is used for the fabrication of lithium-ion MSC by Bao et al. [69]. This is considered as the first prototype of a lithium-ion MSCs. The interdigitated lithium-ion MSC is represented as LTO||AG where "LTO" stands for $Li_4Ti_5O_{12}$ (negative electrode) and "AG" stands for activated graphene (positive electrode). Here, the RTIL used is [PYR_{14}][TFSI] and the MSC is operated within 0–3 V window delivering an energy density of 53.5 mW h cm^{-3}.

Patterning thin-film ionogel directly onto the electrode from a precursor electrolyte solution by means of in situ polymerization is another interesting

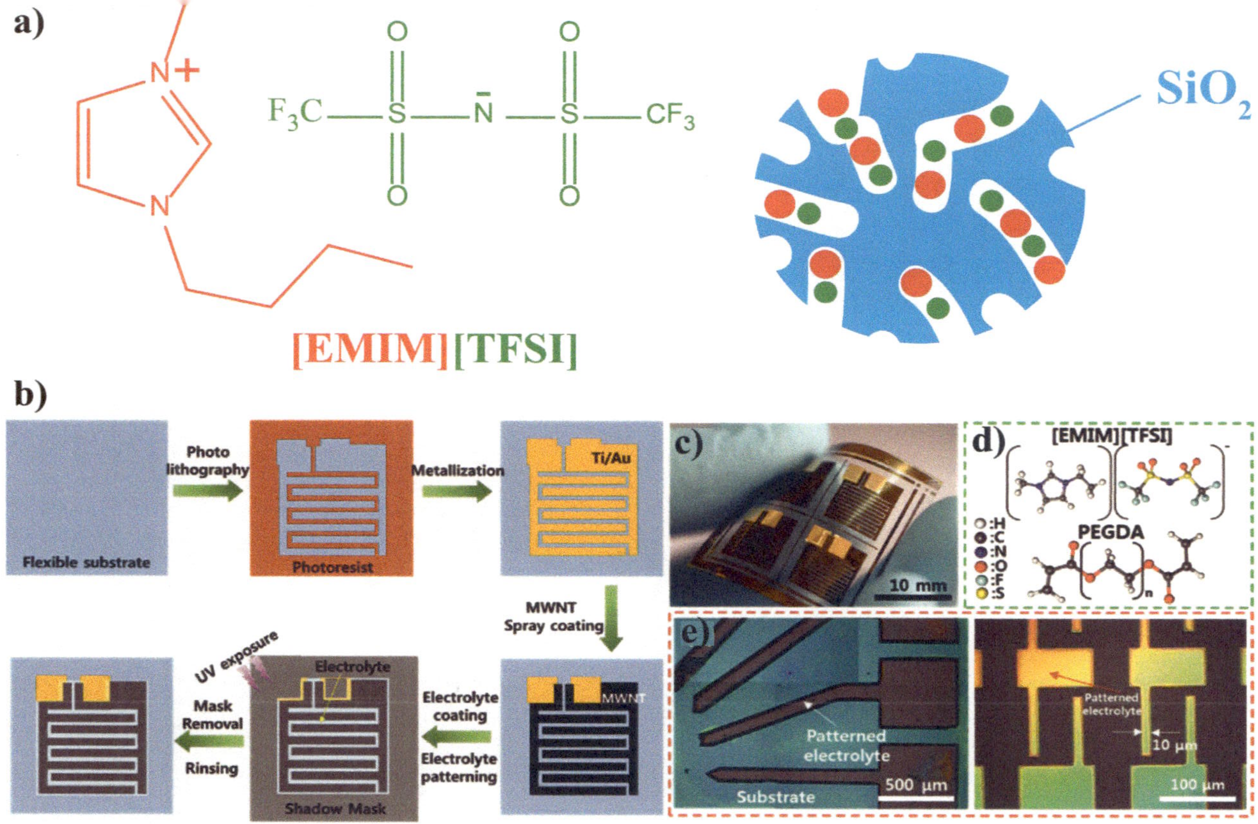

Fig. 3.15 (A) Illustration of fumed SiO_2-based ionogel used for fabrication of LSG-based MSC device [67], and chemical structure of the RTIL used is also displayed; (B) steps involved in the in situ fabrication of MSC device by UV cured ionogel electrolyte; (C) optical image of the MSC fabricated on PET substrate; (D) structure of PEGDA polymer matrix and [EMIM][TFSI] RTIL; and (E) optical microscopy images of the various ionogel patterns [37]. *(American Chemical Society; Reprinted with permission from D. Kim, G. Lee, D. Kim, J.S. Ha, Air-stable, high-performance, flexible microsupercapacitor with patterned ionogel electrolyte, ACS Appl. Mater. Interfaces 7 (8) (2015) 4608–4615, https://doi.org/10.1021/am5077843, Copyright (2015). American Chemical Society.)*

approach to design solid-state MSCs. This solution processing strategy ensures electrolyte penetration throughout the electrode and into the interior leading to maximum utilization of electrode material and low Ohmic resistance of the device. Preparation of such an electrolyte by employing a photolithographic process from a mixture of a UV-curable monomer, a photo-initiator, and the desired RTIL is a straightforward technique. For example, a patterned ionogel with 10 μm thickness is obtained from a mixture of poly(ethylene glycol) diacrylate (PEGDA) and [EMIM][TFSI] RTIL in the presence of 2-hydroxy-2-methylpropiophenone (HMPP) photo-initiator [37]. A spray-coated MWCNT-COOH electrode is first completely coated with this monomer solution. After this process, the electrode is covered with a shadow mask and subjected to UV irradiation to initiate the photo-polymerization process. This is the first report on the fabrication of MSC device by means of in situ PE generation strategy (Fig. 15B–D).

Sol-gel-assisted cocondensation method is developed to immobilize RTIL in a silicon-based scaffold using silicon alkoxides (e.g., tetramethyl orthosilicate (TMOS)) as silica precursor and formic acid as catalyst [220, 221]. Compared to a simple physical mixture of RTIL in a gel polymer matrix, ionogel prepared by condensation method exhibits better mechanical property as well as ionic conductivity (3 mS cm^{-1}). An ionogel electrolyte with [EMIM][TFSI] RTIL is prepared by following the same process and used for fabricating a highly flexible MSC based on CVD-grown vertically aligned CNT as the active material [222]. Addition of dimethyldimethoxysilane (DMDMS) as coprecursor into the TMOS-formic acid sol mixture is found to improve the mechanical property of the ionogel [220]. This electrolyte, when used with silicon nanowire-based solid-state MSC, delivers similar capacitance (60 μF cm^{-2} for 40 μm long nanowire) as that for the nonconfined RTIL. The summary of ionogel-based MSC devices is presented in Table 3.12.

3.5 Conclusions and future perspectives

Futuristic devices such as smart electronics and the Internet of Things (IoTs) demand small footprint, energy-dense, and safe energy storage devices, and microsupercapacitors (MSCs) play a dynamic role in this respect. Even though MSCs have been invented recently and matured by two decades, there have been enormous works toward the selection of a family of electrode materials, materials synthesis, deposition procedures, and device

Table 3.12 Components, fabrication, characteristics, and performance summary of ionogel-based MSC devices.

Ionogel	Electrode material	Substrate	Microelectrode fabrication method	Capacitance (areal/volumetric)	Voltage (V)	Refs.
[EMIM][TFSI-PEGDAE	SWCNT-COOH	Ti/Au current collector deposited PET	Spray coating	5.3 F cm^{-3}	0–2	[37]
[EMIM][TFSI]-fumed SiO_2	CNT	Si/polycarbonate	CVD	430 μF cm^{-2}	0–3	[222]
[EMIM][BF_4]-fumed SiO_2	$Cu(OH)_2$@FeOOH nanotubes	Polyimide	Screen-printing, electroless Cu deposition and immersion	58 mF cm^{-2}	0–1.5	[217]
PVA-[BMIM][BF_4]	VS_2	Au	Mechanical shaping	4760 μF cm^{-2}	0–0.6	[218]
PVDF-HFP-[PYR_{14}][TFSI]	LTO (negative) and AG (positive) electrodes	Nylon membrane	Mask-assisted LbL filtration	28.4 F cm^{-3}	0–3	[69]
[EMIM][TFSI]-fumed SiO_2	Si nanowire	Si	Growth of nanowire by CVD catalyzed by Au	30–60 μF cm^{-2}	0–3.5	[220]
[EMIM][TFSI]-fumed SiO_2	AC	Al	Slurry coating	15 mF cm^{-2}	0–3	[221]

fabrication. However, research is rather limited in fully solid-state systems. Solid-state MSCs can be realized using polymer electrolytes (PEs) or inorganic solid electrolytes (ISEs). The low ionic conductivity, as well as the tedious synthesis methods, pulls back the practicality of ISEs in the context of MSCs. Among the PEs, gel polymer electrolytes (GPEs) exhibit higher ionic conductivity than the solid polymer electrolytes (SPEs). Owing to the presence of a liquid phase in the GPE, electrode/electrolyte interfaces are effectively utilized in addition to providing a solid-like operability and electrode/electrolyte interface behavior close to liquid electrolytes.

Recent reports on the in situ generation of GPEs from the constituent monomers dissolved in an electrolyte by UV-light or thermal exposure although not extensively used for MSC fabrication so far, such processing techniques can drastically improve the electrode/electrolyte interfaces. Since SPEs can also be prepared by means of the same insitu strategy, the application of SPEs in MSCs could be rejuvenated in the upcoming years. With the surge of interest in all-solid-state alkali-ion batteries (e.g., Li-ion and Na-ion batteries), there have been many new Li^+ and Na^+ (in aqueous media, also including H^+ and OH^- conduction) ion conducting ISEs introduced in the recent years and MSCs could largely benefit from this upsurge as well.

The attempts on improving the ionic conductivity of SPEs mainly pertain toward tuning the amorphous character. In order to achieve this, cross-linking of the polymer matrix has been attempted in addition to the synthesis CPEs where inorganic fillers are mixed with the solid polymer matrix. CPEs by combining solid polymer matrices and currently available ISEs are also popular. If the current scenario of SPEs is considered, there have been enormous developments in the past two decades resulting in systems, which can surpass ISEs. These electrolytes are heavily employed for lithium-ion and lithium-metal battery fabrication. As of today, these electrolyte systems are yet to be investigated in MSCs. For the future development of all-solid-state MSCs, there should be attempts to fabricate devices based on SPEs. MSCs can adopt numerous materials that have already been studied so far in lithium and lithium polymer batteries, in particular lithium-based microbatteries. There can be a bright future for research and application of all-solid-state MSCs, once the research approaches and challenges are appropriately addressed.

References

[1] C. Lethien, J. Le Bideau, T. Brousse, Challenges and prospects of 3D micro-supercapacitors for powering the internet of things, Energ. Environ. Sci. 12 (1) (2019) 96–115, https://doi.org/10.1039/C8EE02029A.
[2] L. Xia, L. Yu, D. Hu, G.Z. Chen, Electrolytes for electrochemical energy storage, Mater. Chem. Front. 1 (4) (2017) 584–618, https://doi.org/10.1039/C6QM00169F.
[3] V. Palomares, P. Serras, I. Villaluenga, K.B. Hueso, J. Carretero-González, T. Rojo, Na-ion batteries, recent advances and present challenges to become low cost energy storage systems, Energ. Environ. Sci. 5 (3) (2012) 5884–5901, https://doi.org/10.1039/C2EE02781J.
[4] D. Aurbach, Y. Talyosef, B. Markovsky, E. Markevich, E. Zinigrad, L. Asraf, J.S. Gnanaraj, H.-J. Kim, Design of electrolyte solutions for Li and Li-ion batteries: a review, Electrochim. Acta 50 (2) (2004) 247–254, https://doi.org/10.1016/j.electacta.2004.01.090.
[5] K. Xu, Nonaqueous liquid electrolytes for lithium-based rechargeable batteries, Chem. Rev. 104 (10) (2004) 4303–4418, https://doi.org/10.1021/cr030203g.
[6] J. Owen, 21 - Ionic Conductivity, in: G. Allen, J.C. Bevington (Eds.), Comprehensive Polymer Science and Supplements, Pergamon, Amsterdam, 1989, pp. 669–686.
[7] K. Xu, Electrolytes and interphases in Li-ion batteries and beyond, Chem. Rev. 114 (23) (2014) 11503–11618, https://doi.org/10.1021/cr500003w.
[8] M. Amereller, T. Schedlbauer, D. Moosbauer, C. Schreiner, C. Stock, F. Wudy, S. Zugmann, H. Hammer, A. Maurer, R.M. Gschwind, H.D. Wiemhöfer, M. Winter, H.J. Gores, Electrolytes for lithium and lithium ion batteries: from synthesis of novel lithium borates and ionic liquids to development of novel measurement methods, Prog. Solid State Chem. 42 (4) (2014) 39–56, https://doi.org/10.1016/j.progsolidstchem.2014.04.001.
[9] S. Nowak, M. Winter, Chemical analysis for a better understanding of aging and degradation mechanisms of non-aqueous electrolytes for lithium ion batteries: method development, application and lessons learned, J. Electrochem. Soc. 162 (14) (2015) A2500–A2508.
[10] D.P. Dubal, N.R. Chodankar, D.-H. Kim, P. Gomez-Romero, Towards flexible solid-state supercapacitors for smart and wearable electronics, Chem. Soc. Rev. 47 (6) (2018) 2065–2129, https://doi.org/10.1039/C7CS00505A.
[11] P. Murmann, N. von Aspern, P. Janssen, N. Kalinovich, M. Shevchuk, G.-V. Röschenthaler, M. Winter, I. Cekic-Laskovic, Influence of the fluorination degree of organophosphates on flammability and electrochemical performance in lithium ion batteries, J. Electrochem. Soc. 165 (9) (2018) A1935–A1942.
[12] H. Jia, H. Onishi, N. von Aspern, U. Rodehorst, K. Rudolf, B. Billmann, R. Wagner, M. Winter, I. Cekic-Laskovic, A propylene carbonate based gel polymer electrolyte for extended cycle life and improved safety performance of lithium ion batteries, J. Power Sources 397 (2018) 343–351.
[13] C. Zhong, Y. Deng, W. Hu, J. Qiao, L. Zhang, J. Zhang, A review of electrolyte materials and compositions for electrochemical supercapacitors, Chem. Soc. Rev. 44 (21) (2015) 7484–7539, https://doi.org/10.1039/C5CS00303B.
[14] M. Nazri, Liquid electrolytes: some theoretical and practical aspects, in: G.-A. Nazri, G. Pistoia (Eds.), Lithium Batteries: Science and Technology, Springer US, Boston, MA, 2003, pp. 509–529.
[15] M. Beidaghi, Y. Gogotsi, Capacitive energy storage in micro-scale devices: recent advances in design and fabrication of micro-supercapacitors, Energ. Environ. Sci. 7 (3) (2014) 867–884, https://doi.org/10.1039/C3EE43526A.

[16] T. Famprikis, P. Canepa, J.A. Dawson, M.S. Islam, C. Masquelier, Fundamentals of inorganic solid-state electrolytes for batteries, Nat. Mater. 18 (12) (2019) 1278–1291, https://doi.org/10.1038/s41563-019-0431-3.

[17] D. Golodnitsky, E. Strauss, E. Peled, S. Greenbaum, On order and disorder in polymer electrolytes, J. Electrochem. Soc. 162 (14) (2015) A2551–A2566.

[18] D. Bresser, S. Lyonnard, C. Iojoiu, L. Picard, S. Passerini, Decoupling segmental relaxation and ionic conductivity for lithium-ion polymer electrolytes, Mol. Syst. Des. Eng. 4 (4) (2019) 779–792, https://doi.org/10.1039/C9ME00038K.

[19] P.V. Wright, Electrical conductivity in ionic complexes of poly(ethylene oxide), Br. Polym. J. 7 (5) (1975) 319–327, https://doi.org/10.1002/pi.4980070505.

[20] D.E. Fenton, J.M. Parker, P.V. Wright, Complexes of alkali metal ions with poly(ethylene oxide), Polymer 14 (11) (1973) 589, https://doi.org/10.1016/0032-3861(73)90146-8.

[21] Z. Xue, D. He, X. Xie, Poly(ethylene oxide)-based electrolytes for lithium-ion batteries, J. Mater. Chem. A 3 (38) (2015) 19218–19253, https://doi.org/10.1039/C5TA03471J.

[22] X. Cheng, J. Pan, Y. Zhao, M. Liao, H. Peng, Gel polymer electrolytes for electrochemical energy storage, Adv. Energy Mater. 8 (7) (2018) 1702184, https://doi.org/10.1002/aenm.201702184.

[23] S.S. Sekhon, Conductivity behaviour of polymer gel electrolytes: role of polymer, Bull. Mater. Sci. 26 (3) (2003) 321–328, https://doi.org/10.1007/BF02707454.

[24] M. Watanabe, N. Ogata, Ionic conductivity of polymer electrolytes and future applications, Br. Polym. J. 20 (3) (1988) 181–192, https://doi.org/10.1002/pi.4980200304.

[25] S.B. Aziz, T.J. Woo, M.F.Z. Kadir, H.M. Ahmed, A conceptual review on polymer electrolytes and ion transport models, J. Sci.: Adv. Mater. Dev. 3 (1) (2018) 1–17, https://doi.org/10.1016/j.jsamd.2018.01.002.

[26] A. Manuel Stephan, Review on gel polymer electrolytes for lithium batteries, Eur. Polym. J. 42 (1) (2006) 21–42, https://doi.org/10.1016/j.eurpolymj.2005.09.017.

[27] J. Le Bideau, L. Viau, A. Vioux, Ionogels, ionic liquid based hybrid materials, Chem. Soc. Rev. 40 (2) (2011) 907–925, https://doi.org/10.1039/C0CS00059K.

[28] J.C. Bachman, S. Muy, A. Grimaud, H.-H. Chang, N. Pour, S.F. Lux, O. Paschos, F. Maglia, S. Lupart, P. Lamp, L. Giordano, Y. Shao-Horn, Inorganic solid-state electrolytes for lithium batteries: mechanisms and properties governing ion conduction, Chem. Rev. 116 (1) (2016) 140–162, https://doi.org/10.1021/acs.chemrev.5b00563.

[29] C. Cao, Z.-B. Li, X.-L. Wang, X.-B. Zhao, W.-Q. Han, Recent advances in inorganic solid electrolytes for lithium batteries, Front. Energy Res. 2 (2014) 25.

[30] X. Yu, A. Manthiram, Electrode–electrolyte interfaces in lithium-based batteries, Energ. Environ. Sci. 11 (3) (2018) 527–543, https://doi.org/10.1039/C7EE02555F.

[31] P. Knauth, Inorganic solid Li ion conductors: an overview, Solid State Ion. 180 (14) (2009) 911–916, https://doi.org/10.1016/j.ssi.2009.03.022.

[32] P. Yao, H. Yu, Z. Ding, Y. Liu, J. Lu, M. Lavorgna, J. Wu, X. Liu, Review on polymer-based composite electrolytes for lithium batteries, Front. Chem. 7 (2019) 522.

[33] S. Zheng, X. Shi, P. Das, Z.-S. Wu, X. Bao, The road towards planar microbatteries and micro-supercapacitors: from 2D to 3D device geometries, Adv. Mater. 31 (50) (2019) 1900583, https://doi.org/10.1002/adma.201900583.

[34] V. Vijayakumar, B. Anothumakkool, A.T.A. Torris, S.B. Nair, M.V. Badiger, S. Kurungot, An all-solid-state-supercapacitor possessing a non-aqueous gel polymer electrolyte prepared using a UV-assisted in situ polymerization strategy, J. Mater. Chem. A 5 (18) (2017) 8461–8476, https://doi.org/10.1039/C7TA01514C.

[35] B. Anothumakkool, A.T.A. Torris, S. Veeliyath, V. Vijayakumar, M.V. Badiger, S. Kurungot, High-performance flexible solid-state supercapacitor with an extended nanoregime interface through in situ polymer electrolyte generation, ACS Appl. Mater. Interfaces 8 (2) (2016) 1233–1241, https://doi.org/10.1021/acsami.5b09677.
[36] J.R. Nair, C. Gerbaldi, G. Meligrana, R. Bongiovanni, S. Bodoardo, N. Penazzi, P. Reale; Gentili, V. UV-cured methacrylic membranes as novel gel–polymer electrolyte for Li-ion batteries, J. Power Sources 178 (2) (2008) 751–757, https://doi.org/10.1016/j.jpowsour.2007.08.004.
[37] D. Kim, G. Lee, D. Kim, J.S. Ha, Air-stable, high-performance, flexible microsupercapacitor with patterned ionogel electrolyte, ACS Appl. Mater. Interfaces 7 (8) (2015) 4608–4615, https://doi.org/10.1021/am5077843.
[38] N. Yadav, N. Yadav, M.K. Singh, S.A. Hashmi, Nonaqueous, redox-active gel polymer electrolyte for high-performance supercapacitor, Energ. Technol. 7 (9) (2019) 1900132, https://doi.org/10.1002/ente.201900132.
[39] B. Evanko, S.W. Boettcher, S.J. Yoo, G.D. Stucky, Redox-enhanced electrochemical capacitors: status, opportunity, and best practices for performance evaluation, ACS Energy Lett. 2 (11) (2017) 2581–2590, https://doi.org/10.1021/acsenergylett.7b00828.
[40] D. Kim, G. Lee, D. Kim, J. Yun, S.-S. Lee, J.S. Ha, High performance flexible double-sided micro-supercapacitors with an organic gel electrolyte containing a redox-active additive, Nanoscale 8 (34) (2016) 15611–15620, https://doi.org/10.1039/C6NR04352F.
[41] B. Nagar, D.P. Dubal, L. Pires, A. Merkoçi, P. Gómez-Romero, Design and fabrication of printed paper-based hybrid micro-supercapacitor by using graphene and redox-active electrolyte, ChemSusChem 11 (11) (2018) 1849–1856, https://doi.org/10.1002/cssc.201800426.
[42] L. Liu, H. Zhao, Y. Lei, Advances on three-dimensional electrodes for micro-supercapacitors: a mini-review, InfoMat 1 (1) (2019) 74–84, https://doi.org/10.1002/inf2.12007.
[43] H. Zhang, Y. Cao, M.O.L. Chee, P. Dong, M. Ye, J. Shen, Recent advances in micro-supercapacitors, Nanoscale 11 (13) (2019) 5807–5821, https://doi.org/10.1039/C9NR01090D.
[44] C. Liu, Z. Zhang, Electrochemical materials design for micro-supercapacitors, Supercapacitor Des. Appl. 19 (2016).
[45] S.R. Gowda, A.L.M. Reddy, M.M. Shaijumon, X. Zhan, L. Ci, P.M. Ajayan, Conformal coating of thin polymer electrolyte layer on nanostructured electrode materials for three-dimensional battery applications, Nano Lett. 11 (1) (2011) 101–106, https://doi.org/10.1021/nl102919m.
[46] B. Mendoza-Sánchez, Y. Gogotsi, Synthesis of two-dimensional materials for capacitive energy storage, Adv. Mater. 28 (29) (2016) 6104–6135, https://doi.org/10.1002/adma.201506133.
[47] Y.-T. Chen, C.-W. Ma, C.-M. Chang, Y.-J. Yang, Micromachined planar supercapacitor with interdigital buckypaper electrodes, Micromachines 9 (5) (2018) 242.
[48] J.J. Yoo, K. Balakrishnan, J. Huang, V. Meunier, B.G. Sumpter, A. Srivastava, M. Conway, A.L. Mohana Reddy, J. Yu, R. Vajtai, P.M. Ajayan, Ultrathin planar graphene supercapacitors, Nano Lett. 11 (4) (2011) 1423–1427, https://doi.org/10.1021/nl200225j.
[49] K.U. Laszczyk, D.N. Futaba, K. Kobashi, K. Hata, T. Yamada, A. Sekiguchi, The limitation of electrode shape on the operational speed of a carbon nanotube based micro-supercapacitor, Sustain. Energy Fuels 1 (6) (2017) 1282–1286, https://doi.org/10.1039/C7SE00101K.

[50] Z. Lin, P.-L. Taberna, P. Simon, Electrochemical double layer capacitors: what is next beyond the corner? Curr. Opin. Electrochem. 6 (1) (2017) 115–119, https://doi.org/10.1016/j.coelec.2017.10.013.
[51] A.C. Forse, J.M. Griffin, C. Merlet, J. Carretero-Gonzalez, A.-R.O. Raji, N.M. Trease, C.P. Grey, Direct observation of ion dynamics in supercapacitor electrodes using in situ diffusion NMR spectroscopy, Nat. Energy *2* (3) (**2017**) 16216, https://doi.org/10.1038/nenergy.2016.216.
[52] N.A. Kyeremateng, T. Brousse, D. Pech, Microsupercapacitors as miniaturized energy-storage components for on-chip electronics, Nat. Nanotechnol. 12 (1) (2017) 7–15, https://doi.org/10.1038/nnano.2016.196.
[53] P. Simon, Y. Gogotsi, Capacitive energy storage in nanostructured carbon–electrolyte systems, Acc. Chem. Res. 46 (5) (2013) 1094–1103, https://doi.org/10.1021/ar200306b.
[54] K. Nie, Y. Hong, J. Qiu, Q. Li, X. Yu, H. Li, L. Chen, Interfaces between cathode and electrolyte in solid state lithium batteries: challenges and perspectives, Front. Chem. 6 (2018).
[55] J. Kasnatscheew, R. Wagner, M. Winter, I. Cekic-Laskovic, Interfaces and materials in lithium ion batteries: challenges for theoretical electrochemistry, Top. Curr. Chem. 376 (3) (2018) 16, https://doi.org/10.1007/s41061-018-0196-1.
[56] N. Kazemi, D.L. Danilov, L. Haverkate, N.J. Dudney, S. Unnikrishnan, P.H.L. Notten, Modeling of all-solid-state thin-film Li-ion batteries: accuracy improvement, Solid State Ion. 334 (2019) 111–116, https://doi.org/10.1016/j.ssi.2019.02.003.
[57] M. Yu, X. Feng, Thin-film electrode-based supercapacitors, Joule **3** (2019).
[58] S. Jankowsky, M.M. Hiller, O. Fromm, M. Winter, H.D. Wiemhöfer, Enhanced lithium-ion transport in polyphosphazene based gel polymer electrolytes, Electrochim. Acta 155 (2015) 364–371, https://doi.org/10.1016/j.electacta.2014.12.123.
[59] K. Borzutzki, J. Thienenkamp, M. Diehl, M. Winter, G. Brunklaus, Fluorinated polysulfonamide based single ion conducting room temperature applicable gel-type polymer electrolytes for lithium ion batteries, J. Mater. Chem. A 7 (1) (2019) 188–201, https://doi.org/10.1039/C8TA08391F.
[60] L. Porcarelli, C. Gerbaldi, F. Bella, J.R. Nair, Super soft all-ethylene oxide polymer electrolyte for safe all-solid lithium batteries, Sci. Rep. 6 (2016) 19892, https://doi.org/10.1038/srep19892. https://www.nature.com/articles/srep19892#supplementary-information.
[61] L. Porcarelli, A.S. Shaplov, F. Bella, J.R. Nair, D. Mecerreyes, C. Gerbaldi, Single-ion conducting polymer electrolytes for lithium metal polymer batteries that operate at ambient temperature, ACS Energy Lett. 1 (4) (2016) 678–682, https://doi.org/10.1021/acsenergylett.6b00216.
[62] H.-K. Kim, T.-Y. Seong, J.-H. Lim, W. Ii Cho, Y. Soo Yoon, Electrochemical and structural properties of radio frequency sputtered cobalt oxide electrodes for thin-film supercapacitors, J. Power Sources 102 (1) (2001) 167–171, https://doi.org/10.1016/S0378-7753(01)00864-3.
[63] H.-K. Kim, S.-H. Cho, Y.-W. Ok, T.-Y. Seong, Y.S. Yoon, All solid-state rechargeable thin-film microsupercapacitor fabricated with tungsten cosputtered ruthenium oxide electrodes, J. Vacuum Sci. Technol. B 21 (3) (2003) 949–952.
[64] J.-H. Sung, S.-J. Kim, K.-H. Lee, Fabrication of all-solid-state electrochemical microcapacitors, J. Power Sources 133 (2) (2004) 312–319, https://doi.org/10.1016/j.jpowsour.2004.02.003.
[65] J.-H. Sung, S.-J. Kim, S.-H. Jeong, E.-H. Kim, K.-H. Lee, Flexible microsupercapacitors, J. Power Sources 162 (2) (2006) 1467–1470.

[66] H. Xing, X. Wang, C. Shen, S. Li, Fabrication and tests of a three-dimensional micro-supercapacitor using SU-8 photoresist as the separator, Micro Nano Lett. 7 (12) (2012) 1166–1169.
[67] M.F. El-Kady, R.B. Kaner, Scalable fabrication of high-power graphene micro-supercapacitors for flexible and on-chip energy storage, Nat. Commun. 4 (1) (2013) 1475, https://doi.org/10.1038/ncomms2446.
[68] C.-H. Chang, B. Hsia, J.P. Alper, S. Wang, L.E. Luna, C. Carraro, S.-Y. Lu, R. Maboudian, High-temperature all solid-state microsupercapacitors based on SiC nanowire electrode and YSZ electrolyte, ACS Appl. Mater. Interfaces 7 (48) (2015) 26658–26665, https://doi.org/10.1021/acsami.5b08423.
[69] S. Zheng, J. Ma, Z.-S. Wu, F. Zhou, Y.-B. He, F. Kang, H.-M. Cheng, X. Bao, All-solid-state flexible planar lithium ion micro-capacitors, Energ. Environ. Sci. 11 (8) (2018) 2001–2009, https://doi.org/10.1039/C8EE00855H.
[70] G. Sun, H. Yang, G. Zhang, J. Gao, X. Jin, Y. Zhao, L. Jiang, L. Qu, A capacity recoverable zinc-ion micro-supercapacitor, Energ. Environ. Sci. 11 (12) (2018) 3367–3374, https://doi.org/10.1039/C8EE02567C.
[71] B. Neudecker, N. Dudney, J. Bates, "Lithium-free" thin-film battery with in situ plated Li anode, J. Electrochem. Soc. 147 (2) (2000) 517–523.
[72] H.-K. Kim, Y.-W. Ok, T.-Y. Seong, E.J. Jeon, W.i. Cho, Y.S. Yoon, Microstructures and electrochemical properties of Pt-doped amorphous V 2 O 5 cathode films for thin film microbatteries, J. Vac. Sci. Technol. A 19 (5) (2001) 2549–2553.
[73] J.H. Lim, D.J. Choi, H.-K. Kim, W.I. Cho, Y.S. Yoon, Thin film supercapacitors using a sputtered RuO2 electrode, J. Electrochem. Soc. 148 (3) (2001) A275–A278.
[74] H.-K. Kim, T.-Y. Seong, J.H. Lim, Y.-W. Ok, W.i. Cho, Y.H. Shin, Y.S. Yoon, Correlation between the microstructures and the cycling performance of RuO 2 electrodes for thin-film microsupercapacitors, J. Vacuum Sci. Technol. B 20 (5) (2002) 1827–1832.
[75] M.-J. Lee, J.S. Kim, S.H. Choi, J.J. Lee, S.H. Kim, S.H. Jee, Y.S. Yoon, Characteristics of thin film supercapacitor with ruthenium oxide electrode and Ta2O5 + x solid oxide thin film electrolyte, J. Electroceram. 17 (2) (2006) 639–643, https://doi.org/10.1007/s10832-006-7777-z.
[76] T. Göhlert, P.F. Siles, T. Päßler, R. Sommer, S. Baunack, S. Oswald, O.G. Schmidt, Ultra-thin all-solid-state micro-supercapacitors with exceptional performance and device flexibility, Nano Energy 33 (2017) 387–392, https://doi.org/10.1016/j.nanoen.2017.01.054.
[77] J. Wang, F. Li, F. Zhu, O.G. Schmidt, Recent progress in micro-supercapacitor design, integration, and functionalization, Small Methods 3 (8) (2019) 1800367, https://doi.org/10.1002/smtd.201800367.
[78] V. Sallaz, S. Oukassi, S. Poulet, F. Voiron, M. Bedjaoui, R. Salot, D. Berardan, On-Chip Microsupercapacitors Using Ultrathin Lipon Film, Meeting Abstracts, The Electrochemical Society, 2019, p. 161.
[79] A. Manthiram, X. Yu, S. Wang, Lithium battery chemistries enabled by solid-state electrolytes, Nat. Rev. Mater. 2 (4) (2017) 16103, https://doi.org/10.1038/natrevmats.2016.103.
[80] Y. Wang, S. Song, C. Xu, N. Hu, J. Molenda, L. Lu, Development of solid-state electrolytes for sodium-ion battery—a short review, Nano Mater. Sci. 1 (2) (2019) 91–100, https://doi.org/10.1016/j.nanoms.2019.02.007.
[81] J.R. Nair, L. Imholt, G. Brunklaus, M. Winter, Lithium metal polymer electrolyte batteries: opportunities and challenges, Electrochem. Soc. Interface 28 (2) (2019) 55–61.

[82] H. Gao, A. Virya, K. Lian, Proton conducting H5BW12O40 electrolyte for solid supercapacitors, J. Mater. Chem. A 3 (43) (2015) 21511–21517, https://doi.org/10.1039/C5TA06685A.
[83] X. Wu, H. Cai, Q. Wu, W. Yan, Synthesis and high proton conductive performance of vanadium-substituted Dawson structure heteropoly acid H8P2W16V2O62·20H2O, Mater. Lett. 181 (2016) 1–3, https://doi.org/10.1016/j.matlet.2016.06.012.
[84] Z. Xie, H. Wu, Q. Wu, L. Ai, Synthesis and performance of solid proton conductor molybdovanadosilicic acid, RSC Adv. 8 (25) (2018) 13984–13988, https://doi.org/10.1039/C8RA02390E.
[85] R. Ramachandran, F. Wang, Electrochemical capacitor performance: influence of aqueous electrolytes, in: Supercapacitors-Theoretical and Practical Solutions, InTechOpen, 2017.
[86] J. Lawton, S. Tiano, D. Donnelly, S. Flanagan, T. Arruda, The effect of sulfuric acid concentration on the physical and electrochemical properties of vanadyl solutions, Batteries 4 (3) (2018) 40.
[87] K. Kwon, J.W. Evans, Viscosity changes of Li battery electrolytes and their long-term effect on the frequency of EQCM electrodes, Electrochem. Solid St. 5 (3) (2002) A59–A61.
[88] I. Heng, C.W. Lai, J.C. Juan, Organic electrolytes for supercapacitors, Supercapacitor Technol. Mater. Process. Architect. *61* (**2019**) 1.
[89] Y. Yokoyama, T. Fukutsuka, K. Miyazaki, T. Abe, Origin of the electrochemical stability of aqueous concentrated electrolyte solutions, J. Electrochem. Soc. 165 (14) (2018) A3299–A3303.
[90] Z. Cheng, D. Yida, H. Wenbin, S. Daoming, H. Xiaopeng, Q. Jinli, Z. Jiujun, Electrolytes for electrochemical supercapacitors, in: Electrolytes for Electrochemical Supercapacitors, Chapter Electrolytes for Electrochemical Supercapacitors, CRC Press, 2016.
[91] M. Létiche, K. Brousse, A. Demortière, P. Huang, B. Daffos, S. Pinaud, M. Respaud, B. Chaudret, P. Roussel, L. Buchaillot, P.L. Taberna, P. Simon, C. Lethien, Sputtered titanium carbide thick film for high areal energy on chip carbon-based micro-supercapacitors, Adv. Funct. Mater. 27 (20) (2017) 1606813, https://doi.org/10.1002/adfm.201606813.
[92] L. Wei, N. Nitta, G. Yushin, Lithographically patterned thin activated carbon films as a new technology platform for on-chip devices, ACS Nano 7 (8) (2013) 6498–6506, https://doi.org/10.1021/nn4028129.
[93] Y. Yang, L. He, C. Tang, P. Hu, X. Hong, M. Yan, Y. Dong, X. Tian, Q. Wei, L. Mai, Improved conductivity and capacitance of interdigital carbon microelectrodes through integration with carbon nanotubes for micro-supercapacitors, Nano Res. 9 (8) (2016) 2510–2519, https://doi.org/10.1007/s12274-016-1137-3.
[94] Y. Gao, Z. Shi, L. Li, 2019. Carbon materials for supercapacitors. In *Novel Carbon Materials and Composites*; pp 137-167.
[95] P. Huang, C. Lethien, S. Pinaud, K. Brousse, R. Laloo, V. Turq, M. Respaud, A. Demortière, B. Daffos, P.L. Taberna, B. Chaudret, Y. Gogotsi, P. Simon, On-chip and freestanding elastic carbon films for micro-supercapacitors, Science 351 (6274) (2016) 691, https://doi.org/10.1126/science.aad3345.
[96] B. Gorska, E. Frackowiak, F. Beguin, Redox active electrolytes in carbon/carbon electrochemical capacitors, Curr. Opin. Electrochem. 9 (2018) 95–105, https://doi.org/10.1016/j.coelec.2018.05.006.
[97] D.M. Anjos, J.K. McDonough, E. Perre, G.M. Brown, S.H. Overbury, Y. Gogotsi, V. Presser, Pseudocapacitance and performance stability of quinone-coated carbon onions, Nano Energy 2 (5) (2013) 702–712, https://doi.org/10.1016/j.nanoen.2013.08.003.

[98] B. Anothumakkool, P.-L. Taberna, B. Daffos, P. Simon, Y. Sayed-Ahmad--Baraza, C. Ewels, T. Brousse, J. Gaubicher, Improved electro-grafting of nitropyrene onto onion-like carbon via in situ electrochemical reduction and polymerization: tailoring redox energy density of the supercapacitor positive electrode, J. Mater. Chem. A 5 (4) (2017) 1488–1494, https://doi.org/10.1039/C6TA08170C.
[99] C. Zhan, Jiang, D.-e., Understanding the pseudocapacitance of RuO2 from joint density functional theory, J. Phys. Condens. Matter 28 (46) (2016) 464004, https://doi.org/10.1088/0953-8984/28/46/464004.
[100] A. Eftekhari, L. Li, Y. Yang, Polyaniline supercapacitors, J. Power Sources 347 (2017) 86–107, https://doi.org/10.1016/j.jpowsour.2017.02.054.
[101] G.A. Snook, P. Kao, A.S. Best, Conducting-polymer-based supercapacitor devices and electrodes, J. Power Sources 196 (1) (2011) 1–12, https://doi.org/10.1016/j.jpowsour.2010.06.084.
[102] S. Makino, Y. Yamauchi, W. Sugimoto, Synthesis of electro-deposited ordered mesoporous RuOx using lyotropic liquid crystal and application toward micro-supercapacitors, J. Power Sources 227 (2013) 153–160, https://doi.org/10.1016/j.jpowsour.2012.11.032.
[103] Z. Liu, S. Liu, R. Dong, S. Yang, H. Lu, A. Narita, X. Feng, K. Müllen, High power in-plane micro-supercapacitors based on mesoporous polyaniline patterned graphene, Small 13 (14) (2017) 1603388, https://doi.org/10.1002/smll.201603388.
[104] J.-H. Sung, S.-J. Kim, K.-H. Lee, Fabrication of microcapacitors using conducting polymer microelectrodes, J. Power Sources 124 (1) (2003) 343–350, https://doi.org/10.1016/S0378-7753(03)00669-4.
[105] M. Boota, C. Chen, M. Bécuwe, L. Miao, Y. Gogotsi, *Noncovalent Functionalization of Quinone Derivatives on 2D Graphene for High Performance Pseudocapacitive Electrodes*, Meeting Abstracts, The Electrochemical Society, 2016, p. 4078.
[106] W.A. Badawy, F.M. Al-Kharafi, J.R. Al-Ajmi, Electrochemical behaviour of cobalt in aqueous solutions of different pH, J. Appl. Electrochem. 30 (6) (2000) 693–704, https://doi.org/10.1023/A:1003893122201.
[107] F. Allebrod, C. Chatzichristodoulou, P.L. Mollerup, M.B. Mogensen, Electrical conductivity measurements of aqueous and immobilized potassium hydroxide, Int. J. Hydrogen Energy 37 (21) (2012) 16505–16514, https://doi.org/10.1016/j.ijhydene.2012.02.088.
[108] R. Ramachandran, M. Saranya, V. Velmurugan, B.P.C. Raghupathy, S.K. Jeong, A.N. Grace, Effect of reducing agent on graphene synthesis and its influence on charge storage towards supercapacitor applications, Appl. Energy 153 (2015) 22–31, https://doi.org/10.1016/j.apenergy.2015.02.091.
[109] C. Zhu, T. Liu, F. Qian, T.Y.-J. Han, E.B. Duoss, J.D. Kuntz, C.M. Spadaccini, M.A. Worsley, Y. Li, Supercapacitors based on three-dimensional hierarchical graphene aerogels with periodic macropores, Nano Lett. 16 (6) (2016) 3448–3456, https://doi.org/10.1021/acs.nanolett.5b04965.
[110] C.Y. Foo, H.N. Lim, M.A. Mahdi, M.H. Wahid, N.M. Huang, Three-dimensional printed electrode and its novel applications in electronic devices, Sci. Rep. 8 (1) (2018) 7399, https://doi.org/10.1038/s41598-018-25861-3.
[111] X. Tang, H. Zhou, Z. Cai, D. Cheng, P. He, P. Xie, D. Zhang, T. Fan, Generalized 3D printing of graphene-based mixed-dimensional hybrid aerogels, ACS Nano 12 (4) (2018) 3502–3511.
[112] N. Kurra, Q. Jiang, H.N. Alshareef, A general strategy for the fabrication of high performance microsupercapacitors, Nano Energy 16 (2015) 1–9, https://doi.org/10.1016/j.nanoen.2015.05.031.

[113] R. Lucio-Porto, S. Bouhtiyya, J.-F. Pierson, A. Morel, F. Capon, P. Boulet, T. Brousse, VN thin films as electrode materials for electrochemical capacitors, Electrochim. Acta 141 (2014) 203–211, https://doi.org/10.1016/j.electacta.2014.07.056.
[114] K. Robert, C. Douard, A. Demortière, F. Blanchard, P. Roussel, T. Brousse, C. Lethien, On chip interdigitated micro-supercapacitors based on sputtered bifunctional vanadium nitride thin films with finely tuned inter- and intracolumnar porosities, Adv. Mater. Technol. 3 (7) (2018) 1800036.
[115] P. Hu, M. Yan, X. Wang, C. Han, L. He, X. Wei, C. Niu, K. Zhao, X. Tian, Q. Wei, Z. Li, L. Mai, Single-nanowire electrochemical probe detection for internally optimized mechanism of porous graphene in electrochemical devices, Nano Lett. 16 (3) (2016) 1523–1529, https://doi.org/10.1021/acs.nanolett.5b03576.
[116] M. Li, M.F. El-Kady, J.Y. Hwang, M.D. Kowal, K. Marsh, H. Wang, Z. Zhao, R.B. Kaner, Embedding hollow Co3O4 nanoboxes into a three-dimensional macroporous graphene framework for high-performance energy storage devices, Nano Res. 11 (5) (2018) 2836–2846, https://doi.org/10.1007/s12274-017-1914-7.
[117] X. Ma, S. Feng, L. He, M. Yan, X. Tian, Y. Li, C. Tang, X. Hong, L. Mai, Rapid, all dry microfabrication of three-dimensional Co3O4/Pt nanonetworks for high-performance microsupercapacitors, Nanoscale 9 (32) (2017) 11765–11772, https://doi.org/10.1039/C7NR01789H.
[118] S. Xu, Y. Dall'Agnese, G. Wei, C. Zhang, Y. Gogotsi, W. Han, Screen-printable microscale hybrid device based on MXene and layered double hydroxide electrodes for powering force sensors, Nano Energy 50 (2018) 479–488, https://doi.org/10.1016/j.nanoen.2018.05.064.
[119] Q. Jiang, N. Kurra, C. Xia, H.N. Alshareef, Hybrid microsupercapacitors with vertically scaled 3D current collectors fabricated using a simple cut-and-transfer strategy, Adv. Energy Mater. 7 (1) (2017) 1601257, https://doi.org/10.1002/aenm.201601257.
[120] X. Dong, Z. Guo, Y. Song, M. Hou, J. Wang, Y. Wang, Y. Xia, Flexible and wire-shaped micro-supercapacitor based on Ni(OH)2-nanowire and ordered mesoporous carbon electrodes, Adv. Funct. Mater. 24 (22) (2014) 3405–3412, https://doi.org/10.1002/adfm.201304001.
[121] K. Fic, G. Lota, M. Meller, E. Frackowiak, Novel insight into neutral medium as electrolyte for high-voltage supercapacitors, Energ. Environ. Sci. 5 (2) (2012) 5842–5850, https://doi.org/10.1039/C1EE02262H.
[122] X. Bu, L. Su, Q. Dou, S. Lei, X. Yan, A low-cost "water-in-salt" electrolyte for a 2.3 V high-rate carbon-based supercapacitor, J. Mater. Chem. A 7 (13) (2019) 7541–7547, https://doi.org/10.1039/C9TA00154A.
[123] X. Zang, C. Shen, M. Sanghadasa, L. Lin, High-voltage supercapacitors based on aqueous electrolytes, ChemElectroChem 6 (4) (2019) 976–988, https://doi.org/10.1002/celc.201801225.
[124] W.S.V. Lee, T. Xiong, G.C. Loh, T.L. Tan, J. Xue, Optimizing electrolyte physiochemical properties toward 2.8 V aqueous supercapacitor, ACS Appl. Energy Mater. 1 (7) (2018) 3070–3076, https://doi.org/10.1021/acsaem.8b00751.
[125] Q.T. Qu, B. Wang, L.C. Yang, Y. Shi, S. Tian, Y.P. Wu, Study on electrochemical performance of activated carbon in aqueous Li2SO4, Na2SO4 and K2SO4 electrolytes, Electrochem. Commun. 10 (10) (2008) 1652–1655, https://doi.org/10.1016/j.elecom.2008.08.020.
[126] J.H. Chae, G.Z. Chen, Influences of ions and temperature on performance of carbon nano-particulates in supercapacitors with neutral aqueous electrolytes, Particuology 15 (2014) 9–17, https://doi.org/10.1016/j.partic.2013.02.008.
[127] C. Xu, C. Wei, B. Li, F. Kang, Z. Guan, Charge storage mechanism of manganese dioxide for capacitor application: effect of the mild electrolytes containing alkaline

and alkaline-earth metal cations, J. Power Sources 196 (18) (2011) 7854–7859, https://doi.org/10.1016/j.jpowsour.2011.04.052.
[128] W. Gao, N. Singh, L. Song, Z. Liu, A.L.M. Reddy, L. Ci, R. Vajtai, Q. Zhang, B. Wei, P.M. Ajayan, Direct laser writing of micro-supercapacitors on hydrated graphite oxide films, Nat. Nanotechnol. 6 (8) (2011) 496–500, https://doi.org/10.1038/nnano.2011.110.
[129] M. Beidaghi, C. Wang, Micro-supercapacitors based on interdigital electrodes of reduced graphene oxide and carbon nanotube composites with ultrahigh power handling performance, Adv. Funct. Mater. 22 (21) (2012) 4501–4510, https://doi.org/10.1002/adfm.201201292.
[130] C. Shen, X. Wang, W. Zhang, F. Kang, A high-performance three-dimensional micro supercapacitor based on self-supporting composite materials, J. Power Sources 196 (23) (2011) 10465–10471.
[131] M.F. El-Kady, M. Ihns, M. Li, J.Y. Hwang, M.F. Mousavi, L. Chaney, A.T. Lech, R.-B. Kaner, Engineering three-dimensional hybrid supercapacitors and microsupercapacitors for high-performance integrated energy storage, Proc. Natl. Acad. Sci. 112 (14) (2015) 4233, https://doi.org/10.1073/pnas.1420398112.
[132] M. Beidaghi, C. Wang, Micro-supercapacitors based on three dimensional interdigital polypyrrole/C-MEMS electrodes, Electrochim. Acta 56 (25) (2011) 9508–9514, https://doi.org/10.1016/j.electacta.2011.08.054.
[133] I. Cekic-Laskovic, N. von Aspern, L. Imholt, S. Kaymaksiz, K. Oldiges, B.R. Rad, M. Winter, Synergistic effect of blended components in nonaqueous electrolytes for lithium ion batteries, Top. Curr. Chem. 375 (2) (2017) 37, https://doi.org/10.1007/s41061-017-0125-8.
[134] R.W. Schmitz, P. Murmann, R. Schmitz, R. Müller, L. Krämer, J. Kasnatscheew, P. Isken, P. Niehoff, S. Nowak, G.-V. Röschenthaler, N. Ignatiev, P. Sartori, S. Passerini, M. Kunze, A. Lex-Balducci, C. Schreiner, I. Cekic-Laskovic, M. Winter, Investigations on novel electrolytes, solvents and SEI additives for use in lithium-ion batteries: systematic electrochemical characterization and detailed analysis by spectroscopic methods, Prog. Solid State Chem. *42* (4) (2014) 65–84, https://doi.org/10.1016/j.progsolidstchem.2014.04.003.
[135] M. Winter, The solid electrolyte interphase–the most important and the least understood solid electrolyte in rechargeable Li batteries, Zeitschrift für physikalische Chemie 223 (10–11) (2009) 1395–1406.
[136] E. Kovalska, C. Kocabas, Organic electrolytes for graphene-based supercapacitor: liquid, gel or solid, Mater. Today Commun. 7 (2016) 155–160, https://doi.org/10.1016/j.mtcomm.2016.04.013.
[137] G.H. Lane, E. Jezek, Electrochemical studies of acetonitrile based supercapacitor electrolytes containing alkali and alkaline earth metal cations, Electrochim. Acta 150 (2014) 173–187, https://doi.org/10.1016/j.electacta.2014.10.112.
[138] V. Ruiz, T. Huynh, S.R. Sivakkumar, A.G. Pandolfo, Ionic liquid–solvent mixtures as supercapacitor electrolytes for extreme temperature operation, RSC Adv. 2 (13) (2012) 5591–5598, https://doi.org/10.1039/C2RA20177A.
[139] S. Park, K. Kim, Tetramethylammonium tetrafluoroborate: the smallest quaternary ammonium tetrafluoroborate salt for use in electrochemical double layer capacitors, J. Power Sources 338 (2017) 129–135, https://doi.org/10.1016/j.jpowsour.2016.10.080.
[140] A. Laheäär, H. Kurig, A. Jänes, E. Lust, LiPF6 based ethylene carbonate–dimethyl carbonate electrolyte for high power density electrical double layer capacitor, Electrochim. Acta 54 (19) (2009) 4587–4594.
[141] P. Huang, M. Heon, D. Pech, M. Brunet, P.-L. Taberna, Y. Gogotsi, S. Lofland, J.D. Hettinger, P. Simon, Micro-supercapacitors from carbide derived carbon (CDC) films on silicon chips, J. Power Sources 225 (2013) 240–244, https://doi.org/10.1016/j.jpowsour.2012.10.020.

[142] D. Pech, M. Brunet, P.-L. Taberna, P. Simon, N. Fabre, F. Mesnilgrente, V. Conédéra, H. Durou, Elaboration of a microstructured inkjet-printed carbon electrochemical capacitor, J. Power Sources 195 (4) (2010) 1266–1269.
[143] D. Pech, M. Brunet, H. Durou, P. Huang, V. Mochalin, Y. Gogotsi, P.-L. Taberna, P. Simon, Ultrahigh-power micrometre-sized supercapacitors based on onion-like carbon, Nat. Nanotechnol. 5 (9) (2010) 651–654, https://doi.org/10.1038/nnano.2010.162.
[144] B. Ricketts, C. Ton-That, Self-discharge of carbon-based supercapacitors with organic electrolytes, J. Power Sources 89 (1) (2000) 64–69.
[145] G. Sun, J. An, C.K. Chua, H. Pang, J. Zhang, P. Chen, Layer-by-layer printing of laminated graphene-based interdigitated microelectrodes for flexible planar microsupercapacitors, Electrochem. Commun. 51 (2015) 33–36.
[146] M.F. El-Kady, V. Strong, S. Dubin, R.B. Kaner, Laser scribing of high-performance and flexible graphene-based electrochemical capacitors, Science 335 (6074) (2012) 1326, https://doi.org/10.1126/science.1216744.
[147] L. Yu, G.Z. Chen, Ionic liquid-based electrolytes for supercapacitor and supercapattery, Front. Chem. 7 (2019).
[148] M. Galiński, A. Lewandowski, I. Stępniak, Ionic liquids as electrolytes, Electrochim. Acta 51 (26) (2006) 5567–5580, https://doi.org/10.1016/j.electacta.2006.03.016.
[149] M. Armand, F. Endres, D.R. MacFarlane, H. Ohno, B. Scrosati, Ionic-liquid materials for the electrochemical challenges of the future, Nat. Mater. 8 (8) (2009) 621–629, https://doi.org/10.1038/nmat2448.
[150] A. Brandt, S. Pohlmann, A. Varzi, A. Balducci, S. Passerini, Ionic liquids in supercapacitors, MRS Bull. 38 (7) (2013) 554–559.
[151] E. Paillard, Q. Zhou, W.A. Henderson, G.B. Appetecchi, M. Montanino, S. Passerini, Electrochemical and physicochemical properties of PY14FSI-based electrolytes with LiFSI, J. Electrochem. Soc. 156 (11) (2009) A891–A895.
[152] S. Pohlmann, A. Balducci, A new conducting salt for high voltage propylene carbonate-based electrochemical double layer capacitors, Electrochim. Acta 110 (2013) 221–227.
[153] A.J.R. Rennie, N. Sanchez-Ramirez, R.M. Torresi, P.J. Hall, Ether-bond-containing ionic liquids as supercapacitor electrolytes, J. Phys. Chem. Lett. 4 (17) (2013) 2970–2974, https://doi.org/10.1021/jz4016553.
[154] S. Li, X. Wang, H. Xing, C. Shen, Micro supercapacitors based on a 3D structure with symmetric graphene or activated carbon electrodes, J. Micromech. Microeng. 23 (11) (2013) 114013.
[155] H. Xiao, Z.-S. Wu, L. Chen, F. Zhou, S. Zheng, W. Ren, H.-M. Cheng, X. Bao, One-step device fabrication of phosphorene and graphene interdigital microsupercapacitors with high energy density, ACS Nano 11 (7) (2017) 7284–7292, https://doi.org/10.1021/acsnano.7b03288.
[156] E. Eustache, C. Douard, A. Demortière, V. De Andrade, M. Brachet, J. Le Bideau, T. Brousse, C. Lethien, High areal energy 3D-interdigitated micro-supercapacitors in aqueous and ionic liquid electrolytes, Adv. Mater. Technol. 2 (10) (2017) 1700126, https://doi.org/10.1002/admt.201700126.
[157] D. Aradilla, P. Gentile, V. Ruiz, P. Gómez-Romero, J. Wimberg, B. Iliev, T.J. Schubert, S. Sadki, G. Bidan, SiNWs-based electrochemical double layer microsupercapacitors with wide voltage window (4 V) and long cycling stability using a protic ionic liquid electrolyte, Adv. Nat. Sci. Nanosci. Nanotechnol. 6 (1) (2014), 015004.
[158] A. Platek, J. Piwek, K. Fic, T. Schubert, P. Gentile, G. Bidan, E. Frackowiak, Electrochemical performance of silicon nanostructures in low-temperature ionic liquids for microelectronic applications, J. Mater. Chem. A 5 (43) (2017) 22708–22716.
[159] R. Lin, P.-L. Taberna, S. Fantini, V. Presser, C.R. Pérez, F. Malbosc, N.L. Rupesinghe, K.B. Teo, Y. Gogotsi, P. Simon, Capacitive energy storage from −50 to 100 C using an ionic liquid electrolyte, J. Phys. Chem. Lett. 2 (19) (2011) 2396–2401.

[160] H. Kataoka, Y. Saito, M. Tabuchi, Y. Wada, T. Sakai, Ionic conduction mechanism of PEO-type polymer electrolytes investigated by the carrier diffusion phenomenon using PGSE-NMR, Macromolecules 35 (16) (2002) 6239–6244, https://doi.org/10.1021/ma012139f.

[161] L. Long, S. Wang, M. Xiao, Y. Meng, Polymer electrolytes for lithium polymer batteries, J. Mater. Chem. A 4 (26) (2016) 10038–10069, https://doi.org/10.1039/C6TA02621D.

[162] A. Arya, A.L. Sharma, Polymer electrolytes for lithium ion batteries: a critical study, Ionics 23 (3) (2017) 497–540, https://doi.org/10.1007/s11581-016-1908-6.

[163] Q. Zhao, X. Liu, S. Stalin, K. Khan, L.A. Archer, Solid-state polymer electrolytes with in-built fast interfacial transport for secondary lithium batteries, Nat. Energy 4 (5) (2019) 365–373, https://doi.org/10.1038/s41560-019-0349-7.

[164] D. Zhou, D. Shanmukaraj, A. Tkacheva, M. Armand, G. Wang, Polymer electrolytes for lithium-based batteries: advances and prospects, Chem 5 (9) (2019) 2326–2352, https://doi.org/10.1016/j.chempr.2019.05.009.

[165] J.R. Nair, I. Shaji, N. Ehteshami, A. Thum, D. Diddens, A. Heuer, M. Winter, Solid polymer electrolytes for lithium metal battery via thermally induced cationic ring-opening polymerization (CROP) with an insight into the reaction mechanism, Chem. Mater. 31 (9) (2019) 3118–3133, https://doi.org/10.1021/acs.chemmater.8b04172.

[166] H. Gao, K. Lian, Proton-conducting polymer electrolytes and their applications in solid supercapacitors: a review, RSC Adv. 4 (62) (2014) 33091–33113, https://doi.org/10.1039/C4RA05151C.

[167] A. Manuel Stephan, K.S. Nahm, Review on composite polymer electrolytes for lithium batteries, Polymer 47 (16) (2006) 5952–5964, https://doi.org/10.1016/j.polymer.2006.05.069.

[168] Y.-J. Lim, Y.-H. An, N.-J. Jo, Polystyrene-Al2O3 composite solid polymer electrolyte for lithium secondary battery, Nanoscale Res. Lett. 7 (1) (2012) 19, https://doi.org/10.1186/1556-276X-7-19.

[169] Y. Zhu, J. Cao, H. Chen, Q. Yu, B. Li, High electrochemical stability of a 3D cross-linked network PEO@nano-SiO2 composite polymer electrolyte for lithium metal batteries, J. Mater. Chem. A 7 (12) (2019) 6832–6839, https://doi.org/10.1039/C9TA00560A.

[170] G. Vijayakumar, S.N. Karthick, A.R. Sathiya Priya, S. Ramalingam, A. Subramania, Effect of nanoscale CeO2 on PVDF-HFP-based nanocomposite porous polymer electrolytes for Li-ion batteries, J. Solid State Electrochem. 12 (9) (2008) 1135–1141, https://doi.org/10.1007/s10008-007-0460-8.

[171] X. Liang, D. Han, Y. Wang, L. Lan, J. Mao, Preparation and performance study of a PVDF–LATP ceramic composite polymer electrolyte membrane for solid-state batteries, RSC Adv. 8 (71) (2018) 40498–40504, https://doi.org/10.1039/C8RA08436J.

[172] Y. Li, W. Zhang, Q. Dou, K.W. Wong, K.M. Ng, Li7La3Zr2O12 ceramic nanofiber-incorporated composite polymer electrolytes for lithium metal batteries, J. Mater. Chem. A 7 (7) (2019) 3391–3398, https://doi.org/10.1039/C8TA11449H.

[173] T. Placke, R. Kloepsch, S. Dühnen, M. Winter, Lithium ion, lithium metal, and alternative rechargeable battery technologies: the odyssey for high energy density, J. Solid State Electrochem. 21 (7) (2017) 1939–1964, https://doi.org/10.1007/s10008-017-3610-7.

[174] J. Betz, G. Bieker, P. Meister, T. Placke, M. Winter, R. Schmuch, Theoretical versus practical energy: a plea for more transparency in the energy calculation of different rechargeable battery systems, Adv. Energy Mater. 9 (6) (2019) 1803170, https://doi.org/10.1002/aenm.201803170.

[175] R. Schmuch, R. Wagner, G. Hörpel, T. Placke, M. Winter, Performance and cost of materials for lithium-based rechargeable automotive batteries, Nat. Energy 3 (4) (2018) 267–278, https://doi.org/10.1038/s41560-018-0107-2.

[176] M. Kaempgen, C.K. Chan, J. Ma, Y. Cui, G. Gruner, Printable thin film supercapacitors using single-walled carbon nanotubes, Nano Lett. 9 (5) (2009) 1872–1876, https://doi.org/10.1021/nl8038579.

[177] V. Vijayakumar, M. Ghosh, A.T.A. Torris, M.K.N. Chandran, S.B. Nair, M.V. Badiger, S. Kurungot, Water-in-acid gel polymer electrolyte realized through a phosphoric acid-enriched polyelectrolyte matrix toward solid-state supercapacitors, ACS Sustain. Chem. Eng. 6 (10) (2018) 12630–12640, https://doi.org/10.1021/acssuschemeng.8b01175.

[178] Z. Zuo, Y. Fu, A. Manthiram, Novel blend membranes based on acid-base interactions for fuel cells, Polymers 4 (4) (2012) 1627–1644.

[179] F. Santos, J.P. Tafur, J. Abad, A.J. Fernández Romero, Structural modifications and ionic transport of PVA-KOH hydrogels applied in Zn/Air batteries, J. Electroanal. Chem. 850 (2019) 113380, https://doi.org/10.1016/j.jelechem.2019.113380.

[180] J. Li, J. Qiao, K. Lian, Hydroxide ion conducting polymer electrolytes and their applications in solid supercapacitors: a review, Energy Storage Mater. (2019), https://doi.org/10.1016/j.ensm.2019.08.012.

[181] R. Basnayake, G.R. Peterson, D.J. Casadonte, C. Korzeniewski, Hydration and interfacial water in nafion membrane probed by transmission infrared spectroscopy, J. Phys. Chem. B. 110 (47) (2006) 23938–23943, https://doi.org/10.1021/jp064121i.

[182] K.-D. Kreuer, Proton conductivity: materials and applications, Chem. Mater. 8 (3) (1996) 610–641, https://doi.org/10.1021/cm950192a.

[183] S.H. Kang, I.G. Kim, B.-N. Kim, J.H. Sul, Y.S. Kim, I.-K. You, Facile fabrication of flexible in-plane graphene micro-supercapacitor via flash reduction, ETRI J. 40 (2) (2018) 275–282, https://doi.org/10.4218/etrij.2017-0242.

[184] Z.S. Wu, K. Parvez, X. Feng, K. Müllen, Graphene-based in-plane micro-supercapacitors with high power and energy densities, Nat. Commun. 4 (1) (2013) 2487, https://doi.org/10.1038/ncomms3487.

[185] K. Ariga, J.P. Hill, Q. Ji, Layer-by-layer assembly as a versatile bottom-up nanofabrication technique for exploratory research and realistic application, Phys. Chem. Chem. Phys. 9 (19) (2007) 2319–2340, https://doi.org/10.1039/B700410A.

[186] P. Berndt, K. Kurihara, T. Kunitake, Adsorption of poly(styrenesulfonate) onto an ammonium monolayer on mica: a surface forces study, Langmuir 8 (10) (1992) 2486–2490, https://doi.org/10.1021/la00046a022.

[187] S. Lochmann, J. Grothe, K. Eckhardt, D. Leistenschneider, L. Borchardt, S. Kaskel, Nanoimprint lithography of nanoporous carbon materials for micro-supercapacitor architectures, Nanoscale 10 (21) (2018) 10109–10115, https://doi.org/10.1039/C8NR01535J.

[188] L. Li, E.B. Secor, K.-S. Chen, J. Zhu, X. Liu, T.Z. Gao, J.-W.T. Seo, Y. Zhao, M.C. Hersam, High-performance solid-state supercapacitors and microsupercapacitors derived from printable graphene inks, Adv. Energy Mater. 6 (20) (2016) 1600909, https://doi.org/10.1002/aenm.201600909.

[189] J. Pu, X. Wang, T. Zhang, S. Li, J. Liu, K. Komvopoulos, High-energy-density, all-solid-state microsupercapacitors with three-dimensional interdigital electrodes of carbon/polymer electrolyte composite, Nanotechnology 27 (4) (2015), https://doi.org/10.1088/0957-4484/27/4/045701, 045701.

[190] N. Kurra, M.K. Hota, H.N. Alshareef, Conducting polymer micro-supercapacitors for flexible energy storage and Ac line-filtering, Nano Energy 13 (2015) 500–508, https://doi.org/10.1016/j.nanoen.2015.03.018.

[191] H.U. Lee, S.W. Kim, Pen lithography for flexible microsupercapacitors with layer-by-layer assembled graphene flake/PEDOT nanocomposite electrodes, J. Mater. Chem. A 5 (26) (2017) 13581–13590, https://doi.org/10.1039/C7TA02936E.

[192] W. Yu, H. Zhou, B.Q. Li, S. Ding, 3D printing of carbon nanotubes-based micro-supercapacitors, ACS Appl. Mater. Interfaces 9 (5) (2017) 4597–4604, https://doi.org/10.1021/acsami.6b13904.

[193] X. Yun, B. Lu, Z. Xiong, B. Jia, B. Tang, H. Mao, T. Zhang, X. Wang, Direct 3D printing of a graphene oxide hydrogel for fabrication of a high areal specific capacitance microsupercapacitor, RSC Adv. 9 (50) (2019) 29384–29395, https://doi.org/10.1039/C9RA04882K.
[194] Y. Liu, B. Zhang, Q. Xu, Y. Hou, S. Seyedin, S. Qin, G.G. Wallace, S. Beirne, J.M. Razal, J. Chen, Development of graphene oxide/polyaniline inks for high performance flexible microsupercapacitors via extrusion printing, Adv. Funct. Mater. 28 (21) (2018) 1706592, https://doi.org/10.1002/adfm.201706592.
[195] Y. Wang, Y. Shi, C.X. Zhao, J.I. Wong, X.W. Sun, H.Y. Yang, Printed all-solid flexible microsupercapacitors: towards the general route for high energy storage devices, Nanotechnology 25 (9) (2014), 094010.
[196] A. Ferris, S. Garbarino, D. Guay, D. Pech, 3D RuO2 microsupercapacitors with remarkable areal energy, Adv. Mater. 27 (42) (2015) 6625–6629.
[197] Y. Wang, Y.-Z. Zhang, D. Dubbink, J.E. ten Elshof, Inkjet printing of δ-MnO2 nanosheets for flexible solid-state micro-supercapacitor, Nano Energy 49 (2018) 481–488, https://doi.org/10.1016/j.nanoen.2018.05.002.
[198] A. Morag, R. Jelinek, Flexible asymmetric microsupercapacitors from freestanding hollow nickel microfiber electrodes, Adv. Electron. Mater. 5 (1) (2019) 1800584, https://doi.org/10.1002/aelm.201800584.
[199] P. Giannakou, M.G. Masteghin, R.C.T. Slade, S.J. Hinder, M. Shkunov, Energy storage on demand: ultra-high-rate and high-energy-density inkjet-printed NiO micro-supercapacitors, J. Mater. Chem. A (2019), https://doi.org/10.1039/C9TA07878A.
[200] S. Zheng, W. Lei, J. Qin, Z.-S. Wu, F. Zhou, S. Wang, X. Shi, C. Sun, Y. Chen, X. Bao, All-solid-state high-energy planar asymmetric supercapacitors based on all-in-one monolithic film using boron nitride nanosheets as separator, Energy Storage Mater. 10 (2018) 24–31, https://doi.org/10.1016/j.ensm.2017.08.002.
[201] Y. Yue, Z. Yang, N. Liu, W. Liu, H. Zhang, Y. Ma, C. Yang, J. Su, L. Li, F. Long, Z. Zou, Y. Gao, A flexible integrated system containing a microsupercapacitor, a photodetector, and a wireless charging coil, ACS Nano 10 (12) (2016) 11249–11257, https://doi.org/10.1021/acsnano.6b06326.
[202] W. Sun, X. Chen, Fabrication and tests of a novel three dimensional micro supercapacitor, Microelectron. Eng. 86 (4) (2009) 1307–1310, https://doi.org/10.1016/j.mee.2008.12.010.
[203] B.-S. Lee, S. Cui, X. Xing, H. Liu, X. Yue, V. Petrova, H.-D. Lim, R. Chen, P. Liu, Dendrite suppression membranes for rechargeable zinc batteries, ACS Appl. Mater. Interfaces 10 (45) (2018) 38928–38935, https://doi.org/10.1021/acsami.8b14022.
[204] M. Ghosh, V. Vijayakumar, S. Kurungot, Dendrite growth suppression by Zn2+-integrated nafion ionomer membranes: beyond porous separators toward aqueous Zn/V2O5 batteries with extended cycle life, Energ. Technol. 7 (9) (2019) 1900442, https://doi.org/10.1002/ente.201900442.
[205] H. Cao, F. Wan, L. Zhang, X. Dai, S. Huang, L. Liu, Z. Niu, Highly compressible zinc-ion batteries with stable performance, J. Mater. Chem. A 7 (19) (2019) 11734–11741, https://doi.org/10.1039/C9TA02990G.
[206] Y. Xiao, L. Huang, Q. Zhang, S. Xu, Q. Chen, W. Shi, Gravure printing of hybrid MoS2@S-rGO interdigitated electrodes for flexible microsupercapacitors, Appl. Phys. Lett. 107 (1) (2015), https://doi.org/10.1063/1.4926570, 013906.
[207] Y.G. Zhu, Y. Wang, Y. Shi, J.I. Wong, H.Y. Yang, CoO nanoflowers woven by CNT network for high energy density flexible micro-supercapacitor, Nano Energy 3 (2014) 46–54, https://doi.org/10.1016/j.nanoen.2013.10.006.
[208] Q. Xia, W. Zeng, F. Ji, X. Chen, Y. Zhang, F. Ling, W. Hu, L. Fang, S.N. Khisro, M. Zhou, Graphene quantum dot/Co(OH)2 electrode on nanoporous Au–Ag alloy for superior hybrid micro-supercapacitors, J. Mater. Chem. C (2019), https://doi.org/10.1039/C9TC02435B.

[209] X. Feng, J. Ning, D. Wang, J. Zhang, J. Dong, C. Zhang, X. Shen, Y. Hao, All-solid-state planner micro-supercapacitor based on graphene/NiOOH/Ni(OH)2 via mask-free patterning strategy, J. Power Sources 418 (2019) 130–137, https://doi.org/10.1016/j.jpowsour.2019.01.093.

[210] S. Gu, Z. Lou, L. Li, Z. Chen, X. Ma, G. Shen, Fabrication of flexible reduced graphene oxide/Fe2O3 hollow nanospheres based on-chip micro-supercapacitors for integrated photodetecting applications, Nano Res. 9 (2) (2016) 424–434, https://doi.org/10.1007/s12274-015-0923-7.

[211] J. Xu, G. Shen, A flexible integrated photodetector system driven by on-chip micro-supercapacitors, Nano Energy 13 (2015) 131–139, https://doi.org/10.1016/j.nanoen.2015.02.027.

[212] G. Lee, D. Kim, D. Kim, S. Oh, J. Yun, J. Kim, S.-S. Lee, J.S. Ha, Fabrication of a stretchable and patchable array of high performance micro-supercapacitors using a non-aqueous solvent based gel electrolyte, Energ. Environ. Sci. 8 (6) (2015) 1764–1774, https://doi.org/10.1039/C5EE00670H.

[213] J. Yun, Y. Lim, H. Lee, G. Lee, H. Park, S.Y. Hong, S.W. Jin, Y.H. Lee, S.-S. Lee, J.S. Ha, A patterned graphene/ZnO UV sensor driven by integrated asymmetric micro-supercapacitors on a liquid metal patterned foldable paper, Adv. Funct. Mater. 27 (30) (2017) 1700135, https://doi.org/10.1002/adfm.201700135.

[214] B. Rupp, M. Schmuck, A. Balducci, M. Winter, W. Kern, Polymer electrolyte for lithium batteries based on photochemically crosslinked poly(ethylene oxide) and ionic liquid, Eur. Polym. J. 44 (9) (2008) 2986–2990, https://doi.org/10.1016/j.eurpolymj.2008.06.022.

[215] M.-A. Néouze, J. Le Bideau, P. Gaveau, S. Bellayer, A. Vioux, Ionogels, new materials arising from the confinement of ionic liquids within silica-derived networks, Chem. Mater. 18 (17) (2006) 3931–3936, https://doi.org/10.1021/cm060656c.

[216] S. Wang, B. Hsia, C. Carraro, R. Maboudian, High-performance all solid-state micro-supercapacitor based on patterned photoresist-derived porous carbon electrodes and an ionogel electrolyte, J. Mater. Chem. A 2 (21) (2014) 7997–8002, https://doi.org/10.1039/C4TA00570H.

[217] J.-Q. Xie, Y.-Q. Ji, J.-H. Kang, J.-L. Sheng, D.-S. Mao, X.-Z. Fu, R. Sun, C.-P. Wong, In situ growth of Cu (OH) 2@ FeOOH nanotube arrays on catalytically deposited Cu current collector patterns for high-performance flexible in-plane micro-sized energy storage devices, Energ. Environ. Sci. 12 (1) (2019) 194–205.

[218] J. Feng, X. Sun, C. Wu, L. Peng, C. Lin, S. Hu, J. Yang, Y. Xie, Metallic few-layered VS2 ultrathin nanosheets: high two-dimensional conductivity for in-plane supercapacitors, J. Am. Chem. Soc. 133 (44) (2011) 17832–17838.

[219] D. Kim, G. Shin, Y.J. Kang, W. Kim, J.S. Ha, Fabrication of a stretchable solid-state micro-supercapacitor array, ACS Nano 7 (9) (2013) 7975–7982, https://doi.org/10.1021/nn403068d.

[220] M. Brachet, D. Gaboriau, P. Gentile, S. Fantini, G. Bidan, S. Sadki, T. Brousse, J. Le Bideau, Solder-reflow resistant solid-state micro-supercapacitors based on ionogels, J. Mater. Chem. A 4 (30) (2016) 11835–11843, https://doi.org/10.1039/C6TA03142K.

[221] M. Brachet, T. Brousse, J. Le Bideau, All solid-state symmetrical activated carbon electrochemical double layer capacitors designed with ionogel electrolyte, ECS Electrochem. Lett. 3 (11) (2014) A112–A115.

[222] B. Hsia, J. Marschewski, S. Wang, J.B. In, C. Carraro, D. Poulikakos, C.P. Grigoropoulos, R. Maboudian, Highly flexible, all solid-state micro-supercapacitors from vertically aligned carbon nanotubes, Nanotechnology 25 (5) (2014), https://doi.org/10.1088/0957-4484/25/5/055401, 055401.

CHAPTER 4

Characterization of microsupercapacitors

Mihir Kumar Jha and Chandramouli Subramaniam
Department of Chemistry, Indian Institute of Technology Bombay, Mumbai, Maharashtra, India

4.1 Introduction

At the fundamental level, a supercapacitor consists of two electrodes sandwiching a dielectric that stores electrical energy in the form of an electrochemical double layer at the electrode-electrolyte interface. The critical elements that govern the performance of the microsupercapacitors (MSCs) are (a) choice of materials for the electrode and electrolyte, (b) the choice of substrate and current collector, and (c) the design and form-factor of the MSC device. The seamless integration of all these components through appropriate processing techniques forms an important step in realizing superior performance from the final device. The design and fabrication of the device are two key domains where the MSCs differ widely from conventional supercapacitors. Furthermore, bulk supercapacitors are primarily meant for charge-storage applications due to the high capacitance (~100–150 F/cm^3) [1]; MSCs are mainly envisaged for AC line-filtering applications, interfacing with portable energy harvesters and powering wearable devices, on account of their ultrafast frequency response and smaller capacitance (70–100 F/cm^3) [2].

Sandwich-geometry and roll-to-roll processing form the two predominant approaches for fabrication of conventional supercapacitors [3]. In contrast, microsupercapacitors have been fabricated through two main approaches that differ in the processing routes. The top-down approach consists of employing a combination of chemical and lithographic processes to achieve concurrent formation and integration of the current collector and the active electrode material. Such an approach has been demonstrated in fabricating carbide-derived carbon-based interdigitated microsupercapacitor on a Si-SiO_2 substrate [4]. An alternative bottom-up strategy consists of coating the active electrode material on top of an appropriate current collector. This approach offers greater versatility in the choice of electrode

Microsupercapacitors
https://doi.org/10.1016/B978-0-08-102888-9.00002-1

materials, current collectors, and substrates and therefore has been more prevalent in literature [5–8]. Recent years have also witnessed an explosion of research activities focusing on MSC for wearable electronics and electronic textiles. Such activities primarily aim to develop novel means of fabricating MSCs utilizing biocompatible, abundant, and synergistic substrates such as cotton yarns and textiles. Accordingly, a wide variety of electrode materials ranging from graphene, carbon onions, carbon nanotubes, MXenes, activated carbon, carbide-derived carbons, laser-scribed carbons, and metal oxides have been utilized for realizing MSCs on both hard substrates such as Si/SiO_2, Ni foil, stainless steel and soft, pliable substrates such as PET, PI, Ni foam, Ecoflex, etc. [1,2,4–10].

Catering to both these approaches, the treatment of the characterization techniques, in this chapter, would encompass the characterization of the electrode material and the electrolyte with particular emphasis on the electrochemical characterization techniques that would provide insights into the performance of the MSCs. Therefore gaining fundamental insights into the properties of the electrode, the electrolyte, and their interface mandates a multipronged approach.

Accordingly, this chapter would therefore detail major characterization techniques related to (a) morphological insights, (b) chemical, and (c) physical and elemental properties of the electrode materials. Further, the means to understand the electrochemical properties of electrolyte and thereby tune the electrode-electrolyte interface in MSCs would be simultaneously described. Following this, critical parameters to judge the design of the device and its form factor and the various routes available to assess the electrochemical performance of the device would be discussed in detail. Finally, catering to the exponential increase in miniaturized portable electronics, this chapter would also provide a glimpse into various mechanical aspects of the MSCs. The flow of this chapter, encompassing all these aspects, is schematically depicted in Fig. 4.1.

4.2 Electrode and material characterization techniques

4.2.1 Electron microscopy

The use of the nanomaterials has dominated the landscape of both macro and microsupercapacitors due to their high specific surface area, surface-governed functionality, and electrical conductivity. However, a qualitative estimation of its morphology is still elusive using conventional light-microscopy-based techniques due to diffraction limit. This stipulates that

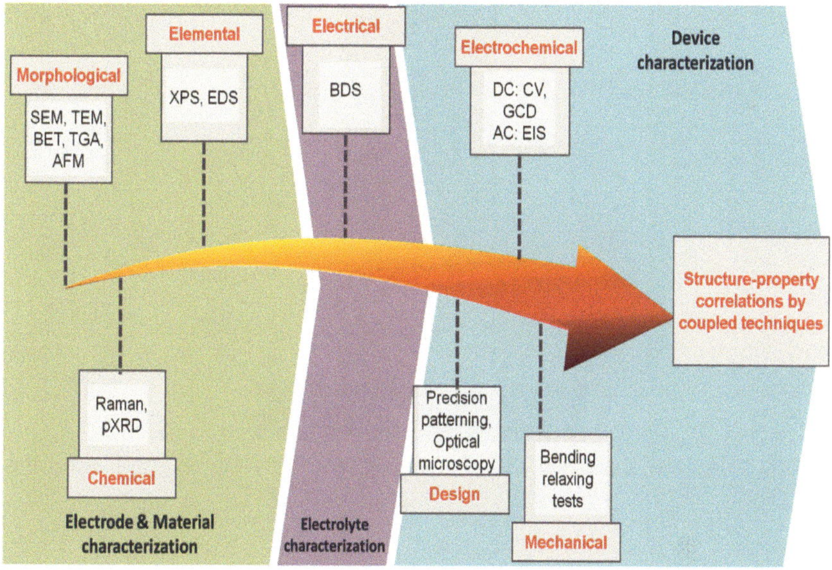

Fig. 4.1 Characterization roadmap for a microsupercapacitor. Insights that each technique offers are given in red text. Abbreviations: SEM, scanning electron microscopy; TEM, transmission electron microscopy; BET, N_2 adsorption isotherm; TGA, thermogravimetric analysis; AFM, atomic force microscopy; pXRD, powder X-ray diffraction; XPS, X-ray photoelectron spectroscopy; EDS, energy-dispersive X-ray spectroscopy; BDS, broadband dielectric spectroscopy; CV, cyclic voltammetry; GCD, galvanostatic charge-discharge cycling; EIS, electrochemical impedance spectroscopy.

the resolving power (δ) of any technique, which is defined as the shortest distance between two resolvable, distinguishable points of the image, is directly dependent on the average wavelength of electromagnetic radiation used to probe it and is given by

$$\delta = \frac{\lambda}{2nsin\theta}$$

where λ is the wavelength of the incident light, θ is the maximum half-angle subtended by the objective of the microscope (Fig. 4.2), and n is the refractive index of the imaging medium ($n = 1$ for air and 1.33 for water).

Accordingly, lowering the wavelength of the electromagnetic radiation is the most-effective manner to enhance the spatial resolution of imaging. This necessitates electron-microscopy-based imaging to derive a quantitative estimate of chemical constitution and monodispersity in dimension, besides providing qualitative information on the morphology and crystallinity of nanomaterials. The two main techniques that are prevalent in this

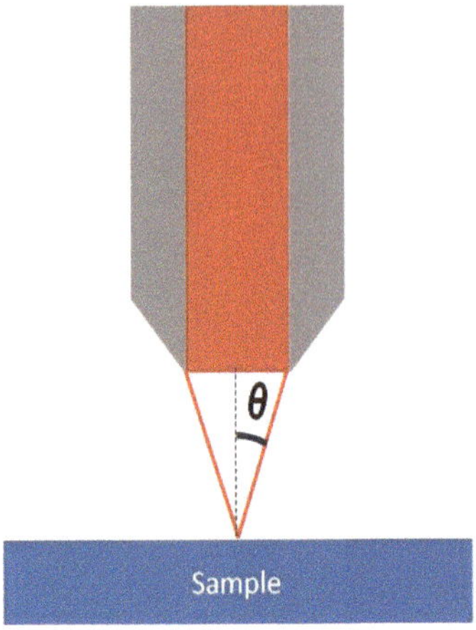

Fig. 4.2 Schematic illustration of half-angle subtended by the objective on to the sample.

context, namely, (a) secondary electron microscopy and (b) transmission electron microscopy, will be discussed in this context.

The scanning electron microscope works on the principle of utilizing the secondary or back-scattered electrons emitted from the sample for direct three-dimensional visualization of the morphology and the structure of the electrode material. The emission of secondary electron depends on the nature of the material, energy of the electron beam, and its angle of incidence. "Soft" nanocarbons such as carbon nanotubes and graphene undergo rapid graphitization and structural changes when exposed to electron beam of higher energy. Therefore it is advisable to utilize an accelerating voltage between 1 and 5 kV for analyzing such samples. In contrast, hard samples such as hard carbons and MXenes can withstand the higher accelerating voltage beyond 5 kV. An anatomic picture of an SEM would look like an electron gun passing through a series of lenses before being incident on the specimen stage (Fig. 4.3) [11]. This is then followed by collection of signals by the secondary electron detector that is subsequently processed to obtain a digital image.

The production of an uninterrupted electron emission is generally facilitated by field emission (FE). This is possible when the applied field is strong enough to overcome the work function of the material leading to emission of electrons with wavelength λ given by

$$\lambda = \frac{h}{\sqrt{2m_o eV}}$$

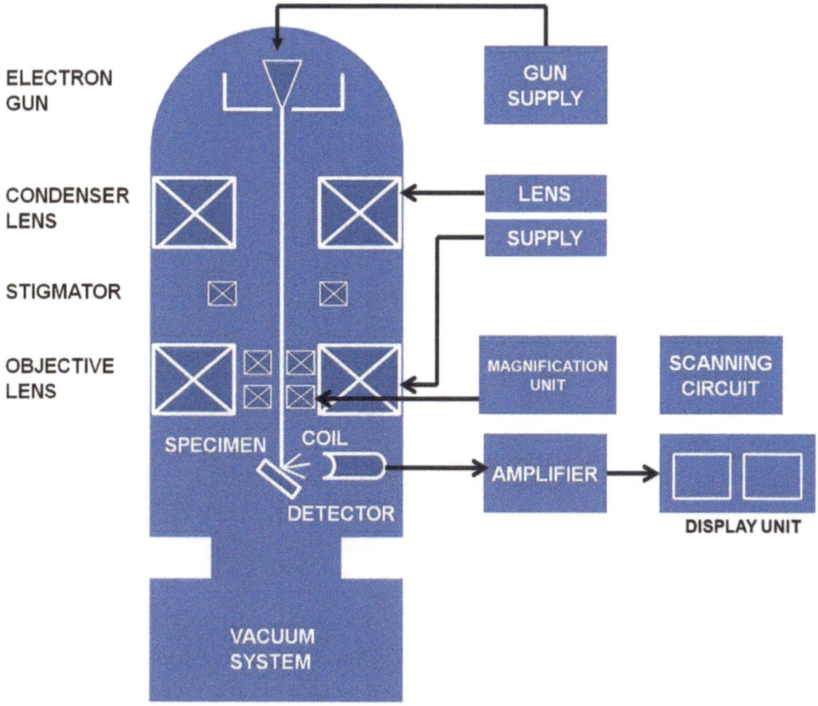

Fig. 4.3 Block diagram showing the basic anatomy of scanning electron microscope. *Reproduced with permission from A.C. Reimschuessel, Scanning electron microscopy—part I, J. Chem. Educ. 49 (1972) A413. https://doi.org/10.1021/ed049pa413.*

where h is Planck's constant, m_0 is the effective mass of electrons, e the elementary charge of an electron, and V is the accelerating voltage. Thus the wavelength of the electrons approaches subnanometer dimensions beyond an accelerating voltage of 10 kV. The emission of electrons can also be achieved thermionically using a LaB_6 single crystal, although the energy distribution of thermionic electrons is broader and is therefore less preferred over field-emission electron technique. The emitted electrons are controlled through a series of electromagnetic lenses that utilize Lorentz force to achieve focusing and defocusing of the electrons. The viewing angle can be controlled by rotation or tilt of the stage, which is manipulated using piezo-resistive motors that are capable of providing highly controlled movements in the x, y, and z directions. Importantly, the SEM chamber is kept in high vacuum (10^{-3}–10^{-4} Pa) to increase the mean-free path of the electrons. It is for the same reason that the samples for SEM need to be thoroughly degassed and dried before loading inside the SEM chamber.

Signals from secondary electrons are detected across the secondary-electron detector. The signals thus collected are finally imaged across an LED screen to obtain the SEM image.

In line with SEM, transmission electron microscopy (TEM) captures information on properties that are intrinsic the material. This includes crystal structure, morphology, atomic arrangement, stress inside a crystal/structure, etc. Imaging using TEM involves transmission of the electron beam through the material. Such a technique is contrary to knocking off electrons in SEM. However, the overall design anatomy of a typical TEM instrument is identical to SEM, comprising.

(a) an electron gun,
(b) a series of electromagnetic lenses (condenser and objective lens), and
(c) specimen stage, detector, and a projection screen.

Despite anatomical similarities, higher resolutions can be obtained in TEM than in SEM. This is because of larger accelerating voltages (60–300 kV) in TEM (for SEM ~5–30 kV). As a result, the maximum magnification obtained in TEM is about 50 times higher than that in SEM. Moreover, sample preparations are comparatively more tedious for TEM imaging. The contrast in a typical TEM image is borne out of differences in intensities of transmitted and diffracted electron beams. Although TEM seems to be a preferential imaging technique, depth of focus and imaging areas are comparatively much larger in SEM. An additional advantage of TEM lies in its ability to examine crystal lattices. The position of atoms in a crystal lattice is confirmed by a simple diffraction pattern obtained from TEM. The diffraction pattern is basically a Fourier transform of the periodicity of atoms in the crystal lattice. In summary, both SEM and TEM are employed in tandem to investigate structure and morphology of materials.

The importance of electron microscopy-based imaging is testified by the exponentially increasing interest in the development of morphologically unique nanostructures as electrode materials for MSCs. Electron microscopy also provides critical fundamental insights for deriving a structure-morphology-performance correlation for MSCs. Generally, the morphology of the electrode material and the porous network of the electrolyte merits closer examination prior to fabrication of MSCs. SEM has been employed as a confirmatory tool to establish nanostructuring in electrode materials. Additionally, morphological information like particle size, presence of grain boundaries, pore-network, etc., are characterized using SEM. TEM is employed in order to probe even finer details of electrode materials. This includes characterizing monolayers of graphene, MoS_2, and other 2D nanostructures. However, the uniformity of the nanomaterials in terms of both morphology and dimensions can be estimated using both SEM and TEM. The monodispersity index is estimated from a collection of SEM or TEM images and is given as a histogram of particle size distribution.

Importantly, the lattice planes of a material can be visualized through diffraction spots of electron beam in TEM. This technique, referred to as selective area electron diffraction (SAED), provides spatially localized information on the crystallinity of the material. Further, the interplanar distance estimated from SAED is an useful indicator of the propensity for formation of electrochemical double layer and intercalation of ions of the electrolyte. Liu et al. characterized RGO/MnO_2/Ag NW hybrid film, which is utilized as the electrode material for MSCs, using TEM [9]. As observed in Fig. 4.4, TEM exhibits diffraction planes signifying crystalline phase. The power of electron microscopy is also illustrated by Lei and coworkers, through a step-by-step approach to follow the fabrication of a honeycomb nanoporous alumina scaffold (Fig. 4.5) [10]. The porosity and density of packing of these nanostructures can be estimated from the images. Further, such imaging also establishes the defect-free structure of the final material (Fig. 4.5).

Since SEM involves simple raster scanning of the sample to generate an image, a powder sample or solid sample devoid of any moisture can be

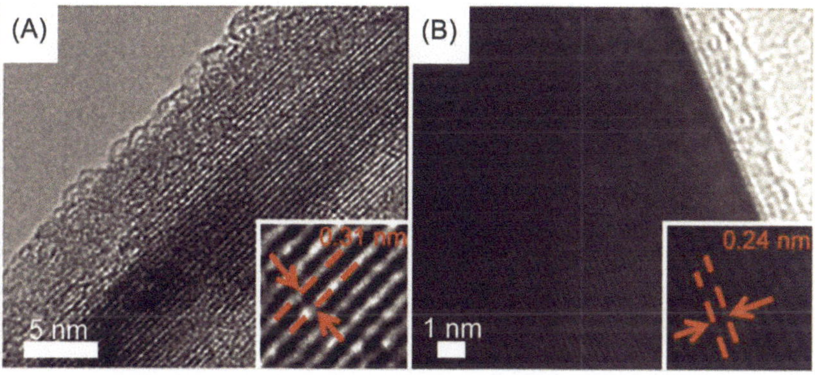

Fig. 4.4 High-resolution TEM images of a graphene/manganese oxide/silver nanowire hybrid showing the region of (A) MnO_2 flakes and (B) Ag nanowire. *Reproduced with permission from W. Liu, C. Lu, X. Wang, R.Y. Tay, B.K. Tay, High-performance microsupercapacitors based on two-dimensional graphene/manganese dioxide/silver nanowire ternary hybrid film, ACS Nano 9 (2015) 1528–1542. https://doi.org/10.1021/nn5060442.*

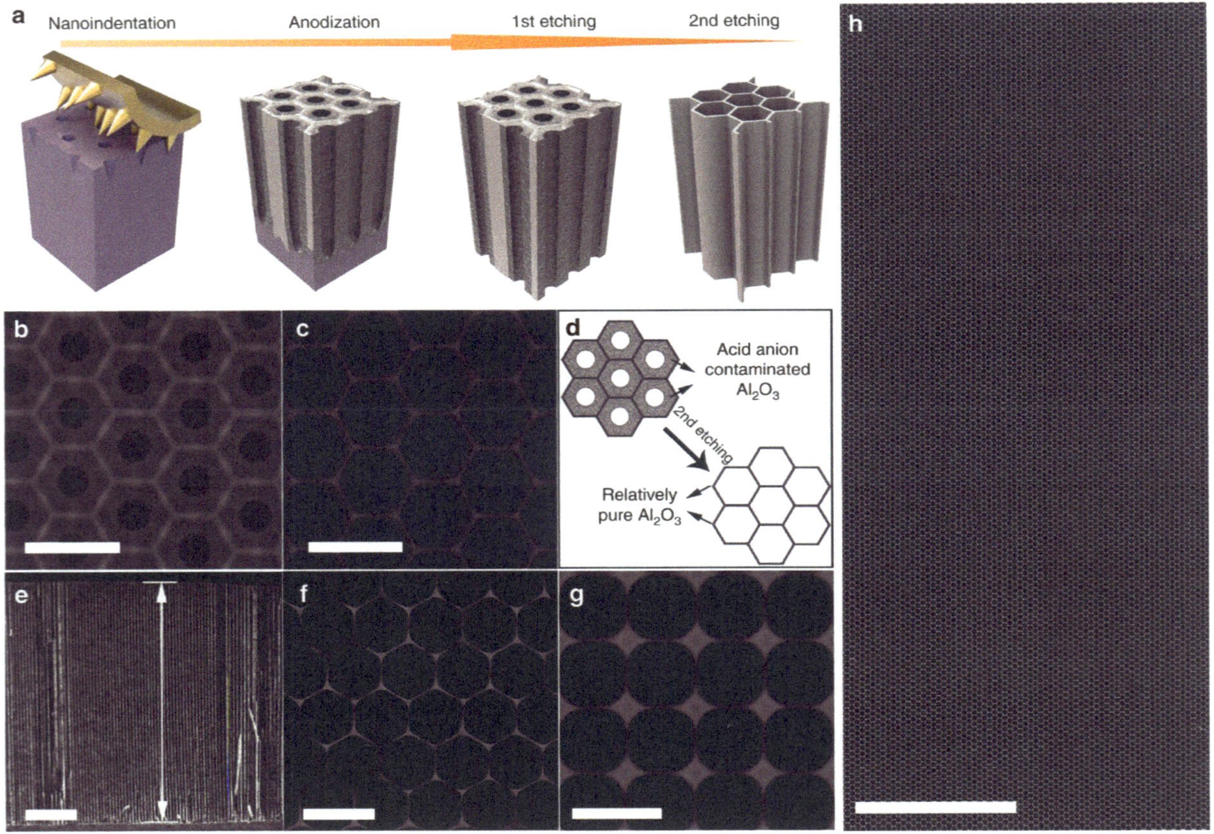

Fig. 4.5 (A) Fabrication process of honeycomb alumina nanoscaffold. (B) SEM image of nanoporous alumina after anodization process revealing the double-layer cell structure. (C) SEM image of HAN with hexagonal cell arrangement and 400-nm intercell spacing showing the structure of HAN with ultrathin wall of 16 ± 2 nm. (D) Schematic of the evolution of the HAN by the second etching process.

(Continued)

employed. Different kinds of SEM instruments cater to the need of imaging varying samples. For instance, the porous morphology of a solid/gel electrolyte can be imaged using a cryo-SEM, where the sample is cooled down to subzero temperatures to restrict segmental polymer motions or any gel-like fluidity that might be present. Similarly, a biomass that does not require very high-resolution imaging can be imaged using an environmental SEM (eSEM). Ultrahigh resolution imaging using SEM can be accomplished using focused ion beam (FIB-SEM). Even a trace amount of moisture in the sample can lead to significant charging leading to deterioration of image quality. This is handled by complete degassing of the sample prior to imaging. While the maximum resolution achieved in SEM is ~1 nm, TEM can be utilized for imaging up to about 0.1 nm. In contrast to SEM, TEM requires the electron beam to transmit through the sample. Care must be exercised to maintain the deaggregated state of the sample during sample preparation.

Sample preparation constitutes an important step in TEM imaging. Electron-beam-induced structural modifications are a common source of erroneous analysis in electron microscopy. The effect of structural changes on the electronic conductivity of carbon nanotubes has been observed [12,13]. The common practice is to drop-cast dispersion of the material onto Cu-TEM grids, followed by drying before imaging. Alternately, capillarity-induced wicking of the sample from the dispersion can also be employed to prevent aggregation of the sample on the grid, during drying. In both these cases, the dispersions are usually prepared in a low-boiling solvent such as ethanol or isopropanol by ultrasonication. Conventionally, the dispersion

Fig. 4.5, cont'd (E) Cross-sectional view SEM image of an HAN indicating the cell depth to be about 25 μm. (F) SEM image of HAN with hexagonal cell arrangement and 800 nm intercell spacing. (G) SEM image of HAN with square cell arrangement and 400 nm intercell spacing. (H) Large-scale SEM image showing the uniform, stable, and defect-free structure of the HAN with hexagonal cell arrangement and 400 nm intercell spacing. Scale bar: 500 nm (B), 500 nm (C), 5 μm (E), 1 μm (F), 500 nm (G), 10 μm (H). *Reproduced with permission from Z. Lei, L. Liu, H. Zhao, F. Liang, S. Chang, L. Li, Y. Zhang, Z. Lin, J. Kröger, Y. Lei, Nanoelectrode design from microminiaturized honeycomb monolith with ultrathin and stiff nanoscaffold for high-energy micro-supercapacitors, Nat. Commun. 11 (2020). https://doi.org/10.1038/s41467-019-14170-6.*

is ultrasonicated at moderate frequencies, prior to drop-casting on Cu grids, to ensure deagglomeration of the material, while retaining its morphology. Such sample preparation techniques find importance in TEM of conductive inks that are employed to fabricate electrodes for MSCs [14]. Samples prone to electron-beam-induced damage such as soft-tissues, biomaterials, and biomass-derived carbons can be fixed in a polymeric matrix and microtomed to obtain thin sections that are scooped onto the Cu grid. Similarly, carbon-film-coated Cu girds can also be used for imaging materials that are prone to electron-beam-induced damage such as biomaterials and graphene flakes.

4.2.2 Surface area

The specific surface area of a material is estimated by the BET analysis. BET adsorption theory was proposed in 1938 by Brunauer, Emmett, and Teller and is based on certain fundamental assumptions [15]. These assumptions are as follows:

(a) Adsorption occurs in multilayers. The adsorbent molecules interact with the adsorbate through van der Waals forces (physisorption).

(b) Adsorption at one position does not affect adsorption at neighboring positions indicating absence of any kind of intramolecular interaction between adsorbent molecules. Additionally, this confirms that the surface energy of the adsorbent does not play any role in the entire adsorption process.

(c) The energy of adsorption of the first monolayer is assumed to be constant (E_1). The adsorption energy for the subsequent layers is identical to the energy of liquefaction (E_L) of the gas.

The second and third assumptions have however been a subject of disagreement for several researchers [16]. Based on the above assumptions, the relation between the total volume of gas adsorbed (v_{total}) at a given pressure (p) is given by

$$\frac{p}{v_{total}(p_0 - p)} = \frac{1}{v_{mono}c} + \frac{c-1}{v_{mono}c}\left(\frac{p}{p_0}\right)$$

where v_{mono} is the volume adsorbed over a monolayer at pressure p and p_0 is the pressure required to condense the gas-to-liquid state. Thus p/p_0 is the relative pressure. "c" in the above equation is a constant and is equal to $e^{\frac{E_1-E_L}{RT}}$. A plot of $\frac{p}{v_{\text{total}}(p_0-p)}$ vs $\left(\frac{p}{p_0}\right)$ is a straight line with slope $\frac{c-1}{v_{\text{mono}}c}$ and intercept $\frac{1}{v_{\text{mono}}c}$. The surface area is therefore given by

$$\text{Surface area} = \frac{1}{slope + intercept}$$

The surface area is estimated by temperature and pressure-controlled adsorption and desorption of an inert gas through the sample to obtain the BET isotherm. Depending on the sample and intended application, any gas can be employed with N_2, CO_2, and air being the most common ones. There are five different kinds of adsorption isotherms encountered routinely with porous materials. Generally, electrode materials meant for applications in MSC exhibit type I and type II isotherms. While the type I isotherm is concave toward the relative pressure (X axis) and represents mono-molecular adsorption, the type II isotherm deals with multilayer adsorption phenomena. Therefore type II isotherm deviates significantly from Langmuir adsorption model. The initial steep rise followed by the knee-point in the type II isotherm is characteristic of pore-filling in micro and mesocavities of the adsorbate.

It is very important to note that the specific surface area (SSA), pore volume, and porosity obtained through N_2 adosprtion isotherms can only provide a rough qualitative estimate of the efficacy of the material as electrode for supercapacitor. Direct correlation between the SSA and the capacitance is strongly discouraged as capacitance is a complex function of not only the SSA and the porosity, but also the type and geometry of pores present in the material. This is experimentally verified by Shi and coworkers who established that the capacitance does not vary linearly with SSA for nanoporous supercapacitive electrode materials (Fig. 4.6) [17]. While the adsorption isotherm estimates the total area available for adsorption of N_2, the electrochemical double layer formed in MSCs is primarily formed by the solvated ions and polarized solvent molecules and hence constitutes a completely different structural organization. Nevertheless, estimate of SSA is a useful technique to assess the possible contribution from the micropores and mesopores of the electrode material for further structural and morphological refinement.

The accessible electrochemical surface area of the electrode material is a key factor controlling the performance of MSCs. While the specific surface area (SSA) of a graphene electrode can be as high as 2600 m^2/g, carbon nanotubes (SSA ~1200 m^2/g) are sometimes preferred for MSCs due to their ease of synthesis and handling. Similarly, the high surface area of MXenes leads to large specific and volumetric capacitance [18–20]. As is discussed at the beginning, nanostructuring of materials is preferred due to the

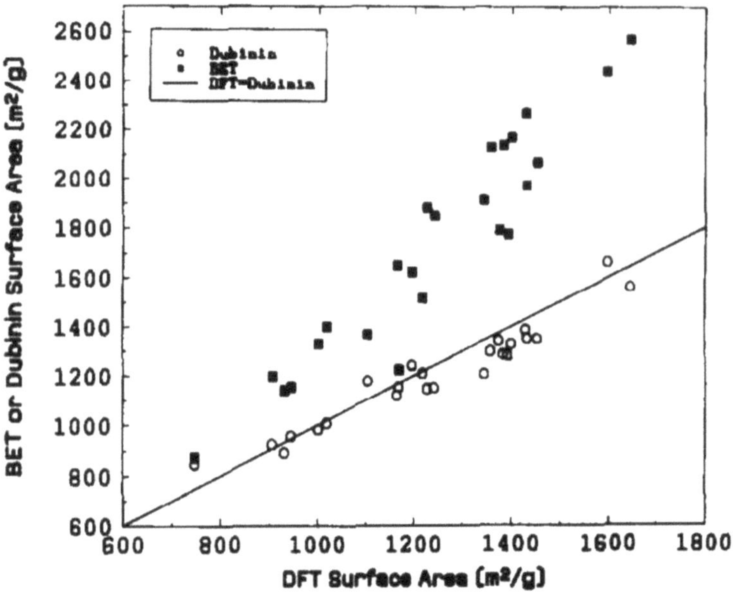

Fig. 4.6 Variation of specific double-layer capacitance with total DFT surface area for activated carbon microbeads and carbons fibers. *Reproduced with permission from H. Shi, Activated carbons and double layer capacitance, Electrochim. Acta 41 (1996) 1633–1639. https://doi.org/10.1016/0013-4686(95)00416-5.*

increase in surface-to-volume ratio and specific surface area. This can generally be achieved either by chemical etching or by thermal treatment of the electrode material [21–23].

As discussed earlier, the specific surface area of the electrode material cannot be used as the sole determinant driving higher capacitance. The pore-size distribution has an equally important contribution. A model to correlate parameters relating to capacitance, surface area, and pore volumes in activated carbon material was proposed [17]. Materials with high SSA may not result in superior capacitance due to a blocked pore network and inaccessibility of the overall surface area for double-layer formation. Pores can be divided into three kinds depending on their size—macro (>50 nm), *meso* (2–50 nm), and micro (<2 nm). The initial hike in volume adsorbed in a typical BET adsorption isotherm (Fig. 4.7) is due to rapid infiltration of N_2 on to the micro and mesopores of the sample. Such pore structures can be fine-tuned in activated carbon samples by varying the activation temperature. It is important to note that the SSA estimated from such N_2 adsorption isotherms, while being a good qualitative measure of the propensity for

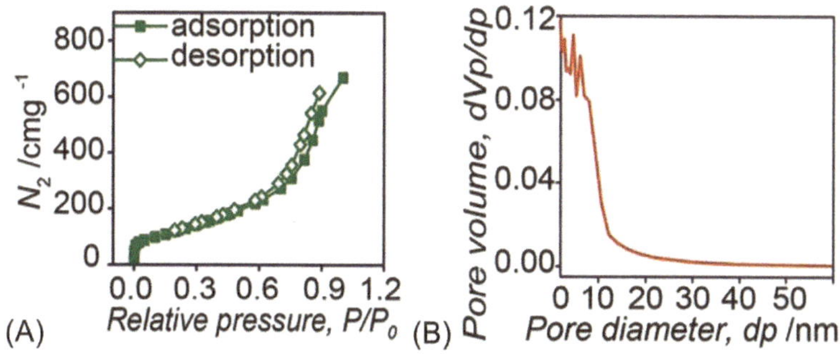

Fig. 4.7 Surface area of SWCNT film employed to fabricate MSC: (A) BET adsorption isotherm and (B) pore size distribution. *Reproduced with permission from K.U. Laszczyk, K. Kobashi, S. Sakurai, A. Sekiguchi, D.N. Futaba, T. Yamada, K. Hata, Lithographically integrated microsupercapacitors for compact, high performance, and designable energy circuits, Adv. Energy Mater. 5 (2015) 6–11. https://doi.org/10.1002/aenm.201500741.*

EDL formation, it cannot provide a direct quantitative measure of the capacitance of the electrode material. This is because access to the pores would be hindered by the solvation of ions. This would result in an extra energy requirement to partially/fully strip the ions of their solvation sphere before adsorption into the micropores. A theoretical model based on an electric wire-in cylinder for micropores has been initially proposed [24]. Such a model accounts for the shape of micro and mesopores to decipher the extent of double-layer formation. Accordingly, for cylindrical micropores, the double-layer capacitance C of wire-in cylinder model is given by

$$C = \frac{2\pi\varepsilon_r\varepsilon_0 L}{\ln(b/a)}$$

where a and b are the radii of the inner and outer radius, respectively, and L is the pore length. Interestingly, the wire-in cylinder model can also be viewed as a double-cylinder model, where the capacitance is

$$C = \frac{2\pi\varepsilon_r\varepsilon_0 L}{\ln(b/a_0)}$$

where a_0 is the radius of the innercylinder.

Unlike the case of SSA, a direct correlation of pore accessibility and the role of micropores in the final capacitance of the device have been established by Patrice Simon and coworkers [25]. Carbide-derived carbons, which is the electrode material, are synthesized using a simple chlorination reaction of titanium carbide, wherein amorphous carbon is obtained.

$$\mathrm{TiC} + 2\mathrm{C1}_2 \rightarrow \mathrm{TiC1}_4 + \mathrm{C}$$

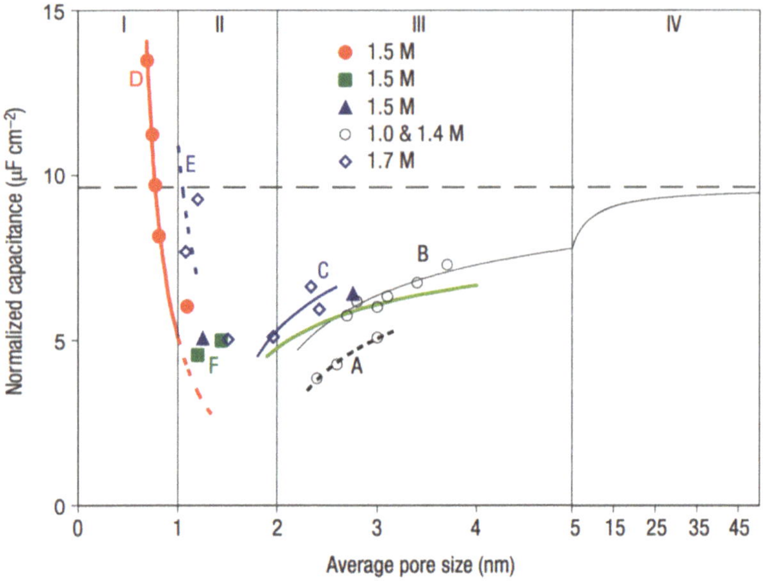

Fig. 4.8 Variation of normalized specific capacitance with pore size for different carbon sample. The legends indicate concentration of the liquid electrolyte employed. *Reproduced with permission from P. Simon, Y. Gogotsi, Materials for electrochemical capacitors, Nat. Mater. 7 (2008) 845–854. https://doi.org/10.1142/9789814317665_0021.*

Pores of different sizes are obtained by varying the chlorination temperatures. As observed in Fig. 4.8, the capacitance gradually drops with decreasing pore size. Ion solvation leads to obstruction of pore accessibility and results in such a drop in capacitance. Interestingly, capacitance increases sharply when the pore size approaches to less than 1 nm. This is because of the prevalence of micropores in the sample that leads to enhanced capacitance. Such direct evidence clearly establishes the importance of *meso* and micropores. Different mechanisms have been proposed to explain charge storage in subnanometer pores as well [23]. Therefore the pore-size distribution and solvation of ions are crucial parameters that drive enhanced performance. Hence, a BET adsorption isotherm must always be accompanied by a pore-size distribution plot in order to provide any conclusive evidence about the performance of an electrode material. BJH (Barrett, Joyner, and Halenda) and DFT (density functional theory) methods are employed to estimate pore-size distributions of samples. While the BJH method is based on the model of simple pore filling, inconsistency or nonuniformity of the electrode material is taken into account in the DFT method (REF).

4.2.3 Thermogravimetric analysis

Thermogravimetric analysis (TGA), as the name suggests, is the technique of monitoring the weight of the material while raising its temperature. The environment in which the heating is carried out can either by inert (He, He, Ar, N_2) or reactive (Air, O_2, H_2, CO_2). The TGA of electrode as well as electrolyte materials employed in MSCs is crucial to determine (a) the filler content of the electrode material, for example, the quantity of carbon black added, (b) the presence of moisture, (c) overall thermal stability of the sample, (d) presence of impurities, and (e) thermal stability of the material.

While the temperature range for a thermogravimetric measurement is determined by the chemical constitution of the sample, a slower heating rate is recommended to precisely capture all the features of a thermogram. A typical TGA profile of a COF composite employed as an MSC electrode looks like the one given in Fig. 4.9A [26]. Herein, TGA has been recorded to substantiate claims of high thermal stability of the electrode material. High stability is attributed to a weight loss of $<10\%$ even after heating to 500 °C.

Often, the first derivative of the TGA curve is plotted to obtain the exact value of temperature at the time of weight loss. This technique is called differential thermal analysis (DTA) and is frequently overlayed on the TGA plot. Fig. 4.9B shows DTA curve of sewage sludge overlayed on TGA [27]. Multiple slopes of the TGA curve are divided into three zones in

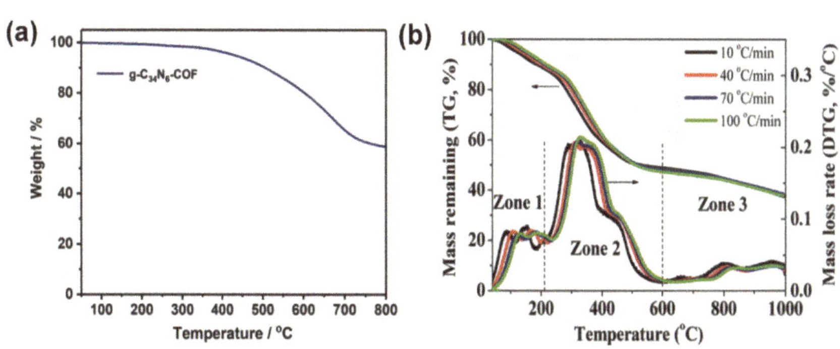

Fig. 4.9 (A) TGA profile of a COF composite employed as an electrode material for fabricating MSC. (B) TGA and DTA profiles of sewage sludge at different heating rates. *Reproduced with permission from J. Xu, Y. He, S. Bi, M. Wang, P. Yang, D. Wu, J. Wang, F. Zhang, An olefin-linked covalent organic framework as a flexible thin-film electrode for a high-performance micro-supercapacitor, Angew. Chem., Int. Ed. 58 (2019) 12065–12069. https://doi.org/10.1002/anie.201905713; T. Kan, V. Strezov, T. Evans, Effect of the heating rate on the thermochemical behavior and biofuel properties of sewage sludge pyrolysis, Energy Fuels 30 (2016) 1564–1570. https://doi.org/10.1021/acs.energyfuels.5b02232.*

the DTA curve. Each zone is indicative of thermal decomposition and pyrolysis of different constituents of sludge. Interestingly, increased heating rate leads to gradual shift of the TGA and DTA toward higher temperatures. This is attributed to lower thermal conductivity of the sludge that results in greater time requirement for the material to take up the heat.

4.2.4 Atomic force microscopy

Atomic force microscopy (AFM) is a very powerful surface-specific microscopic technique capable of imaging nanoscopic features. The operation of an AFM is based on the principle of interactive forces between the surface atoms of the material and a sharp tip mounted on a cantilever. The following processes occur as the tip of the cantilever approaches the surface of the material:

(a) attractive forces between the surface atoms and the cantilever tip dominate resulting in the cantilever tip being drawn toward the surface. This is mainly driven by attractive interactions.

(b) beyond a critical distance of separation between the tip and the surface, repulsive forces dominate, as seen from Fig. 4.10.

AFM can be conducted in either a constant-height mode or a constant-force mode. The precise movements of the cantilever, as is scans the surface of the sample, are captured by bouncing a laser off the surface of the cantilever and capturing its reflection in a four-segmented, position-sensitive photodiode (PSPD). Thus the movement of the cantilever during a raster scan is a dedicated reproduction of the surface morphology at the nanoscale level.

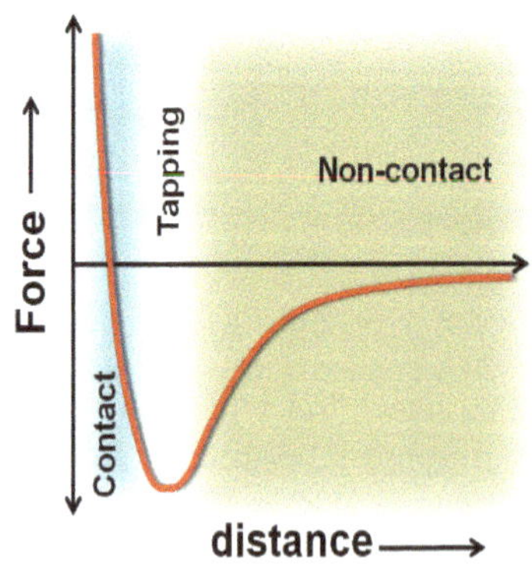

Fig. 4.10 Variation of force with decreasing distance during gradual approach of an AFM tip toward the sample.

The three main modes of operation are classified based on the force domain of interaction between the tip and the sample, as

(a) *contact*: where the tip is in contact with the surface,

(b) *noncontact*: where the nature of the sample does not require the tip to be in contact with the surface, and

(c) *tapping*: this is basically an intermittent mode of operation between the contact and noncontact modes.

Besides the topological information acquired during a scan, the phase difference between multiple oscillations of the cantilever in tapping mode can also be utilized for acquiring physical information of the surface. Interestingly, the versatility of fundamental insights obtained using an AFM enables intricate and specialized techniques such as conducting AFM, Piezoresponse force microscopy, Kelvin probe microscopy, etc.

The applications of AFM in characterizing a microsupercapacitor include (a) probing the topography of the coating of electrode material, (b) estimating thickness of electrode material coating, and (c) establishing uniformity of the electrode film/surface (Figs. 4.11 and 4.12). Additionally, AFM can also provide a detailed insight on the precision of lithographic patterning on substrates. The power of AFM allows it to even manipulate atomistic features on a surface; however, such applications have not been reported on microsupercapacitors as yet. This leaves a scope of fascination toward tuning miniaturized features on the electrode surface to achieve higher performance.

4.2.5 Raman

Since its discovery by Sir C.V. Raman in 1928, Raman spectroscopy has been employed as a primary tool for investigating structural insights about materials for electrodes. A typical Raman spectrum provides the bond vibrational frequencies that carry the spectroscopic fingerprint of the bond, orientation of the overall structure and symmetry of associated bond vibrations. Specifically, the sp^2-sp^2 carbon framework prevalent in graphene and carbon nanotubes (CNTs) is probed using Raman. Among all the phonon modes for sp^2 carbons, the singly degenerate A, and doubly degenerate E_1 and E_2 are Raman active for carbon nanotubes, while the E_{2g} mode is active for graphite. Pure monolayer graphene exhibits two distinct Raman peaks—the G-band at $\sim$1530 cm^{-1} and the G′-band $\sim$2700 cm^{-1} (Fig. 4.13). Importantly, the ratios of the peak intensities of G and G′-band can be employed to estimate the number of graphene layers. Similarly, the presence

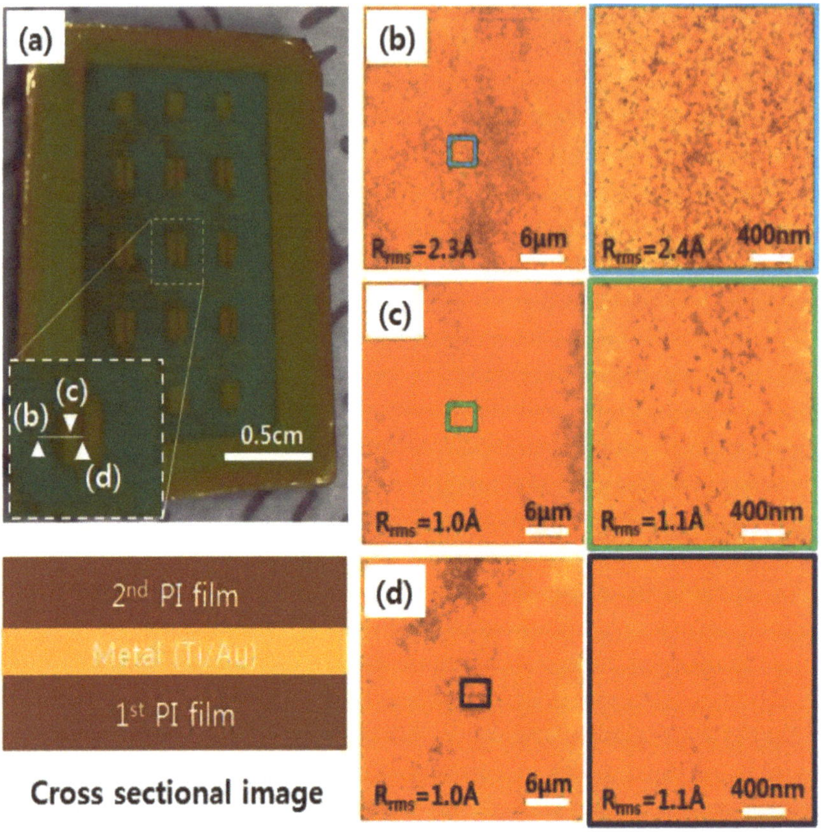

Fig. 4.11 (A) Optical image of MSC array. (B)–(D) AFM images at specific locations of the MSC array, as indicated in (A). The remaining panels indicate high-resolution AFM images of regions square-marked in (B)–(D), respectively. *Reproduced with permission from D. Kim, G. Shin, Y.J. Kang, W. Kim, J.S. Ha, Fabrication of a stretchable solid-state micro-supercapacitor array, ACS Nano 7 (2013) 7975–7982. https://doi.org/10.1021/nn403068d.*

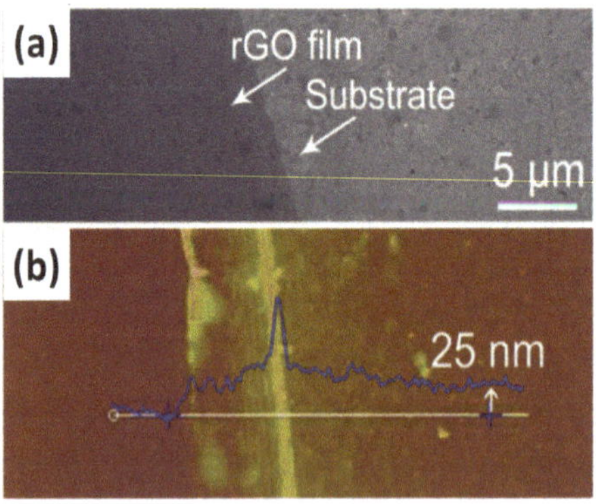

Fig. 4.12 (A) SEM image of an rGO film electrode coated on a substrate. (B) AFM image of the rGO film on the substrate. The height profile overlayed on the AFM image shows the average height of the coating. *Reproduced with permission from Z. Niu, L. Zhang, L. Liu, B. Zhu, H. Dong, X. Chen, All-solid-state flexible ultrathin micro-supercapacitors based on graphene, Adv. Mater. 25 (2013) 4035–4042. https://doi.org/10.1002/adma.201301332.*

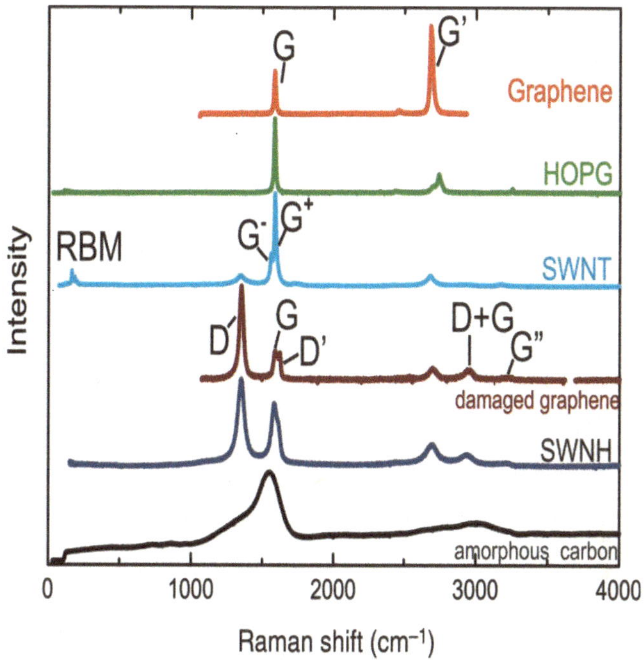

Fig. 4.13 Comparative Raman spectra of various carbon-based nanostructures including graphene, HOPG, SWNT, etc. *Reproduced with permission from R. Saito, M. Hofmann, G. Dresselhaus, A. Jorio, M.S. Dresselhaus, Raman spectroscopy of graphene and carbon nanotubes, Adv. Phys. 60 (2011) 413–550. https://doi.org/10.1080/00018732.2011.582251.*

of any structural defects, sp^3 carbons, or Stone-Wales defects is characterized; an additional defect-induced peak called the D-band appears at $\sim$1340 cm^{-1}, in addition to the symmetry allowed G-band. Such a spectrum forms the basis for estimating the defect density (n_D) of graphene as given by [28]

$$n_D = \frac{(1.8 \pm 0.5) \times 10^{22}}{\lambda_L^4} \left(\frac{I_D}{I_G} \right)$$

where I_D and I_G are the respective intensities of D and G-band in the Raman spectrum. λ_L denotes laser wavelength.

The diameter of a single-walled carbon nanotube, d, is given by

$$d = \frac{248}{\omega_{RBM}}$$

where ω_{RBM} is the Raman shift of the peak corresponding to radial breathing mode (RBM). RBM is characteristic of single-walled carbon nanotubes and appears in the range 100–300 cm^{-1}. Multiwalled nanotubes do not exhibit RBM peak.

On amorphotization of graphene, the D-band position shifts indicating substantial changes in its interplanar spacing. Consequently, the position of the D-band in graphenic systems is a measure of their defect densities.

Experimentally, Raman spectrum is recorded by focusing a monochromatic laser onto the sample surface. The scattered photons are then passed through a beam splitter and a filter to cut off the incident wavelength, before being dispersed over a charge-coupled device (CCD) detector. The most common laser wavelengths are 532, 633, and 785 nm, although lower wavelengths are employed too. The choice of excitation wavelength is dictated by the optical properties of the sample with the general consideration being to avoid the wavelength that results in emission from the sample. This is due to the fact that the Raman scattering is an inelastic scattering process and is thus significantly weaker than elastic Rayleigh scattering (10^6 time stronger) and absorption emission (10^9 times stronger) processes. With an experimental laser system of varying wavelengths, the energy associated with the incident laser can match that of phonons of the sample molecule. This results in enhanced Raman features, in what is called as resonance Raman spectroscopy. An important advantage of Raman is that it does not require any kind of sample preparation unlike other microscopic tools like SEM and TEM. Therefore structural aspects of pristine electrode materials can easily be probed before fabricating devices.

Confocal Raman spectromicroscopy constitutes a significant advancement of conventional Raman spectral acquisition and combines the strength of both microscopy and spectroscopy. Herein the signal consisting of inelastically scattered photons is focused through a pin-hole detector before being analyzed. This results in spectral information from a spot whose lateral resolution is controlled by the laser spot size and diffraction limit while the z-resolution is usually of the order of 10 nm. Such spectromicroscopy has found lot of applications in in-situ/operand techniques where localized morphological and structural impact during charge-discharge processed can be examined. For example, in-situ monitoring of the O—H bond was carried out to comprehend the structural evolution and charge storage in nickel hydroxide-based electrode materials using resonance Raman spectroscopy (Fig. 4.14) [29]. Furthermore, dispersion of D-band in the Raman spectra of carbonaceous electrode materials with varying laser energies has

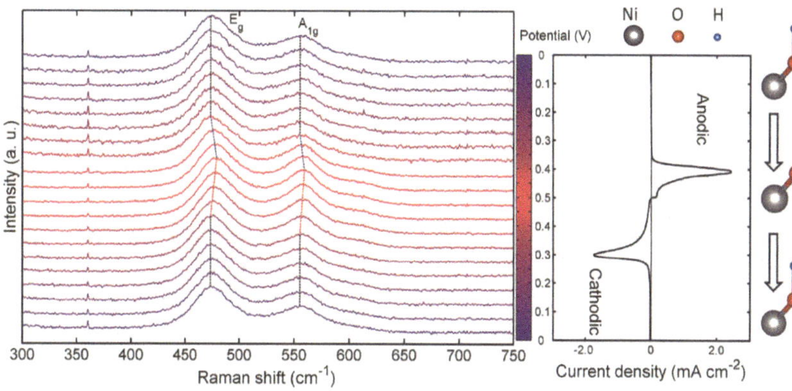

Fig. 4.14 Probing the charge storage mechanism of nickel hydroxide-based electrode material using Raman spectroscopy. *Reproduced with permission from D. Chen, X. Xiong, B. Zhao, M.A. Mahmoud, M.A. El-Sayed, M. Liu, Probing structural evolution and charge storage mechanism of NiO2 Hx electrode materials using in operando resonance Raman spectroscopy, Adv. Sci. 3 (2015) 1–6. https://doi.org/10.1002/advs.201500433.*

been examined by Matthews et al. [30] The peak intensities are often mapped into a two- or three-dimensional Raman image that provides clearer insights into the distribution of chemical groups across a specific area and the role of such distribution in determining the performance of the material. In this context, Raman spectral mapping was demonstrated to understand the extent and uniformity of 2D peaks in a laser-irradiated graphene electrode material (Fig. 4.15) [31]. Such spectral maps can also be utilized to generate spectral line profiles, similar to what is generated in SEM and TEM.

In another example, Reddy and coworkers have elaborated an operando Raman spectroelectrochemical experiment to investigate Na-ion intercalation on to nongraphitizable hard carbons (Fig. 4.16) [32]. The position of the Raman G-band is monitored throughout the intercalation process by varying the potential across the electrodes. The variation in G-band intensity has been correlated to the discharge profile of the electrochemical system.

Recently, the combination of Raman spectral mapping and electrochemical measurements has been demonstrated to provide both spatial and spectral information of the electrode-electrolyte interface (Fig. 4.17). The temperature-dependent infiltration of the PVA-H_2SO_4 solid electrolyte into the pores of a CNT-yarn electrode was established through such operand spectro-electrochemical investigation. Such a mapping and spectral profiling clearly demonstrates the diffused nature of the electrode-electrolyte interface

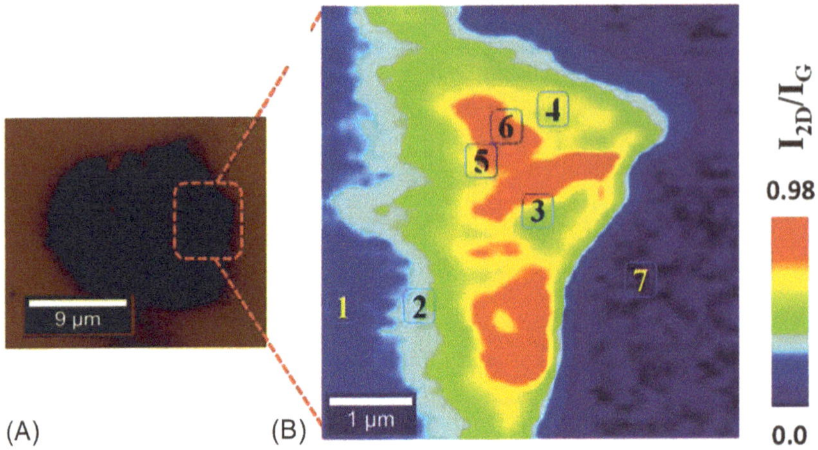

Fig. 4.15 (A) Optical image of laser-irradiated graphene. (B) Raman mapping of the ratio of 2D to G-band for the laser-irradiated graphene sample. *Reproduced with permission from N. Kamboj, T. Purkait, M. Das, S. Sarkar, K.S. Hazra, R.S. Dey, Ultralong cycle life and outstanding capacitive performance of a 10.8 v metal free micro-supercapacitor with highly conducting and robust laser-irradiated graphene for an integrated storage device, Energy Environ. Sci. 12 (2019) 2507–2517. https://doi.org/10.1039/c9ee01458f.*

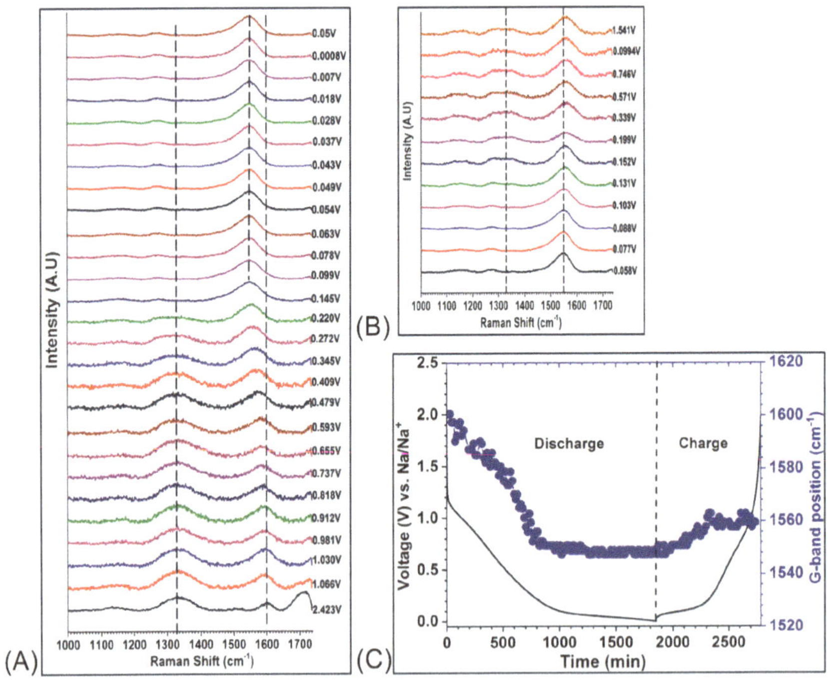

Fig. 4.16 Probing Na-ion intercalation using operando Raman electrochemistry during (A) discharge from OCV to 0 V. (B) Charge from 0 to 2 V. (C) Evolution of G-band during charging and discharging. *Reproduced with permission from M. Anji Reddy, M. Helen, A. Groß, M. Fichtner, H. Euchner, Insight into sodium insertion and the storage mechanism in hard carbon, ACS Energy Lett. 3 (2018) 2851–2857. https://doi.org/10.1021/acsenergylett.8b01761.*

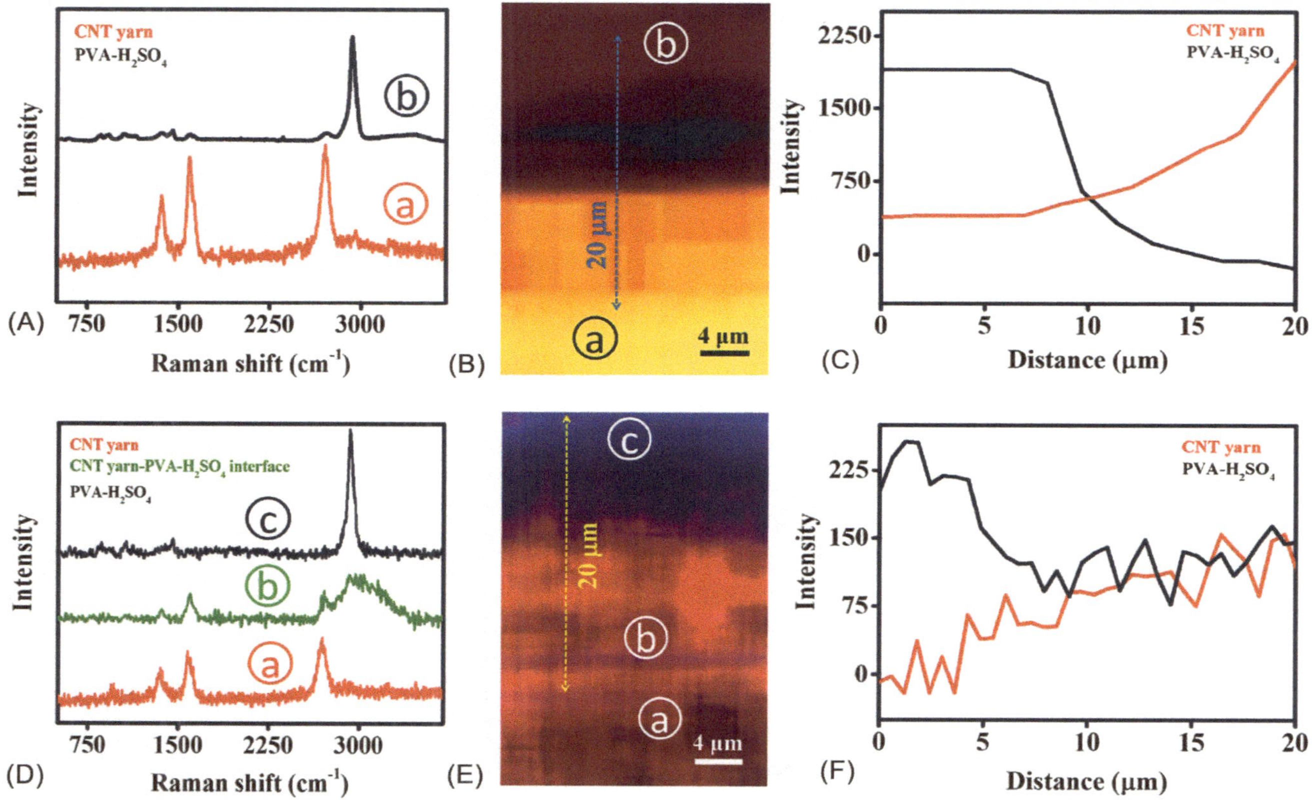

Fig. 4.17 Probing electrode-electrolyte interface through micro-Raman spectro-electrochemical profiling. (A) Raman spectra of locations "a" and "b" (as shown in (B)) representing CNT-yarn and PVA-H_2SO_4 electrolyte, respectively, (B) corresponding Raman spectral mapping indicating the interface of CNT-yarn and the PVA-H_2SO_4 electrolyte. (C) Comparative Raman line profiles exhibiting the variation of intensities of characteristic peaks of PVA-H_2SO_4 and CNT-yarn across a fixed distance of 20 μm through the interface, as depicted by a line in the Raman image. (D) Raman spectra of locations "a," "b," and "c" (as shown in (E)) representing CNT-yarn, diffused layer, and PVA-H_2SO_4 electrolyte,

(Continued)

achieved by thermal annealing at 80 °C, that is critical to achieve ultrafast frequency response in coaxial solid-state microsupercapacitors [33,34].

4.2.6 Powder XRD

Powder XRD (pXRD) is a rapid analytical method employed to detect the crystallinity of the specimen sample. This is actually a complementary technique to single-crystal XRD for powdered samples, like the electrode material. M. von Laue discovered that atomic lattice spacing is identical to the wavelength of X-rays, indicating that crystalline materials can act as three-dimensional diffraction grating to X-rays. A typical pXRD instrument consists of

(a) *X-ray generator*: X-rays are usually generated by bombarding the target material with accelerated electrons such that innershell electrons of the target are significantly knocked off. This produces characteristic X-rays. An electron beam can be obtained by heating a filament, for example, in a cathode ray tube. Copper $K\alpha$ (wavelength = 1.54 Å) is the most common target material employed to obtain X-rays.

(b) *Specimen stage*: This is the component where the sample is mounted. In modern sophisticated pXRD setups, the specimen stage is connected to a humidity chamber and temperature stage. This opens up scope to record XRD spectrum at different humidity and temperature conditions.

(c) *Detector*: When angle of incidence of X-rays satisfies Bragg's equation, constructive interference occurs and the detector provides a signal, confirming the presence of a diffraction plane. Bragg's equation is given by

$$n\lambda = 2dsin\theta$$

where λ and θ are the wavelength and angle of incidence of X-rays, respectively. d is the interplanar spacing and n refers to the order of diffraction.

The signal is obtained as a characteristic XRD peak of the given sample. Moreover, the interplanar spacing of the material can henceforth be

Fig. 4.17, cont'd respectively, after thermal annealing to 80 °C. (E) Corresponding Raman map of the interface. (F) Comparative Raman line profiles for characteristic peaks of PVA-H_2SO_4 and CNT-yarn. *Reproduced with permission from M.K. Jha, R. Ball, R. Seelaboyina, C. Subramaniam, All solid-state coaxial supercapacitor with ultrahigh scan rate operability of 250000 mV/s by thermal engineering of the electrode–electrolyte interface, ACS Appl. Energy Mater. (2020) acsaem.9b02528. https://doi.org/10.1021/acsaem.9b02528.*

estimated. The pXRD spectrum can therefore be employed to examine the nature (crystalline/amorphous) of the sample material. Additionally, unit cell dimensions of a crystalline substance can also be extrapolated from the characteristic peaks.

The use of pXRD in microsupercapacitor research is not limited to characterizing the crystallinity of the electrode material but also finds substantive application in postmortem analysis of the MSC, establishing the temperature-dependent stability of the electrode material and phase transition of the material during cycling, to name a few. For example, Feng and coworkers employed XRD to investigate the energy storage mechanism of a Zn-ion hybrid MSC (Fig. 4.18) [35]. The change in relative intensities of (0 0 2) peak of Zn-nanosheet-based anode material is examined at the 50th charge-discharge cycle. The (0 0 2) peak intensities at different potentials are then correlated with the time of the measurement to obtain a charge-discharge cycle from 0.5 to 1.5 V. Therefore the change in crystal orientation on consecutive cycling can be concluded. In another similar experiment, a phase transition during consecutive cycling from V_2CT_X to V_2O_5 is confirmed using pXRD of the electrode material at 1000th, 10,000th, and 18,000th discharge cycles (Fig. 4.19) [36]. A separate peak corresponding to the (1 0 0) phase of V_2O_5 appears and increases in intensity gradually during cycling. pXRD peaks of carbon materials extensively employed in MSCs are usually

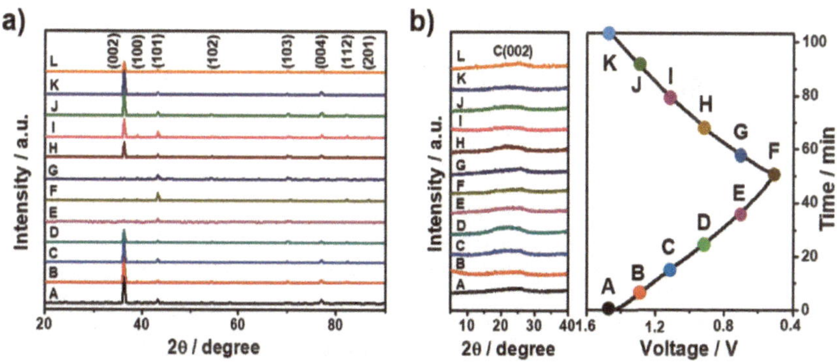

Fig. 4.18 Investigating the discharge characteristics of Zn-nanosheet anode-based MSC using pXRD technique. (A) pXRD spectra of the material at different potential steps as shown in (B). (B) pXRD spectra magnified in the region $2\theta = 10°–40°$. Correlated voltage versus time curve. *Reproduced with permission from P. Zhang, Y. Li, G. Wang, F. Wang, S. Yang, F. Zhu, X. Zhuang, O.G. Schmidt, X. Feng, Zn-ion hybrid micro-supercapacitors with ultrahigh areal energy density and long-term durability, Adv. Mater. 31 (2019) 1–6. https://doi.org/10.1002/adma.201806005.*

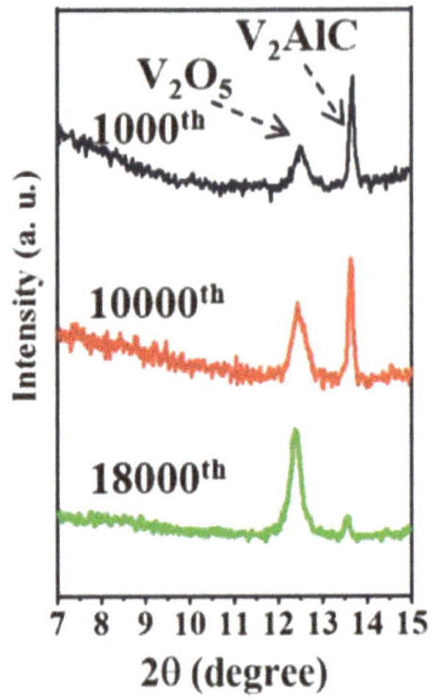

Fig. 4.19 pXRD spectra of a vanadium oxide-based electrode material at different discharge cycles indicating a phase transition. *Reproduced with permission from X. Li, M. Li, Q. Yang, H. Li, H. Xu, Z. Chai, K. Chen, Z. Liu, Z. Tang, L. Ma, Z. Huang, B. Dong, X. Yin, Q. Huang, C. Zhi, Phase transition induced unusual electrochemical performance of V2CTX MXene for aqueous zinc hybrid-ion battery, ACS Nano 14 (2020) 541–551. https://doi.org/10.1021/acsnano.9b06866.*

broad ($2\theta = 20°–25°$ for graphitic (0 0 2) plane), indicating their amorphous nature. However, sharp XRD peaks are obtained for carbon composites comprising a nanostructured metal salt along.

4.2.7 X-ray photoelectron spectroscopy

X-ray photoelectron spectroscopy (XPS) is a powerful surface-sensitive technique that provides chemical and quantitative estimation of elements present in a sample. Further, XPS can also be utilized for investigating empirical formulae of materials, prevalence of surface contamination, and uniformity of elemental composition at the surface (XPS mapping). XPS belongs to the larger group of techniques that rely on utilizing X-rays for chemical analysis, which is called electron spectroscopy for chemical analysis (ESCA).

The kinetic energy of electrons (E_k) emitted during the interaction of monochromatic X-rays (frequency ν) with the sample is correlated to the binding energy (BE) of the electrons as given by

$$E_k = h\nu - BE - \phi$$

where ϕ is the work function and h is Planck's constant.

Usually, a wide spectrum of binding energy ranging from 50 to 1200 eV can be investigated using a standard XPS instrument. The entire setup is kept in ultrahigh vacuum to assist longer path length of the ejected electrons and also to ensure noncontamination of the sample surface that can give rise to additional unwanted peaks. Since the inelastic mean free path of electrons is very less, XPS is restricted to the surface electrons and is a surface-sensitive technique. Following are some noteworthy points concerning accurate deciphering of the XPS spectrum:

(a) The binding energy (BE) axis is plotted in a negative scale on the *X* axis. This indicates that the BE ranges from higher to lower values as we move from left to right through the *X* axis.

(b) The background in XPS spectra is usually stepped. This is because inelastic scattering of some ejected electrons leads to their lower kinetic energy, which contributes to peaks at higher binding energies.

(c) The splitting of peaks in the XPS spectrum depends on spin-orbit coupling. Therefore an Al 2p would evidently be split into a doublet consisting of $2p_{1/2}$ and $2p_{3/2}$ peaks. The higher binding energies correspond to lower values of total angular momentum quantum number, indicating that $2p_{1/2}$ peak will have higher BE than $2p_{3/2}$.

(d) The relative intensity ratio of XPS peaks depends on the degeneracy of the levels. For example, a 3d electronic level would split into a doublet consisting of $3d_{3/2}$ and $3d_{5/2}$. The degeneracy can be estimated as

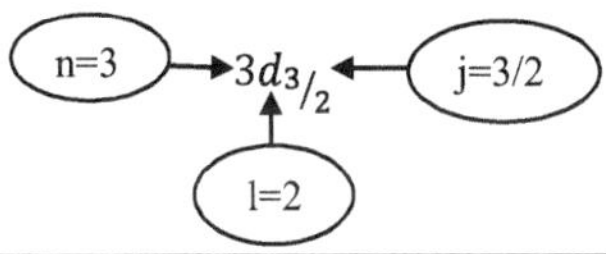

degeneracy, g= 2j+1= 4

for j=5/2, g= 6

Hence, degeneracy ratio= 2:3

Intensity ratio= 2:3

(e) The shifts in position of XPS peaks known as chemical shift depend on the electronegativity of the atoms and the chemical environment around it. Accordingly, a higher positive oxidation state leads to higher BE. For instance, Mn^{4+} will have higher BE than Mn^{2+}.

Applications of XPS or ESCA in characterizing MSCs are manifold. Generally, XPS offers confirmation on the elemental composition of the material. Although EDS serves identical purpose, XPS is way more reliable since it is quantitative in nature. The oxidation number of the element and its electronic environment can also be predicted using XPS. For example, in functionalized graphene oxide samples, the extent of functionalization is estimated using XPS [37]. The C 1 s XPS spectrum would have peaks corresponding to C–O, C–O–C, C=O bonds, thereby confirming the chemical modification of graphene. Rogers and coworkers employed XPS to investigate the variation of electronic state of the constituent elements with gradual charge-discharge cycling of a microsupercapacitor (Fig. 4.20) [38]. Therefore not only does XPS bring out elemental analysis, the technique can also be employed to provide insights on the mechanism of energy storage. This in turn opens up scope for enhancing the overall performance of the device.

4.3 Electrolyte characterization techniques

4.3.1 Ionic conductivity measurement using BDS

The electrolyte forms an important part of the MSCs, governing the potential window for operation, interfacing with the electrode, and minimizing the self-discharge of the device. Liquid electrolytes, whether aqueous or organic, offer high ionic mobility and ion transference number due to the inherent nature of its physical state that facilitates superior movement of ions. The ionic conductivity for liquid electrolytes is governed by Debye Huckel's theory for strong electrolytes and ranges from 0.5 to 1 S/cm at 298 K.

The fluidity of liquid electrolytes and the presence of dissolved gases that trigger Faradic reactions are two main challenges while employing liquid electrolytes for MSCs. These challenges witnessed the development of ion-gel and solid-electrolyte systems with improved functionality for miniaturized and portable MSCs. Herein, the ionic conductor in a liquid electrolyte (NaOH, KOH, H_2SO_4, etc.) is channeled into a solid matrix, which is often a polymer or a dielectric, to obtain ion-immobilized self-standing compartments. The synergistic interaction ensures that the matrix provides the mechanical robustness and the channels for transport of the ionic components. Solid electrolytes offer an ionic conductivity in the order of milli-Siemens/cm (10^{-3} S/cm). While the lowered ionic conductivity in solid electrolytes, as compared to liquid counterparts, is compensated by its mechanical robustness, the cumulative performance in MSCs using solid electrolytes is severely restricted. Therefore energy density of MSCs can be as high as 100 W h/kg with liquid electrolytes. However, the energy density of similar MSCs with solid-state electrolytes drops to 50–60 W h/kg [39]. Gel electrolytes that exhibit liquid-like ionic conductivity and

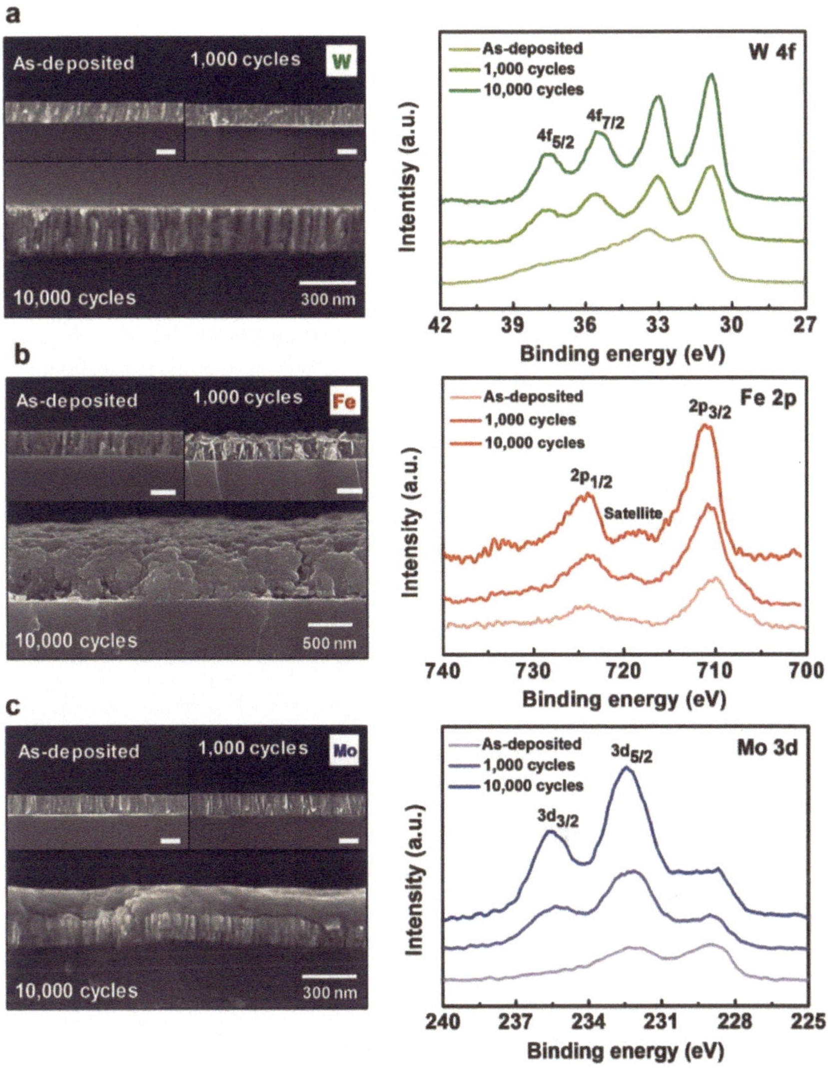

Fig. 4.20 XPS analysis of the electrode material after 10,000 cycles. SEM images after 10,000 cycles for as-deposited films of (A) W, (B) Fe, and (C) Mo. The corresponding initial XPS spectra at specific regions characteristic to these elements along with the ones after 1000 and 10,000 cycles are provided. *Reproduced with permission from G. Lee, S.K. Kang, S.M. Won, P. Gutruf, Y.R. Jeong, J. Koo, S.S. Lee, J.A. Rogers, J.S. Ha, Fully biodegradable microsupercapacitor for power storage in transient electronics, Adv. Energy Mater. 7 (2017) 1–12. https://doi.org/10.1002/aenm.201700157.*

solid-like mechanical robustness can be a suitable replacement that combines the advantages of both liquid and solid electrolytes. Several aqueous-based gel electrolytes (PVA-H_2SO_4, PVA-H_3PO_4, PVA-KOH, etc.) have previously been employed for fabrication of MSCs; however, they are limited in their operational potential windows (maximum till 1.1 V). This is overcome by employing ionic liquids as the ionic conductors. Ionic liquids are greener alternatives to enhanced ionic mobility and offer wider electrochemical potential window resulting in enhanced energy density. The resulting gels are called ion gels that are obtained by immobilization of ionic liquids inside a polymer matrix. Importantly, ionic conductivity of ion gels is $\sim 10^{-2}$ to 10^{-4} S/cm.

Higher ionic conductivity can lead to self-discharge leading to higher leakage current. Contrarily, lower conductivity results in larger iR_{drop} and decreased Coulombic efficiency. Therefore it is best to maintain an optimum level of ionic conductivity (10^{-2} to 10^{-3} S/cm) in the solid electrolyte. The ionic conductivity of solid/gel electrolytes is estimated by the broadband dielectric spectroscopy (BDS) instrument that utilizes an alternating current (AC) and measures the interaction of electromagnetic waves with matter across a broad range of frequencies (10^{-6} to 10^{12} Hz) and thereby provides information on the dynamics of charge carriers present in the sample and their polarizability. Therefore it is crucial both from a fundamental as well as application point of view. A typical BDS instrument comprises.

(a) *Sample cell*: This usually consists of two parallel metal plates that sandwiches the sample. The sample cell is connected to a temperature control setup and a cryostat to enable measurements across a wide range of temperatures (−150 to 400 °C). Often, a piezoresistive sensor control also forms a part of the sample cell to investigate rheological attributes of the material.

(b) *Impedance analyzer*: This setup measures the variation of Re(Z) and −Img(Z), as well as phase angle based on the applied AC bias. Such measurements give out information on conducting ion-diffusion in the electrolyte.

Since measurements are carried out in the sandwich-cell setup, it is essential to maintain sufficient physical contact between the sample and metal plate. As a result, a thin film of the electrolyte is required. This forms a major drawback in characterizing highly viscose or dilute solid solution-based electrolytes. Ionic conductivities of the order of 10^{-2} to 10^{-3} S/cm are considered to be suitable for electrolytes in MSCs.

4.4 Electrochemical characterization techniques

4.4.1 DC techniques

The primary electrochemical measurement carried out for estimating performance parameters of a microsupercapacitor is cyclic voltammetry (CV) [40–42]. Cyclic voltammetry, as the name suggests, involves a linear, time-dependent change in potential of the electrochemical system while simultaneously measuring the current flowing through the system. The current-voltage curve in a typical CV experiment obtained for a supercapacitor is generally rectangular and symmetric in nature (Fig. 4.21). The area of the rectangle can be employed to estimate the capacitance of the system. Mathematically, capacitance,

$$C = \frac{\int IdV}{\nu \Delta V}$$

where the numerator is obtained by integrating a cyclic voltammogram and is actually the area of the rectangle, ΔV is the electrochemical potential window of the voltammogram, and ν is the scan rate.

This directly follows from the electrical double-layer mechanism of charge storage, wherein the response of the system to a linearly varying applied potential corresponds to the build-up of the electrochemical double layer at the electrode-electrolyte interface. This results in a symmetric and rectangular shape in case of perfectly non-Faradaic systems such as supercapacitors (Fig. 4.21).

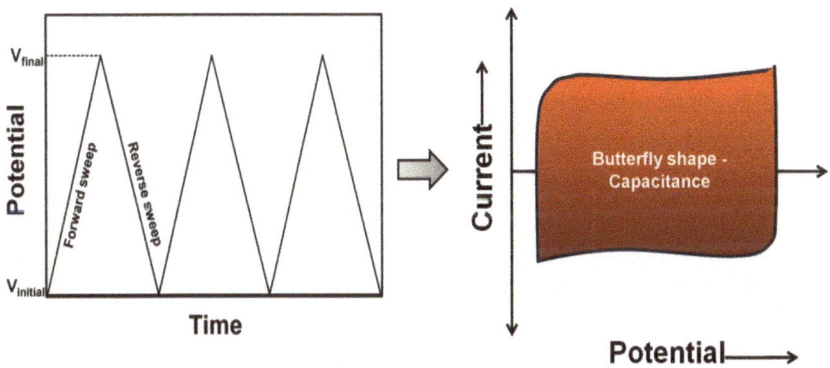

Fig. 4.21 Variation of potential with time in a typical CV experiment. The corresponding butterfly-shaped CV curve of MSC is shown aside.

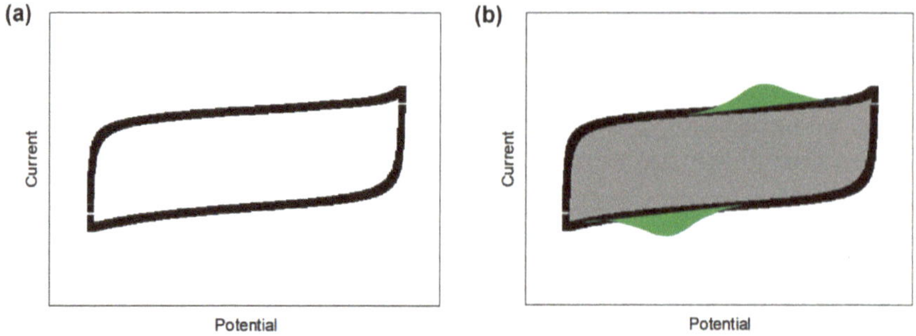

Fig. 4.22 (A) A typical CV of a microsupercapacitor (MSC). (B) Representative CV of a pseudocapacitor. The *green-shaded region* represents the Faradaic component and the *gray-shaded region* represents the capacitive component.

Further, the magnitude of such capacitive current varies linearly with the scan rate of the system, due to the decrease in the thickness of the electrochemical double layer. However, the presence of a peak in the voltammogram is indicative of Faradic redox process occurring at the interface. Such redox processes are facilitated by metal oxides, sulfides, spinels present in the electrode material [21,43–46]. Such surface-mediated Faradic processes therefore lead to "pseudocapacitance." The nature of the cyclic voltammograms exhibits a departure from the ideal rectangular shape with clearly identifiable electron-transfer processes identified as redox peaks (Fig. 4.22). Nanocarbons including graphene, amorphous carbon, and carbon nanotubes have been suitably modified by laser engraving techniques to fabricate microsupercapacitors with enhanced performance [47,48].

MSCs can be characterized by both two- and three-electrode measurements. While a two-electrode measurement is essential to estimate the performance of the device and determine its utility, a three-electrode measurement provides the potential distribution of the active material and can provide information about the mechanism of charge storage. Two-electrode measurements in MSCs are primarily restricted to symmetric electrodes, due to challenges in fabrication of asymmetric supercapacitive electrodes. A three-electrode setup utilizes working electrode that carries the active material, counter electrode for measuring current, and the reference electrode for controlling the applied potential. The working electrode contains the active electrode material. While potential applied to the working electrode is with respect to the reference electrode, total current is measured between the working and counter electrodes. Since a small counter electrode can lead to higher current density passing through it, high area surfaces (like Pt mesh) are employed as

Table 4.1 Comparison of capacitance (F/g) by CV and GCD using a three-electrode and two-electrode cell using PPy and PANI as the active electrode material.

ECP in the composite electrode	Three-electrode cell		Two-electrode cell	
	CV	Galvanostatic discharge	CV	Galvanostatic discharge
PPy	506	495	192	200
PANI	670	650	344	360

Reproduced with permission from V. Khomenko, E. Frackowiak, Determination of the specific capacitance of conducting polymer/nanotubes composite electrodes using different cell configurations, Electrochim. Acta 50 (2005) 2499–2506. doi:10.1016/j.electacta.2004.10.078.

counter electrodes. To avoid potential drop due to solution resistance, a Luggin capillary is often employed to facilitate smaller distance between the working and reference electrode. Fundamentally, the potential applied to a two-electrode cell is half that of the entire range of the potential axis and is identical for both the electrodes. Hence, for a given range of potentials, the working electrode essentially has double the potential applied, when compared to the two-electrode cell. This results in double capacitance. Khomenko et al. estimated capacitance of the same electrode material using different configurations and obtained almost double capacitance with three-electrode cells (Table 4.1) [49]. This shows that while a three-electrode measurement is suited to define Faradaic reactions in the system, it might not be the best possible way to assess the performance of solid-state MSCs.

It is a good practice to provide the current density (normalized to either weight of the electrode material or the area of the electrode) while reporting cyclic voltammograms. In the specific case of a microsupercapacitor, the interdigitated electrodes offer superior performance due to greater accessibility of the electrodes to the electrolyte. It is not advisable to consider weight normalized energy density and power density for MSCs, as the performance of the material does not scale linearly with mass of the electrode material [8]. The areal capacitance of very thin interdigitated electrodes might however be limited, since higher amount of active electrode material is required to accommodate the patterned surface. Despite such limitations, interdigitated electrodes are preferred to conventional planar sandwich geometries. This is because the planar design lacks the competence to scale up, indicating that larger area electrodes stacked in a sandwich-form can suffer from area-dependent potential variations resulting in lowered performance. Additionally, thicker electrodes are not better alternatives since these lead to sluggish ion kinetics and double-layer formation significantly affecting the power density of the MSC. The area of interdigitated fingers can be estimated by controlling the precision of microfabrication.

Fundamentally, there are two different design principles associated with the fabrication of interdigitated architectures for MSCs. The first one involves coating of the active material on a prepatterned substrate. The area between the interdigitated fingers remains in contact to the current collectors and the electrolyte and contribute to double-layer formation. The second category involves patterning on a film of the active material itself. This facilitates greater accessibility of the active material to the electrolyte. The area and depth of patterning that can give a clue about the distance between the electrodes can be provided as inputs during the lithographic process. The capacitance of ultralight carbon-based materials can be normalized to the weight of the active electrode material. If the active material is coated from any dispersion, its weight can be determined using the concentration of the material in the dispersion, its density, and the area of coating. It is therefore essential to choose the correct normalization parameter (area, weight, etc.) for comparing performance of the MSC. In some cases like in a coaxial microsupercapacitor, for example, the volume of the electrode material is taken into account instead of weight or area. This directly follows from the foreseen application of the device being fabricated. For instance, specific current (in units of mA/g) is the parameter required to characterize lightweight microsupercapacitors [23]. Similarly, a device modeled as a wire-in cylinder is expected to occupy a small volume. Hence, it is quintessential to choose the accurate descriptor for reporting the obtained current. Besides current, choosing an appropriate range of scan rates is equally important. The scan rate observes a direct proportionality relationship with the obtained current in a supercapacitor. This is explained by the Randles-Sevcik equation for supercapacitors:

$$i = i_c + i_F = a\upsilon + b\upsilon^{1/2}$$

where i_C is the capacitive current and i_F is the Faradaic current. a and b are arbitrary coefficients. υ is the scan rate. Thus while the capacitive current is directly proportional to the scan rate, the Faradaic current varies as its square root. These are illustrated in Fig. 4.22 where the gray-shaded region corresponds to capacitive current and the green-shaded part is the component of current that arises due to Faradaic processes in the system.

Additionally, Randles-Sevcik analysis allows us to deconvolute the capacitive and noncapacitive segments of the operating mechanism behind energy storage. Such analyses have been utilized to investigate and decouple the effect of redox partners like MoS_2 toward the performance metrics of microsupercapacitors (Fig. 4.23) [50].

Since capacitance is charge per unit potential, the specific capacitance decreases with increasing scan rates. Articles on microsupercapacitors report a wide variety of applied scan rates (Fig. 4.24). While it is imperative to employ smaller scan rates for better CVs, the microsupercapacitor must

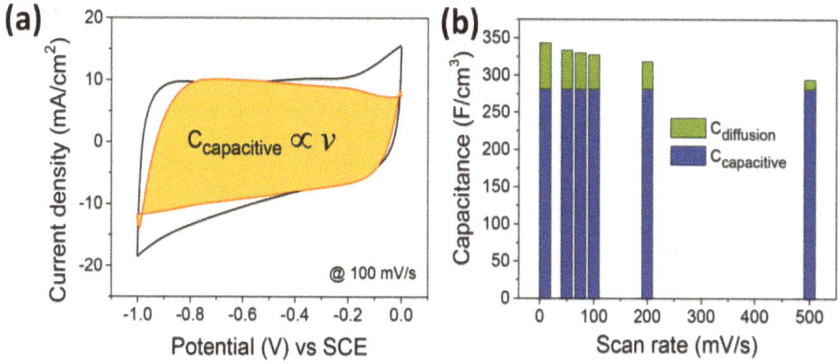

Fig. 4.23 (A) CV for a supercapacitor recorded at 100 mV/s using MoS_2-rGO hybrid material showing the segregation of capacitive and diffusive contributions. (B) Quantitative segregation of capacitive and diffusion-controlled components. *Reproduced with permission from D. Sarkar, D. Das, S. Das, A. Kumar, S. Patil, K.K. Nanda, D.D. Sarma, A. Shukla, Expanding interlayer spacing in MoS2 for realizing an advanced supercapacitor, ACS Energy Lett. 4 (2019) 1602–1609. https://doi.org/10.1021/acsenergylett.9b00983.*

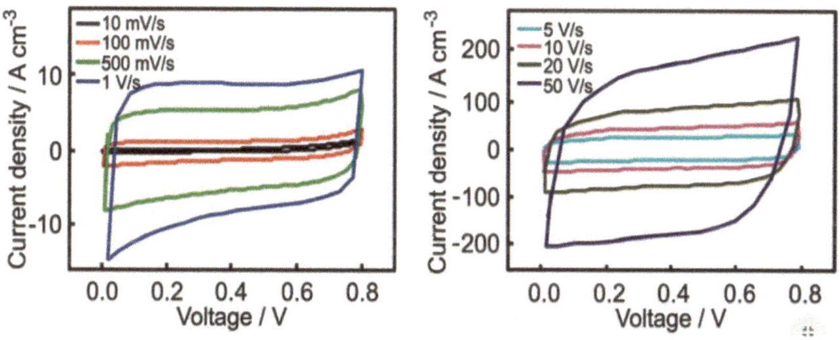

Fig. 4.24 CVs of as-fabricated MSCs recorded at varying scan rates ranging from a low scan rate of 10 mV/s to a high scan rate of 50 V/s. *Reproduced with permission from W. Si, C. Yan, Y. Chen, S. Oswald, L. Han, O.G. Schmidt, On chip, all solid-state and flexible micro-supercapacitors with high performance based on MnOx/au multilayers, Energy Environ. Sci. 6 (2013) 3218–3223. https://doi.org/10.1039/c3ee41286e.*

be operated at the maximum possible scan rate. Cycling the device at higher scan rates results in loss of potential Faradic processes. Therefore higher scan rates (beyond 500 mV/s) should always be employed only after recording the CVs at lower scan rates to ensure the absence of any Faradic behavior. Additionally, the capability of the device to handle higher scan rates and yet retain the EDL mechanism is directly correlated to the relaxation time constant of the device. Another important aspect in a typical CV experiment is the electrochemical potential window. Not only does the potential window depend on the electrode material, the electrolyte employed for the measurement also has a crucial role. Quite expectedly, the operational potential window is the least for aqueous-based electrolytes that cannot be utilized beyond 1.1 V. Alkali polymer electrolytes (PVA-H_2SO_4, PVA-KOH, PVA-H_3PO_4) have been employed to fabricate microsupercapacitors operable at 1 V [7,51,52]. Similarly, organic electrolytes (acetonitrile-based, propylene carbonate-based) are functional to slightly above 2.2 V [53,54]. High potential windows are achieved with ionic-liquid-based liquid as well as solid electrolytes [7,55]. Such high operational potential offers enhanced energy density and therefore opens up avenues for their applicability in microelectronic circuitry (Fig. 4.25).

Galvanostatic charge-discharge (GCD) measurements are carried out in conjunction with CVs to examine performance pertaining to charging and

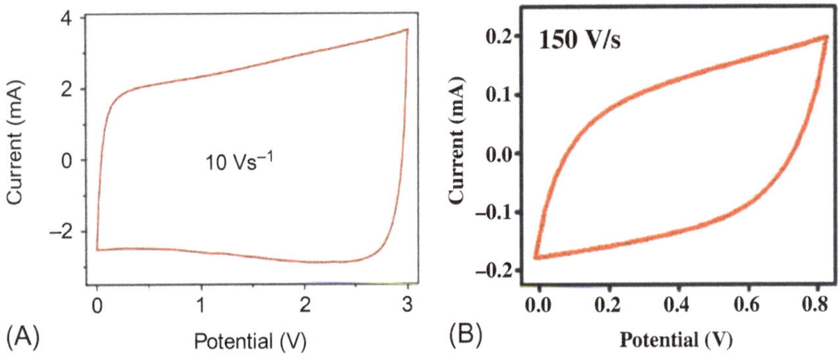

Fig. 4.25 CV of MSC fabricated using (A) carbon onions and (B) CNT-yarn as the active electrode material. *Reproduced with permission from M.K. Jha, R. Ball, R. Seelaboyina, C. Subramaniam, All Solid-State Coaxial Supercapacitor with Ultrahigh Scan Rate Operability of 250 000 mV/s by Thermal Engineering of the Electrode–Electrolyte Interface, ACS Appl. Energy Mater. (2020) acsaem.9b02528. https://doi.org/10.1021/acsaem.9b02528; D. Pech, M. Brunet, H. Durou, P. Huang, V. Mochalin, Y. Gogotsi, P.L. Taberna, P. Simon, Ultrahigh-power micrometre-sized supercapacitors based on onion-like carbon, Nat. Nanotechnol. 5 (2010) 651–654. https://doi.org/10.1038/nnano.2010.162.*

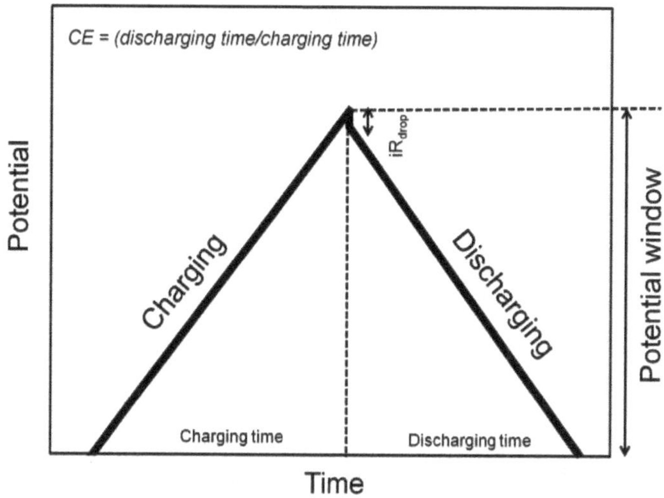

Fig. 4.26 Shape of a typical galvanostatic charge–discharge curve showing iR_{drop}, charging, and discharging curve.

discharging of microsupercapacitors. A closer inspection of GCD curves of supercapacitors reveals three clearly identifiable regions (Fig. 4.26). The uni-gradient nature of the discharging curve is an indication of electrical double-layer mechanism of charge storage. A pseudocapacitive material therefore would exhibit multiple slopes in the GCD curve.

The choice of discharge current density dictates the overall metrics in supercapacitors reflected in their energy and power densities. Higher the discharge current density, lower is the energy density. Being mutually exclusive, power densities are relatively higher at large discharge current densities. Moreover, the specific capacitance of the device also depends on the time of discharge. A microsupercapacitor with superior performance would be operable even at higher discharge current densities with a large time of discharge. Just like in CVs, an enhanced operational potential window of charging and discharging provides advanced performance. Capacitance (in farads) can be estimated from GCD curves using the equation:

$$C = \frac{I\Delta t}{\Delta V}$$

where I is the discharge current, Δt and ΔV are the discharge time and voltage, respectively. The specific capacitance can therefore be obtained from [56]

$$C_{sp} = \frac{4C}{m}$$

where C is the capacitance measured in a two-electrode cell from the equation above, and m denotes total mass of the active material in both the electrodes. The multiplier "4" in the equation adjusts the cell capacitance (C) and combined mass (m) to the capacitance and mass of a single electrode.

Coulombic efficiency (CE) of a microsupercapacitor is derived by dividing the area under discharging curve with that of the charging one (Fig. 4.26). Alternatively, it can be obtained by simply dividing the discharging time to the time of charging for ideal non-Faradic systems. This translates to the fundamental definition of CE, which is essentially a measure of estimating what is obtained from the MSC with respect to what has been fed. Therefore

$$CE = \frac{\textit{time of discharge}}{\textit{time of charging}} = \frac{\int (V - iR)\Delta t_{discharge}}{\int V \Delta t_{charge}}$$

an ideal microsupercapacitor would possess 100% CE. However, it deviates from ideality due to (a) iR_{drop} that arises out of internal resistance in the device configuration and (b) fast discharge facilitated by decreased accessibility to the electrochemical surface area after charging, leading to lowered energy density. Enhanced time of discharge is usually due to change in morphology of the electrode material during charging or gradual decomposition of the electrolyte. This might reflect in an efficiency value of more than 100%, which is meaningless for a microsupercapacitor.

4.4.2 AC techniques

Electrochemical impedance spectroscopy (EIS) is based on measuring the impedance of the system with applied AC potential bias. All DC techniques provide a dynamic, permanent perturbation to the electrode-electrolyte interface and thereby drive permanent changes to surface, interface, and volume of the interface. Such changes are usually accompanied by Faradic redox processes as well. In contrast, AC techniques involve a small, negligible perturbation to the equilibrium state of the system and thereby provide key information of the nearly unperturbed state of the system. The nomenclature of EIS comprises the term "spectroscopy" because the applied potential varies across a range of frequencies. This bears resemblance with other spectroscopic methods wherein frequency of the incident radiation is varied during measurement. EIS of a microsupercapacitive system provides crucial information about the role of ion-diffusion across the electrode-electrolyte interface, prevalence of mass-transfer effects, electrolyte resistance, and the kinetics of double-layer formation. Such insights on to the system pave the way for effective realization of higher performance metrics.

In conventional direct current (DC) measurements, the property of a material to inhibit the flow of electricity, known as resistance (R), is given by Ohm's law,

$$R = \frac{V}{I}$$

where V is the potential difference and I is the current obtained. Resistance is independent of the frequency of potential bias applied. However in practical systems, the frequency of the applied bias cannot be neglected. This is where the concept of impedance comes. Impedance is basically the overall inhibition of current flow. This indicates that impedance is resistance for alternating current (AC) systems. The excitation potential (V) of an AC system can be written as

$$V = V_0 \sin(\omega t)$$

Here V_0 is the amplitude of the AC potential wave propagating with frequency ω at time t. In response to the AC potential excitation, the measured current (I) is

$$I = I_0 \sin(\omega t \pm \varphi)$$

Therefore the obtained current with amplitude I_0 maintains a constant phase difference (φ) with respect to the applied potential (Fig. 4.27).

The phase difference between the potential and current can clearly be visualized. In DC setup, there is no phase difference between potential and current leading to the cancellation of the frequency terms. Thus impedance can be estimated as

$$Z = \frac{V}{I} = \frac{V_0 \sin(\omega t)}{I_0 \sin(\omega t \pm \varphi)}$$

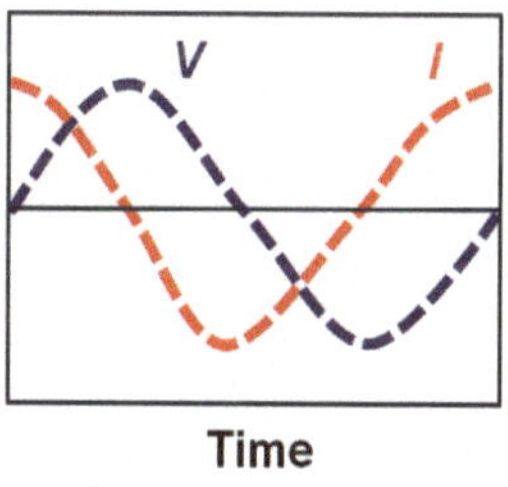

Fig. 4.27 Variation of voltage and current with time, representing an AC sinusoidal potential and current wave.

Impedance is usually expressed as a complex function. Mathematically,

$$V = V_0 \sin(\omega t) = V_0 e^{j\omega t}$$
$$I = I_0 \sin(\omega t \pm \varphi) = I_0 e^{(j\omega t \mp \varphi)}$$

Similarly, impedance, $Z = \frac{V}{I} = \frac{V_0}{I_0} e^{j\varphi}$. Based on practical applications, there can be different electrical elements in a circuit. Most common of them are resistor, capacitor, and inductor. Except the resistance, the impedance for both capacitance and inductance are frequency dependent and are given by

$$Z_C = \frac{1}{j\omega C};\ Z_L = j\omega L$$

The above relations can be derived by the basic definitions of capacitance and inductance and taking their frequency dependence into account. Interestingly, any interface or electrochemical phenomena can be modeled as a combination of several circuit elements comprising resistor, capacitor, inductor, etc. A simple example to begin with involves modeling of an electrode-electrolyte interface as an RC circuit (Fig. 4.28).

The total impedance of such a circuit is obtained by summation of impedances of individual resistor and capacitor. Therefore in this case, impedance, $Z = \frac{1}{j\omega C} + j\omega L$. In a typical EIS experiment, two key parameters are required to be provided as inputs to the potentiostat. These are the potential bias and the range of frequencies. While the usual bias for characterizing a microsupercapacitor can be somewhere between 10 and 100 mV, it would be wiser to employ a broad range of frequency ranging from millihertz to megahertz in a primary EIS experiment. Such a broad range scales ensure minimal loss of vital information about the system. In some cases, for instance, the charge-transfer properties of the system can be shadowed if high-frequency measurements are not carried out. Generally, there are three kinds of data representations that are extracted from EIS—(a) plot of real impedance and imaginary impedance, popularly known as the Nyquist plot; (b) a plot of phase angle with frequency, called Bode plot; and (c) plot of impedance with frequency. Often logarithmic scales are employed to obtain a clear picture of change in these parameters.

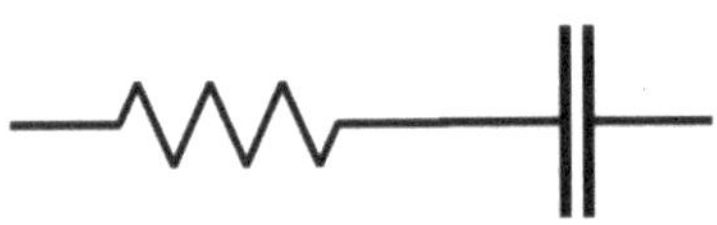

Fig. 4.28 A resistor and capacitor connected in series, RC circuit.

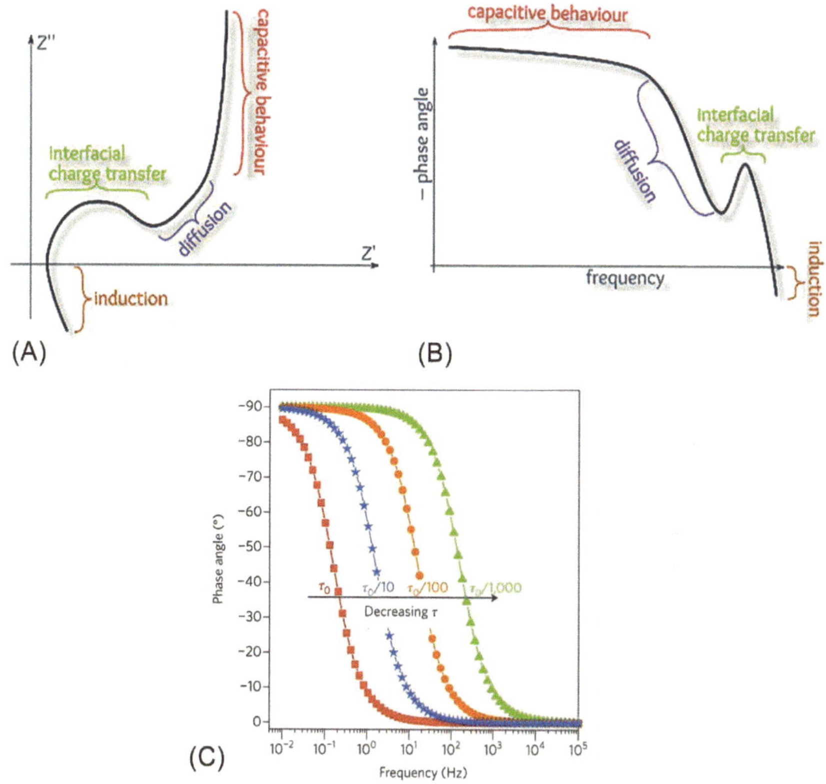

Fig. 4.29 Representative (A) Nyquist plot and (B) Bode plot showing the significance of different regions. (C) Bode plots of MSC devices with different relaxation time constants. *Reproduced from N.A. Kyeremateng, T. Brousse, D. Pech, Microsupercapacitors as miniaturized energy-storage components for on-chip electronics, Nat. Nanotechnol. 12 (2017) 7–15. https://doi.org/10.1038/nnano.2016.196; A. Eftekhari, The mechanism of ultrafast supercapacitors, J. Mater. Chem. A. 6 (2018) 2866–2876. https://doi.org/10.1039/c7ta10013b.*

Fig. 4.29A shows representative Nyquist plot for microsupercapacitors. The initial value of real impedance at zero imaginary impedance signifies the electrolyte resistance of the microsupercapacitor. This can be lowered either by using electrolytes of high ionic conductivity or by employing techniques and designs that facilitate seamless electrode-electrolyte interface [57]. At slightly lower frequencies, lies a semicircular region. The diameter of the semicircle is a measure of the charge-transfer resistance (R_{ct}) of the system. Lower the R_{ct}, better is the capacitor. The R_{ct} value of microsupercapacitors fabricated using transition metal chalcogenide-based electrode materials is relatively lower than their graphene counterparts, indicating the role of interstitial voids in these materials that lead to the formation of efficient electrode-electrolyte interfaces [58]. Since the shape of the semicircle and abscissa of a Nyquist plot determines vital parameters pertaining to the resistance of the system, both the *X* and *Y* scales of Nyquist plot should always be identical for an effective comparison. Finally, the slope of the linear region is indicative of capacitance. Higher slope reflects higher capacitance. The Nyquist plot is usually modeled to an equivalent circuit that consists of electrical elements like a resistor (*R*), capacitor (*C*), and inductor (*L*). Such a circuit is chosen based on its resemblance to the overall microsupercapative system. For example, an electrode-electrolyte interface can simply be modeled using RC circuit.

The variation of phase angle with frequency is represented in a Bode plot (Fig. 4.29b,c). For microsupercapacitive systems, Bode plot is deemed to be a more accurate descriptor than the Nyquist plot [59]. An ideal microsupercapacitor would exhibit a phase angle of $-90°$. High phase angle at a frequency of 120 Hz is critical for microsupercapacitors targeted for AC-line filtering applications. Kim and coworkers have devised a microsupercapacitor with high operational potential window intended for line filtering application [55]. As observed in Fig. 4.29C, the phase angle for the as-fabricated device is close to $-80°$ at 120 Hz. Dielectric relaxation time constant is estimated at the frequency where phase angle equals to $-45°$. The time constant is a measure of kinetics of diffusion and mass-transfer processes occurring in the microsupercapacitor and directly relates to its power delivery rate. MSCs with ultralow relaxation time constants have been reported with carbon onions comprising endohedral pores [7]. Similarly, MSCs operable at ultrahigh scan rates exhibit low time constants indicating fast ion diffusion [33].

While the Nyquist and Bode plots represent crucial parameters of the electrode-electrolyte interface, the overall impedance of the system is sometimes plotted against the frequency of applied AC bias. This illustrates the cumulative contribution of all constant phase elements (i.e., *R*, *L*, and *C*) to overall discharge kinetics of the MSC. Additionally, Warburg component is another such constant phase element with a phase angle of 45°. As observed in Fig. 4.29, the Warburg component exemplifies diffusion-limited regimes across the electrical double layer. A temperature variation across the interface can have significant effect on the Warburg impedance of the MSC.

4.5 Summary

To summarize, this chapter details the various techniques that would be essential to derive a complete understanding of the working of a microsupercapacitor. Equal emphasis has been provided to both the material and device characterization aspects in order to enable a thorough structure-property-performance correlation for MSCs. Although all these techniques discussed are standardized to a large extent in the context of MSCs, the emergence of newer materials, processes, and techniques mandate continuous evolution of these concepts to keep up with the rapid advancements in the field of MSCs.

References

[1] H. Li, Y. Tao, X. Zheng, J. Luo, F. Kang, H.M. Cheng, Q.H. Yang, Ultra-thick graphene bulk supercapacitor electrodes for compact energy storage, Energy Environ. Sci. 9 (2016) 3135–3142, https://doi.org/10.1039/c6ee00941g.

[2] J. Yoo, S. Byun, C.W. Lee, C.Y. Yoo, J. Yu, Precisely geometry controlled microsupercapacitors for ultrahigh areal capacitance, volumetric capacitance, and energy density, Chem. Mater. 30 (2018) 3979–3990, https://doi.org/10.1021/acs.chemmater.7b03786.

[3] M.K. Jha, C. Subramaniam, Design principles for manipulating electrochemical interfaces in solid-state supercapacitors for wearable applications, ACS Omega 6 (12) (2021) 7970–7978, https://doi.org/10.1021/acsomega.1c00172.

[4] J. Chmiola, P.L.T. Celine Largeot, P. Simon, Y. Gogotsi, Monolithic carbide-derived carbon films for micro-supercapacitors, Science (80–) 328 (2010) 480–483, https://doi.org/10.1126/science.1184126.

[5] Z.S. Wu, Y.Z. Tan, S. Zheng, S. Wang, K. Parvez, J. Qin, X. Shi, C. Sun, X. Bao, X. Feng, K. Müllen, Bottom-up fabrication of sulfur-doped graphene films derived from sulfur-annulated nanographene for ultrahigh volumetric capacitance micro-supercapacitors, J. Am. Chem. Soc. 139 (2017) 4506–4512, https://doi.org/10.1021/jacs.7b00805.

[6] X. Li, J. Shao, S.K. Kim, C. Yao, J. Wang, Y.R. Miao, Q. Zheng, P. Sun, R. Zhang, P.-V. Braun, High energy flexible supercapacitors formed via bottom-up infilling of gel electrolytes into thick porous electrodes, Nat. Commun. 9 (2018), https://doi.org/10.1038/s41467-018-04937-8.

[7] N.A. Kyeremateng, T. Brousse, D. Pech, Microsupercapacitors as miniaturized energy-storage components for on-chip electronics, Nat. Nanotechnol. 12 (2017) 7–15, https://doi.org/10.1038/nnano.2016.196.

[8] M. Beidaghi, Y. Gogotsi, Capacitive energy storage in micro-scale devices: recent advances in design and fabrication of micro-supercapacitors, Energy Environ. Sci. 7 (2014) 867–884, https://doi.org/10.1039/c3ee43526a.

[9] W. Liu, C. Lu, X. Wang, R.Y. Tay, B.K. Tay, High-performance microsupercapacitors based on two-dimensional graphene/manganese dioxide/silver nanowire ternary hybrid film, ACS Nano 9 (2015) 1528–1542, https://doi.org/10.1021/nn5060442.

[10] Z. Lei, L. Liu, H. Zhao, F. Liang, S. Chang, L. Li, Y. Zhang, Z. Lin, J. Kröger, Y. Lei, Nanoelectrode design from microminiaturized honeycomb monolith with ultrathin

and stiff nanoscaffold for high-energy micro-supercapacitors, Nat. Commun. 11 (2020), https://doi.org/10.1038/s41467-019-14170-6.
[11] A.C. Reimschuessel, Scanning electron microscopy—part I, J. Chem. Educ. 49 (1972) A413, https://doi.org/10.1021/ed049pa413.
[12] A. Vijayaraghavan, K. Kanzaki, S. Suzuki, Y. Kobayashi, H. Inokawa, Y. Ono, S. Kar, P.M. Ajayan, Metal-semiconductor transition in single-walled carbon nanotubes induced by low-energy electron irradiation, Nano Lett. 5 (2005) 1575–1579, https://doi.org/10.1021/nl0509935.
[13] C. Subramaniam, T.S. Sreeprasad, T. Pradeep, G.V. Pavan Kumar, C. Narayana, T. Yajima, Y. Sugawara, H. Tanaka, T. Ogawa, J. Chakrabarti, Visible fluorescence induced by the metal semiconductor transition in composites of carbon nanotubes with noble metal nanoparticles, Phys. Rev. Lett. 99 (2007) 1–4, https://doi.org/10.1103/PhysRevLett.99.167404.
[14] C.(J.). Zhang, L. McKeon, M.P. Kremer, S.H. Park, O. Ronan, A. Seral-Ascaso, S. Barwich, C. Coileáin, N. McEvoy, H.C. Nerl, B. Anasori, J.N. Coleman, Y. Gogotsi, V. Nicolosi, Additive-free MXene inks and direct printing of micro-supercapacitors, Nat. Commun. 10 (2019) 1–9, https://doi.org/10.1038/s41467-019-09398-1.
[15] S. Brunauer, P.H. Emmett, E. Teller, Adsorption of gases in multimolecular layers, J. Am. Chem. Soc. 60 (1938) 309–319, https://doi.org/10.1021/ja01269a023.
[16] W.G. McMillan, E. Teller, The assumptions of the BET theory, J. Phys. Chem. 55 (1951) 17–20.
[17] H. Shi, Activated carbons and double layer capacitance, Electrochim. Acta 41 (1996) 1633–1639, https://doi.org/10.1016/0013-4686(95)00416-5.
[18] C.J. Zhang, B. Anasori, A. Seral-Ascaso, S.H. Park, N. McEvoy, A. Shmeliov, G.S. Duesberg, J.N. Coleman, Y. Gogotsi, V. Nicolosi, Transparent, flexible, and conductive 2D titanium carbide (MXene) films with high volumetric capacitance, Adv. Mater. 29 (2017) 1–9, https://doi.org/10.1002/adma.201702678.
[19] M.R. Lukatskaya, O. Mashtalir, C.E. Ren, Y. Dall'Agnese, P. Rozier, P.L. Taberna, M. Naguib, P. Simon, M.W. Barsoum, Y. Gogotsi, Cation intercalation and high volumetric capacitance of two-dimensional titanium carbide, Science (80-) 341 (2013) 1502–1505, https://doi.org/10.1126/science.1241488.
[20] M. Ghidiu, M.R. Lukatskaya, M.Q. Zhao, Y. Gogotsi, M.W. Barsoum, Conductive two-dimensional titanium carbide "clay" with high volumetric capacitance, Nature 516 (2015) 78–81, https://doi.org/10.1038/nature13970.
[21] J. Wang, X. Wang, S.W. Lee, Q. Zhang, Enhanced performance of an electric double layer microsupercapacitor based on novel carbon-encapsulated cu nanowire network structure as the electrode, ACS Appl. Mater. Interfaces 11 (2019) 40481–40489, https://doi.org/10.1021/acsami.9b13417.
[22] J. Li, Q. Shi, Y. Shao, C. Hou, Y. Li, Q. Zhang, H. Wang, Cladding nanostructured AgNWs-MoS2 electrode material for high-rate and long-life transparent in plane micro-supercapacitor, Energy Storage Mater. 16 (2019) 212–219, https://doi.org/10.1016/j.ensm.2018.05.013.
[23] P. Simon, Y. Gogotsi, Materials for electrochemical capacitors, Nat. Mater. 7 (2008) 845–854, https://doi.org/10.1142/9789814317665_0021.
[24] J. Huang, B.G. Sumpter, V. Meunier, Theoretical model for nanoporous carbon supercapacitors, Angew. Chem., Int. Ed. 47 (2008) 520–524, https://doi.org/10.1002/anie.200703864.
[25] P. Huang, C. Lethien, S. Pinaud, K. Brousse, R. Laloo, V. Turq, M. Respaud, A. Demortière, B. Daffos, P.L. Taberna, B. Chaudret, Y. Gogotsi, P. Simon, On-chip and freestanding elastic carbon films for micro-supercapacitors, Science (80-) 351 (2016) 691–695.

[26] J. Xu, Y. He, S. Bi, M. Wang, P. Yang, D. Wu, J. Wang, F. Zhang, An olefin-linked covalent organic framework as a flexible thin-film electrode for a high-performance micro-supercapacitor, Angew. Chem., Int. Ed. 58 (2019) 12065–12069, https://doi.org/10.1002/anie.201905713.

[27] T. Kan, V. Strezov, T. Evans, Effect of the heating rate on the thermochemical behavior and biofuel properties of sewage sludge pyrolysis, Energy Fuels 30 (2016) 1564–1570, https://doi.org/10.1021/acs.energyfuels.5b02232.

[28] L.G. Cançado, A. Jorio, E.H.M. Ferreira, F. Stavale, C.A. Achete, R.B. Capaz, M.V.O. Moutinho, A. Lombardo, T.S. Kulmala, A.C. Ferrari, Quantifying defects in graphene via Raman spectroscopy at different excitation energies, Nano Lett. 11 (2011) 3190–3196, https://doi.org/10.1021/nl201432g.

[29] D. Chen, X. Xiong, B. Zhao, M.A. Mahmoud, M.A. El-Sayed, M. Liu, Probing structural evolution and charge storage mechanism of NiO2 Hx electrode materials using in operando resonance Raman spectroscopy, Adv. Sci. 3 (2015) 1–6, https://doi.org/10.1002/advs.201500433.

[30] M.J. Matthews, M.A. Pimenta, G. Dresselhaus, M.S. Dresselhaus, M. Endo, Origin of dispersive effects of the Raman D band in carbon materials, Phys. Rev. B Condens. Matter Mater. Phys. 59 (1999) 6585–6588, https://doi.org/10.1103/physrevb.59.r6585.

[31] N. Kamboj, T. Purkait, M. Das, S. Sarkar, K.S. Hazra, R.S. Dey, Ultralong cycle life and outstanding capacitive performance of a 10.8 v metal free micro-supercapacitor with highly conducting and robust laser-irradiated graphene for an integrated storage device, Energy Environ. Sci. 12 (2019) 2507–2517, https://doi.org/10.1039/c9ee01458f.

[32] M. Anji Reddy, M. Helen, A. Groß, M. Fichtner, H. Euchner, Insight into sodium insertion and the storage mechanism in hard carbon, ACS Energy Lett. 3 (2018) 2851–2857, https://doi.org/10.1021/acsenergylett.8b01761.

[33] M.K. Jha, R. Ball, R. Seelaboyina, C. Subramaniam, All solid-state coaxial supercapacitor with ultrahigh scan rate operability of 250000 mV/s by thermal engineering of the electrode–electrolyte interface, ACS Appl. Energy Mater. (2020), https://doi.org/10.1021/acsaem.9b02528, acsaem.9b02528.

[34] M.K. Jha, T. Jain, C. Subramaniam, Origami of solid-state supercapacitive microjunctions operable at 3 V with high specific energy density for wearable electronics, ACS Appl. Electron. Mater. 2 (2020) 659–669.

[35] P. Zhang, Y. Li, G. Wang, F. Wang, S. Yang, F. Zhu, X. Zhuang, O.G. Schmidt, X. Feng, Zn-ion hybrid micro-supercapacitors with ultrahigh areal energy density and long-term durability, Adv. Mater. 31 (2019) 1–6, https://doi.org/10.1002/adma.201806005.

[36] X. Li, M. Li, Q. Yang, H. Li, H. Xu, Z. Chai, K. Chen, Z. Liu, Z. Tang, L. Ma, Z. Huang, B. Dong, X. Yin, Q. Huang, C. Zhi, Phase transition induced unusual electrochemical performance of V2CTX MXene for aqueous zinc hybrid-ion battery, ACS Nano 14 (2020) 541–551, https://doi.org/10.1021/acsnano.9b06866.

[37] D. Kashyap, P.K. Hee, D. Marshal, Two-step process for programmable removal of oxygen functionality of graphene oxide: functional, structural and electrical characteristics, RSC Adv. 5 (2015) 95657–95665, https://doi.org/10.1039/C5RA18880F.

[38] G. Lee, S.K. Kang, S.M. Won, P. Gutruf, Y.R. Jeong, J. Koo, S.S. Lee, J.A. Rogers, J.S. Ha, Fully biodegradable microsupercapacitor for power storage in transient electronics, Adv. Energy Mater. 7 (2017) 1–12, https://doi.org/10.1002/aenm.201700157.

[39] C. Lethien, J. Le Bideau, T. Brousse, Challenges and prospects of 3D micro-supercapacitors for powering the internet of things, Energy Environ. Sci. 12 (2019) 96–115, https://doi.org/10.1039/c8ee02029a.

[40] N. Elgrishi, K.J. Rountree, B.D. McCarthy, E.S. Rountree, T.T. Eisenhart, J.L. Dempsey, A practical beginner's guide to cyclic voltammetry, J. Chem. Educ. 95 (2018) 197–206, https://doi.org/10.1021/acs.jchemed.7b00361.

[41] L.R. Bard, A.J. Faulkner, Electrochemical Methods, Wiley, 2000.
[42] B.E. Conway, Electrochemical Supercapacitors: Scientific Fundamentals and Technological Applications, Springer US, 1999, https://doi.org/10.1007/978-1-4757-3058-6.
[43] J. Jiang, Y. Li, J. Liu, X. Huang, C. Yuan, X.W. Lou, Recent advances in metal oxide-based electrode architecture design for electrochemical energy storage, Adv. Mater. 24 (2012) 5166–5180, https://doi.org/10.1002/adma.201202146.
[44] P. Chen, G. Shen, Y. Shi, H. Chen, C. Zhou, Preparation and characterization of flexible asymmetric supercapacitors, ACS Nano 4 (2010) 4403–4411, https://doi.org/10.1021/nn100856y.
[45] M. Beidaghi, C. Wang, Micro-supercapacitors based on three dimensional interdigital polypyrrole/C-MEMS electrodes, Electrochim. Acta 56 (2011) 9508–9514, https://doi.org/10.1016/j.electacta.2011.08.054.
[46] C. Shen, X. Wang, S. Li, J. Wang, W. Zhang, F. Kang, A high-energy-density micro supercapacitor of asymmetric MnO 2-carbon configuration by using micro-fabrication technologies, J. Power Sources 234 (2013) 302–309, https://doi.org/10.1016/j.jpowsour.2012.10.101.
[47] M.F. El-Kady, R.B. Kaner, Scalable fabrication of high-power graphene micro-supercapacitors for flexible and on-chip energy storage, Nat. Commun. 4 (2013), https://doi.org/10.1038/ncomms2446.
[48] L. Cao, S. Yang, W. Gao, Z. Liu, Y. Gong, L. Ma, G. Shi, S. Lei, Y. Zhang, S. Zhang, R. Vajtai, P.M. Ajayan, Direct laser-patterned micro-supercapacitors from paintable MoS2 films, Small 9 (2013) 2905–2910, https://doi.org/10.1002/smll.201203164.
[49] V. Khomenko, E. Frackowiak, Determination of the specific capacitance of conducting polymer/nanotubes composite electrodes using different cell configurations, Electrochim. Acta 50 (2005) 2499–2506, https://doi.org/10.1016/j.electacta.2004.10.078.
[50] D. Sarkar, D. Das, S. Das, A. Kumar, S. Patil, K.K. Nanda, D.D. Sarma, A. Shukla, Expanding interlayer spacing in MoS2 for realizing an advanced supercapacitor, ACS Energy Lett. 4 (2019) 1602–1609, https://doi.org/10.1021/acsenergylett.9b00983.
[51] Y. Shao, J. Li, Y. Li, H. Wang, Q. Zhang, R.B. Kaner, Flexible quasi-solid-state planar micro-supercapacitor based on cellular graphene films, Mater. Horizons. 4 (2017) 1145–1150, https://doi.org/10.1039/c7mh00441a.
[52] M.K. Jha, K. Hata, C. Subramaniam, Interwoven carbon nanotube wires for high-performing, mechanically robust, washable, and wearable supercapacitors, ACS Appl. Mater. Interfaces 11 (2019) 18285–18294, https://doi.org/10.1021/acsami.8b22233.
[53] F. Gao, M.T. Wolfer, C.E. Nebel, Highly porous diamond foam as a thin-film micro-supercapacitor material, Carbon N. Y. 80 (2014) 833–840, https://doi.org/10.1016/j.carbon.2014.09.007.
[54] D. Kim, G. Lee, D. Kim, J. Yun, S.S. Lee, J.S. Ha, High performance flexible double-sided micro-supercapacitors with an organic gel electrolyte containing a redox-active additive, Nanoscale 8 (2016) 15611–15620, https://doi.org/10.1039/c6nr04352f.
[55] Y.J. Kang, Y. Yoo, W. Kim, 3-V solid-state flexible supercapacitors with ionic-liquid-based polymer gel electrolyte for AC line filtering, ACS Appl. Mater. Interfaces 8 (2016) 13909–13917, https://doi.org/10.1021/acsami.6b02690.
[56] M.D. Stoller, R.S. Ruoff, Best practice methods for determining an electrode material's performance for ultracapacitors, Energy Environ. Sci. 3 (2010) 1294–1301, https://doi.org/10.1039/c0ee00074d.
[57] E. Barsoukov, J.R. MacDonald, Impedance Spectroscopy: Theory, Experiment, and Applications, Wiley, 2005, https://doi.org/10.1002/0471716243.
[58] B. Anasori, Y. Gogotsi, 2D Metal Carbides and Nitrides (MXenes), Springer, 2019.
[59] A. Lasia, Electrochemical Impedance Spectroscopy and its Applications, Springer, 2014, https://doi.org/10.1007/978-1-4614-8933-7.

PART II

Applications for Microsupercapacitors

CHAPTER 5

Battery/Supercapacitor hybrid energy storage system in vehicle applications

Ziyou Song
Department of Electrical Engineering and Computer Science, University of Michigan, Ann Arbor, MI, United States

5.1 Introduction

Due to the energy crisis and the greenhouse gas increase, electric vehicles (EVs) and hybrid electric vehicles (HEVs) have been extensively investigated and widely adopted as the promising solution [1], given that the electrified vehicles can achieve the high efficiency and the low emissions when compared to the conventional vehicles only including the internal combustion engine (ICE). Among all components, the energy storage system (ESS) is the key part and the bottleneck of EVs and HEVs [2]. Lithium-ion batteries have been widely used as the ESS of the vehicle given their high-energy density (i.e., 100–265 W h/kg) and mature production technology, which makes lithium-ion batteries affordable than the like products [3]. For example, Tesla adopts around 8000 cells (up to models) in one EV to satisfy the long mileage. However, based on the current technologies, the battery-only ESS still has several drawbacks, such as low power density, short lifetime, and high cost, which cannot be ultimately solved [4]. Specifically, the battery used in EVs needs to supply instantaneously varying power under a wide range of temperatures (e.g., −20 to 45°C) [5], and it has been verified that the frequent charge and discharge operations, especially the ones including high current demands, will cause the significant degradation on batteries (i.e., 20% capacity degradation at most for vehicle applications) [6]. To address these problems, many alternative strategies have been proposed and the battery/supercapacitor hybrid energy storage system (HESS) is one of the most noticeable solutions [7]. The simulation results show that the battery lifetime can be significantly extended by incorporating supercapacitor [8], and it is intuitive that the HESS can utilize the advantages of different power sources (i.e., high-power density of supercapacitor and

https://doi.org/10.1016/B978-0-08-102888-9.00007-0

high-energy density of battery), therefore achieves a better performance when compared to the use of either of power sources alone [8].

To effectively use the HESS, the system should be optimized and the controller should be well designed [9]. The HESS optimization mainly involves three aspects: configuration design [10], component sizing [11], and energy management strategy (EMS) optimization [12]. First, the HESS configurations can be classified into three types, which are passive, semiactive, and fully active. The passive HESS connects the battery and the supercapacitor in parallel and the entire HESS is directly coupled to the DC bus [13], as shown in Fig. 5.1A. The passive HESS is low cost; however, its performance is not satisfactory because the supercapacitor voltage is constrained by the battery voltage, which only varies in a narrow range, therefore the supercapacitor cannot be effectively used. On the contrary, the fully active HESS, as shown in Fig. 5.1B, has the best performance, since it adopts two DC/DC converters and therefore the EMS can ensure both the stability of DC bus voltage and the high efficiency of HESS [14]. Nonetheless, the fully active configuration compromises in terms of capital cost, weight volume, and simplicity. As shown in Fig. 5.1C, the semiactive HESS, which only employs one DC/DC converter to decouple either the battery or the supercapacitor with the DC bus, is a good tradeoff between performance and cost [15]. In consequence, this chapter will focus on the typical semiactive HESS configuration shown in Fig. 5.1C.

This chapter will focus on the component sizing and the design of energy management strategy using global optimization methodology, as shown in Fig. 5.2. Specifically, in terms of the component sizing, it generally indicates the size optimization of the supercapacitor, as the battery size is determined by the all-electric mileage requirement of EVs and HEVs [16]. To effectively prolong the battery lifespan by involving the supercapacitor, the supercapacitor size and the EMS should be optimized concurrently to reduce the electricity cost and the battery degradation cost (fuel cost should also be considered for HEVs). It has been validated that supercapacitor size and EMS should be optimized together since they are inherently coupled [17]. According to the existing studies, the EMS for HESS can be classified as online or off-line strategies. For example, the common online strategies include the "all or nothing" strategy [18], the filtration-based strategy [19], the rule-based strategy [20], the model-predictive control strategy [21], and the fuzzy-logic control strategy [22]. These strategies are generally designed based on engineering experience and can be simply implemented in practical controllers, but it is difficult to achieve the optimal performance. In contrast,

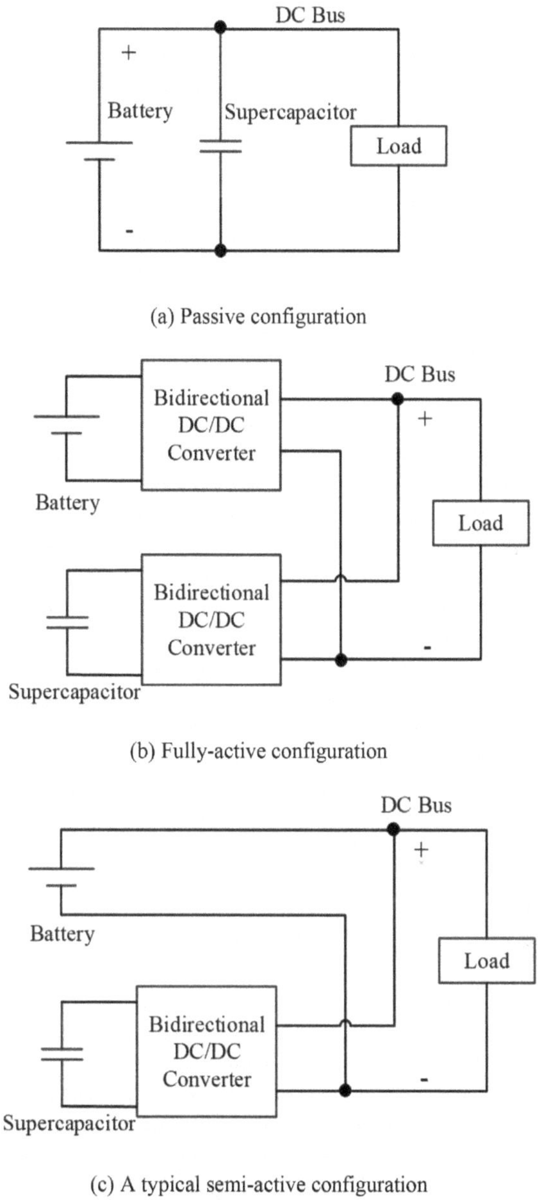

Fig. 5.1 Different HESS configurations. (A) Passive configuration, (B) fully active configuration, and (C) a typical semiactive configuration.

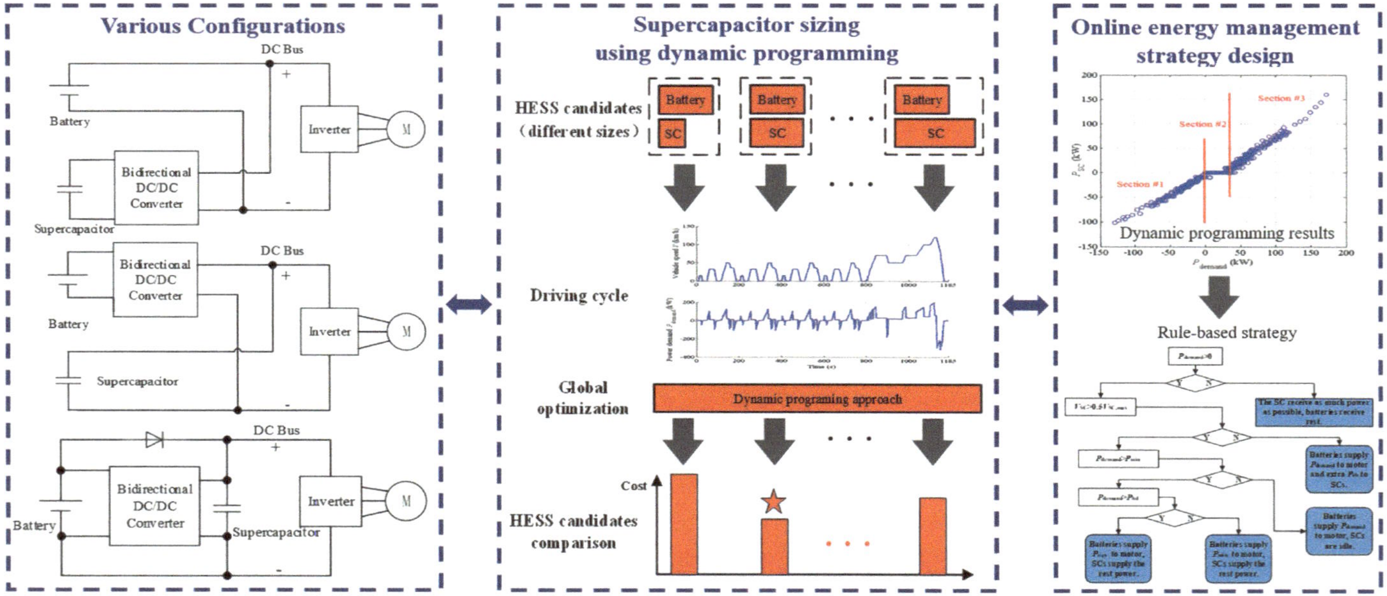

Fig. 5.2 The optimization process of HESS.

the off-line strategies, such as the dynamic programming (DP) approach [17] and the Pontryagin's minimum principle (PMP) [23], can achieve the globally optimal performance, even though it is difficult to implement these strategies online due to the heavy computational burden [24]. The feasible solution for the aforementioned problem can be used to design the online strategies based on the results derived from the off-line strategies. For example, the rule-based strategy can be optimized based on the DP results [17]. Recently, with the development of machine learning techniques, the data-based methods, such as the reinforcement learning-based strategy, have also been investigated as the potential online EMS for HESS [25].

This chapter presents several topics on the optimization of battery/supercapacitor HESS in vehicle applications. In Section 5.2, based on a battery degradation model, the DP approach is used to deal with the integrated design for optimizing the supercapacitor size and the system-level EMS under the typical driving cycle. And a near-optimal rule-based strategy is proposed according to the DP results. In Section 5.3, various temperatures and battery prices are considered and a case study is conducted to quantitatively examine the HESS performance in EV applications in the coming years. This case study is necessary since the HESS is promising and worth using when the battery is expensive; however, the price of lithium batteries has continuously decreased in recent years [26]. In Section 5.4, the two-dimensional PMP is adopted to determine the optimal strategy for the instantaneous power split between the engine, the battery, and the supercapacitor, for a plug-in hybrid electric vehicle (PHEV), considering the fuel cost, the electricity cost, and the battery degradation cost. Finally, in Section 5.5, the microsupercapacitor is briefly introduced to show its potential adoption in vehicle applications, as EVs and HEVs have the critical requirement on the ESS volume and weight. It shows that there will be a huge chance in the vehicle market for the microsupercapacitor, as long as its price is acceptable.

5.2 The optimization of HESS in EVs

In this section, the component sizing and the EMS optimization for a typical semiactive HESS adopted in an electric bus are conducted. The schematic topology of the entire electric powertrain and the corresponding control flowchart are shown in Fig. 5.3. The bidirectional DC/DC converter, which decouples the supercapacitor with the DC bus, is the only actuator in the control loop to ensure the SC power demand P_{SC} tracks the desired

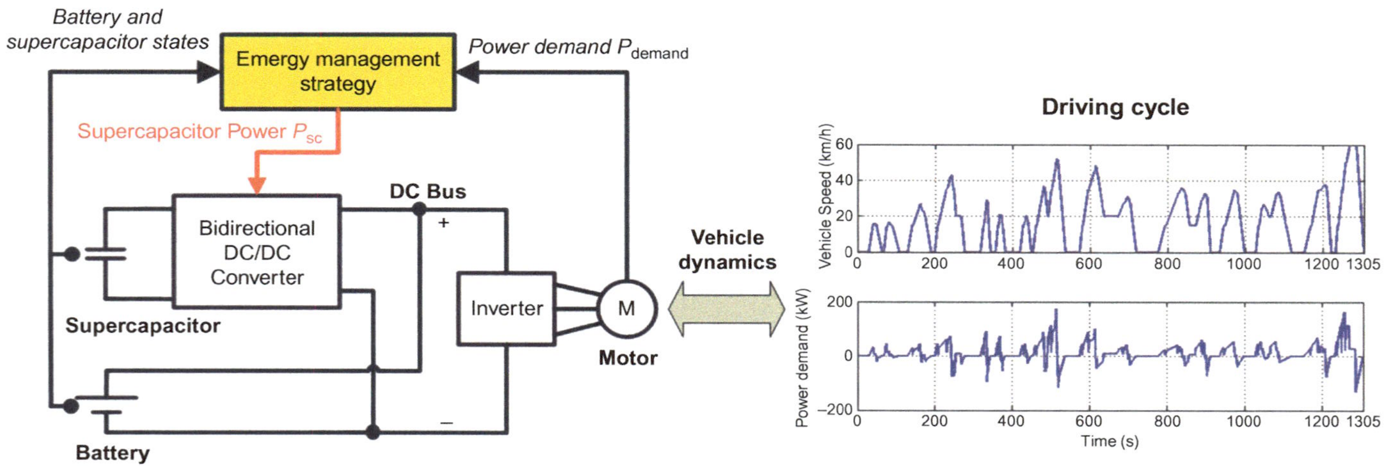

Fig. 5.3 The electric powertrain of the electric bus involving the typical semiactive HESS.

value, which is calculated by the proposed EMS based on the information of supercapacitor, battery, and electric powertrain. The battery is directly connected to the DC bus, thus its power P_{bat} is the difference between the power demand P_{demand} from the inverter and P_{SC}.

The relationship between P_{demand}, P_{SC}, and P_{bat} can be determined as [17].

(1) Traction mode:

$$\begin{aligned} P_{demand} &= P_{bat} + P_{SC}\eta_{DC/DC} \\ &= \left(V_{bat}I_{bat} - I_{bat}^2 R_{bat}\right) + \left(V_{SC}I_{SC} - I_{SC}^2 R_{SC}\right)\eta_{DC/DC}, \end{aligned} \tag{5.1}$$

(2) Regenerative braking mode:

$$\begin{aligned} P_{demand} &= P_{bat} + P_{SC}/\eta_{DC/DC} \\ &= \left(V_{bat}I_{bat} - I_{bat}^2 R_{bat}\right) + \left(V_{SC}I_{SC} - I_{SC}^2 R_{SC}\right)/\eta_{DC/DC}, \end{aligned} \tag{5.2}$$

where V_{bat}/V_{SC} are the open-circuit voltages (OCVs) of the battery/supercapacitor packs, R_{bat} is the battery resistance, R_{SC} is the supercapacitor resistance, I_{bat}/I_{SC} are the currents of the battery/SC (positive for discharging and negative for charging), and $\eta_{DC/DC}$ is the average efficiency of the DC/DC converter. Note that in this work, the charging and discharging resistances are assumed to be the same for both the battery and the supercapacitor. State of charge (SoC) is the crucial state for both battery and supercapacitor and should be carefully considered in the EMS. The SoC dynamics of battery and supercapacitor are given as

$$\begin{cases} SOC_{bat}(t) = SOC_{bat}(0) - \displaystyle\int_0^t \frac{I_{bat}(t)}{C_{bat}} dt, \\ SOC_{SC}(t) = SOC_{SC}(0) - \displaystyle\int_0^t \frac{I_{bat}(t)}{C_{SC} V_{SC,\max}} dt, \end{cases} \tag{5.3}$$

where $SOC_{bat}(t)/SOC_{SC}(t)$ are SoC values of battery/supercapacitor packs at instant t, C_{bat} is the battery capacity, C_{SC} is the supercapacitor capacitance, and $V_{SC,max}$ is the maximum voltage of supercapacitor. Since this chapter emphasizes the energy efficiency optimization and focuses on the HESS responses over a long time interval (more than 20 min for one driving cycle), the *Rint* model shown in Fig. 5.4 is adopted to represent the characteristics of both power sources due to its simplicity and sufficient accuracy [27]. The basic parameters of the battery cell and the SC module adopted in this study are provided in [4].

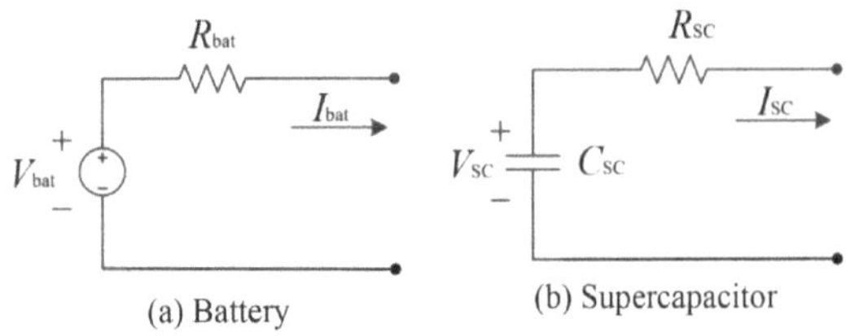

Fig. 5.4 Simplified circuit models of battery and supercapacitor packs. (A) Battery and (B) supercapacitor.

The adoption of supercapacitor in EVs can significantly extend the battery lifetime and therefore the battery degradation model, which can be used in the dynamic process (i.e., nonstandard profiles), is important to the optimization and evaluation of HESS. A dynamic battery degradation model, which can be used at various temperatures, has been proposed in [4], based on a semiempirical model, considering usage time, temperature, and discharge rate [28].

$$Q_{loss} = Ae^{-\left(\frac{E_a + B \cdot C\_Rate}{RT_{bat}}\right)}(A_h)^z, \tag{5.4}$$

where Q_{loss} is the normalized capacity loss, A is the preexponential factor, E_a is the activation energy (J), R is the gas constant (J/(mol K)), A_h is the amp-hour throughput, C_Rate is the battery discharge rate, B is the compensation factor of discharge rate, and z is the time factor [4]. To tailor this model for the dynamic process, we discretize the model as [12]

$$Q_{loss,p+1} - Q_{loss,p} = \Delta A_h z A^{\frac{1}{z}} e^{-\left(\frac{E_a + B \cdot C\_Rate}{zRT_{bat}}\right)} Q_{loss,p}{}^{\frac{z-1}{z}}, \tag{5.5}$$

where $Q_{loss,p}$ and $Q_{loss,p+1}$ are the battery capacity loss at t_p and t_{p+1}, and ΔA_h is the A_h-throughput from t_p to t_{p+1}, which is defined as

$$\Delta A_h = \frac{1}{3600}\int_{t_p}^{t_{p+1}} |I_{bat}| dt. \tag{5.6}$$

To ensure that the model is accurate, especially at subzero temperatures, the battery degradation model was further modified as

$$Q_{loss,p+1} - Q_{loss,p} = \Delta A_h z A^{\frac{1}{z}} e^{-\left(\frac{E_a + B \cdot C\_Rate}{zR(|T_b - T_{bat}| + T_C)}\right)} Q_{loss,p}{}^{\frac{z-1}{z}}, \tag{5.7}$$

where T_b is the optimal temperature for battery to achieve the least degradation and T_C is a constant temperature for compensation. Parameters T_b,

T_C, z, and B are calibrated by the experiments and the detailed information is provided in [11]. The experimental validation for the proposed model under a wide range of temperature is shown in Fig. 5.5. The battery degradation model is finally given as

$$Q_{loss} = 0.0032e^{-\left(\frac{15162-1516C\_Rate}{R(|285.75-T_{bat}|+265)}\right)}(A_h)^{0.849}. \quad (5.8)$$

Based on the dynamic model of the electric bus, the power demand P_{demand} of HESS can be determined. Due to the space limitation, the vehicle dynamic model is not specified in this chapter and the detailed information is provided in [8, 10–12]. The studied electric bus is a 12-m bus with a total weight of 15,500 kg. China Bus Driving Cycle (CBDC) is focused in the optimization and the corresponding speed and power demand profiles are shown in Fig. 5.6.

The amount of battery cells is chosen as 600 given the requirement that the mileage should be more than 100 km at a constant speed (i.e., 50 km/h) on a flat road [11]. The integrated optimization is conducted using DP to determine the best solution of supercapacitor size and EMS, based on a

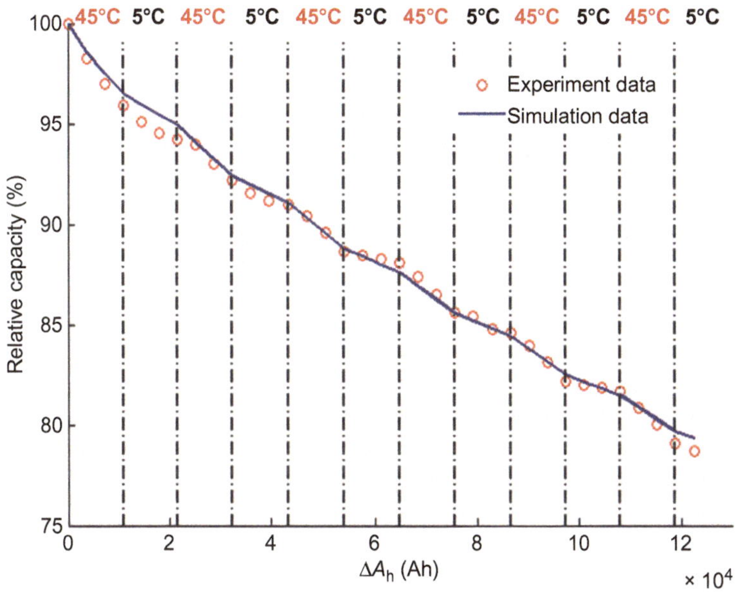

Fig. 5.5 Experimental validation of battery degradation model [4]. *(With permission from Elsevier).*

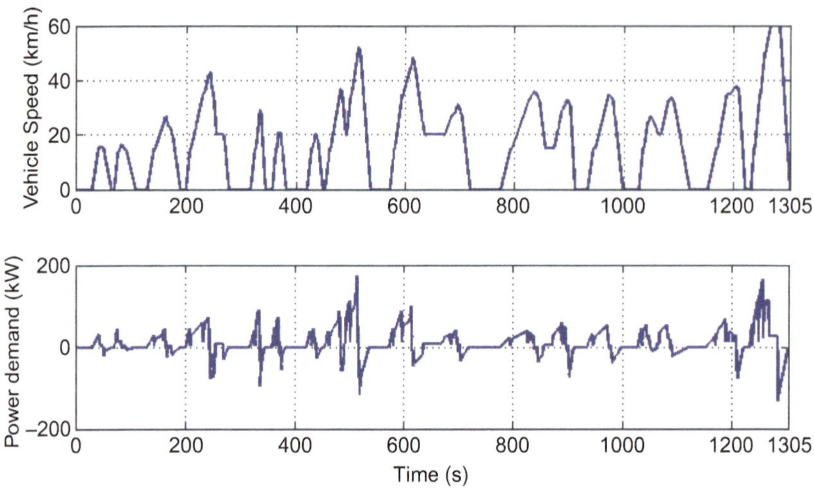

Fig. 5.6 The speed and power demand profiles of CBDC.

preset cost function, which considers the battery degradation cost and the electricity cost. The total operation cost of HESS at any time instant k is defined as follows:

$$\text{Minimize}\left\{\sum_{k=1}^{k=k_{\max}}\frac{[Cost_{\text{bat,loss}}(k)+Cost_{\text{ele}}(k)]}{L_{\text{c}}}\right\}, \tag{5.9}$$

where

$$Cost_{\text{bat,loss}}(k)=\frac{C_{\text{bat}}V_{\text{bat}}p_{\text{bat}}}{0.2\times 1000}\times\left[9.78\times 10^{-4}\frac{|I_{\text{bat}}(k)|T_s}{3600M_{\text{bat}}}e^{-\left(\frac{15162-1516C\_Rate}{0.849R(|285.75-T_{\text{bat}}(k)|+265)}\right)}Q_{\text{loss}}(k-1)^{-0.1779}\right],$$

$$Cost_{\text{ele}}(k)=\frac{p_{\text{ele}}T_s}{3600}\left[P_{\text{SC}}(k)+P_{\text{bat}}(k)\right],$$

where T_s is the sample time and set to 1 s, $Cost_{\text{bat,loss}}$ is the battery degradation cost, p_{bat} and p_{ele} are the battery and electricity prices, respectively, M_{bat} is the number of parallel strings in the battery pack, and L_{c} is the driving distance. The supercapacitor initial voltage is set to 90% of its maximum value and the lower voltage for supercapacitor is set to 50% of the maximum value considering its low efficiency in low voltage regions. The current price of lithium-ion battery pack is about \$257/kW h [29]. In addition, the prices of electricity and supercapacitor are set to \$0.11/kW h and \$15,000/kW h in the following analysis [30]. Note that the electricity price means the cost of electricity per unit (\$/kW h), while the electricity cost means the total cost of electricity in the optimization period (\$). The definition can also be generalized to battery price, battery degradation cost, fuel price, and fuel cost, which will be used in the sequel.

First, DP approach is used for different sizes of supercapacitor to obtain the optimal design and the detailed information of DP is provided in [17]. In the sizing process, the series and parallel numbers of supercapacitor modules are set to vary from 1 to 6 and 15 to 29, respectively. Note that the temperature, which significantly influences the battery degradation, is fixed at 20 °C in this section, and the case study considering various battery prices and temperatures will be conducted in the next section. As shown in Fig. 5.7, the "knee point" can be selected as the optimal size, as when the capital cost reaches the "knee point," the operation cost starts to decrease less fast and almost linearly. As shown in Fig. 5.7, the optimal supercapacitor size can be determined as 50 (2 in parallel and 25 in series).

Based on the optimal supercapacitor size derived in the above analysis, the DP result is investigated to propose the online EMS, since the DP approach cannot be directly used in practical applications. As shown in Fig. 5.8A, the power demand can be optimally split between battery and supercapacitor and the voltage of supercapacitor fluctuates in a wide range. It reveals that the supercapacitor is effectively used to tolerate high-power

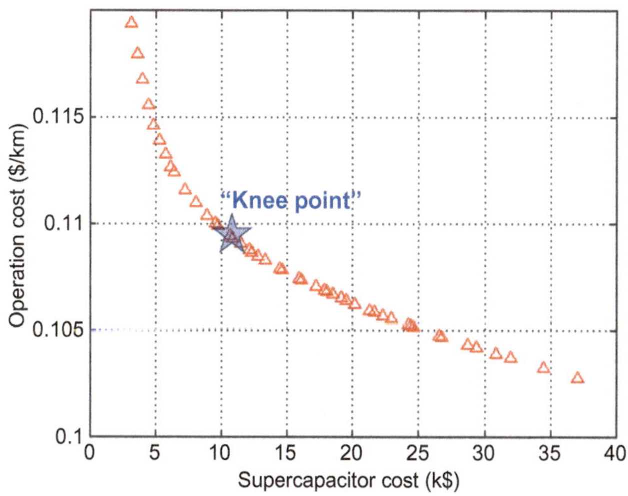

Fig. 5.7 The supercapacitor sizing results for the studied electric bus.

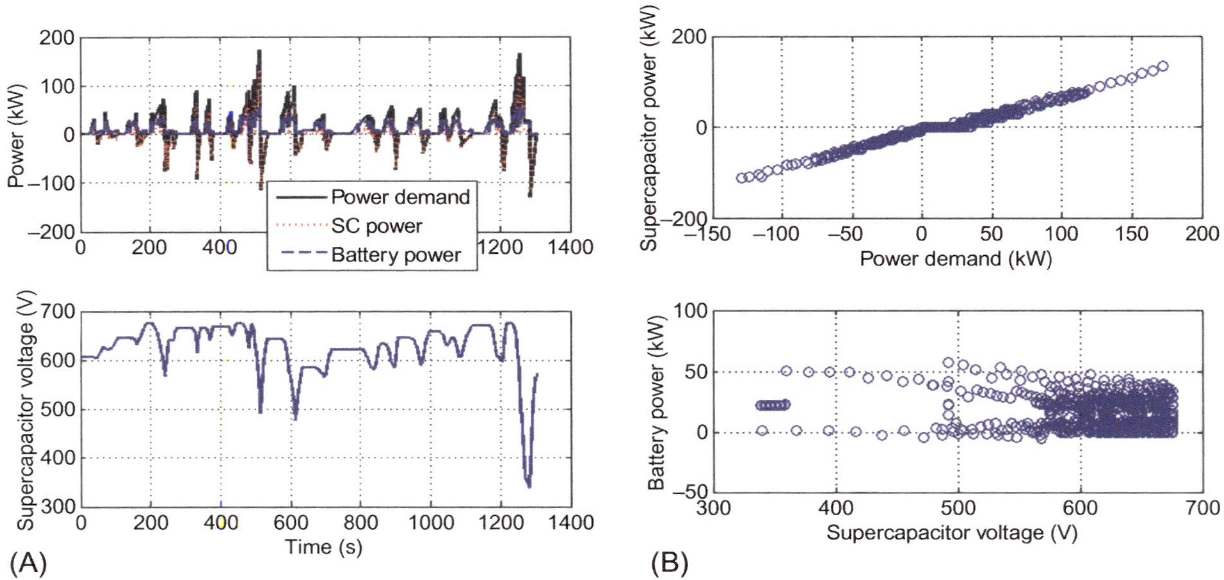

Fig. 5.8 The DP results under CBDC. (A) Power split results and (B) the piecewise linear relationship.

demand and the battery can be therefore protected. As shown in Fig. 5.8B, the relationship between P_{demand} and P_{SC} is piecewise linear in three separate sections. Section 5.1 represents the DP results under the regenerative braking condition (i.e., $P_{demand} < 0$), and it shows that the supercapacitor absorbs all regenerative energy to avoid charging battery. Note that the slope of the linearized relationship in Section 5.1 is slightly below 1 because of the power loss of the DC/DC converter. Sections 5.2 and 5.3 show the DP results under the traction condition (i.e., $P_{demand} \geq 0$), and we clearly find the rule that the supercapacitor does not support power to load when P_{demand} is less than a threshold denoted as P_{min}. Moreover, when P_{demand} exceeds P_{min}, the battery supplies the constant power P_{min} and the supercapacitor supplies the rest of the power ($P_{demand} - P_{min}$). This rule is extracted from the DP results, indicating the battery power concentrates around 0 and P_{min} regardless of the supercapacitor voltage, as shown in Fig. 5.8B. Similarly, the slope of the linearized relationship in Section 5.3 is slightly more than 1 also caused by the power loss of the DC/DC converter.

By summarizing all the rules extracted from the DP results, a rule-based EMS can be designed and implemented online, as depicted in Fig. 5.9. In the rule-based EMS, the protection rules, especially the ones for the

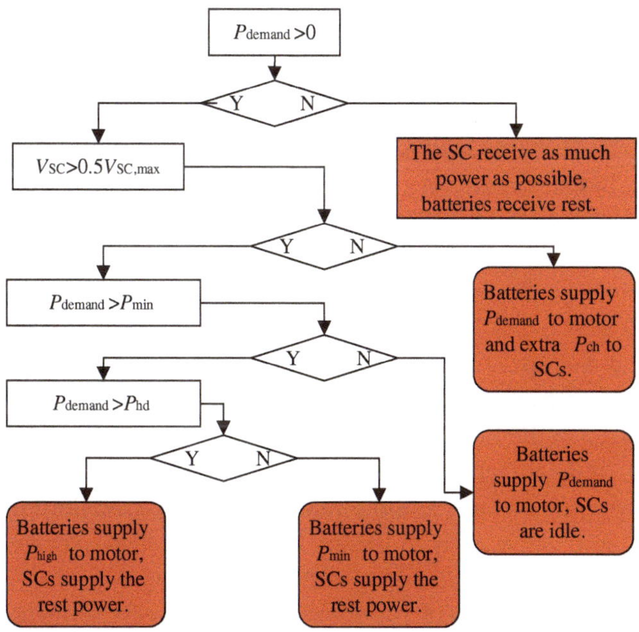

Fig. 5.9 The rule-based EMS for online applications.

supercapacitor, are also considered. According to the values of P_{demand}, the proposed rule-based EMS can directly split the power and assign the demands to both energy sources, and then it is adopted in the simulation. Note that P_{hd} is another threshold, which is chosen to avoid fast discharge of supercapacitor, given that the energy density of supercapacitor is low and therefore cannot continuously supply high power. Specifically, when P_{demand} is higher than P_{hd}, the output power of battery is increased from P_{min} to P_{high} to supply more power. When compared to the battery energy storage system (BESS) without supercapacitor, the operation cost can be reduced by up to 20% by the HESS during the 10-year operation period [29]. Therefore it is validated that the HESS is effective, even though the capital cost increases due to the adoption of supercapacitor. Considering the issue of consumer acceptance, the initial price and the HESS performance should be carefully balanced based on the sizing result shown in Fig. 5.7.

5.3 A case study of HESS in electric buses

It has been shown that the battery degradation cost and the electricity cost should be considered concurrently in the HESS optimization, as shown in Eq. (5.9). The battery degradation cost is linearly proportional to the price, thus the continuous decline in battery price may influence the analysis shown in the previous section due to the corresponding decrease in battery degradation cost. In this section, the influence of various temperatures and battery prices on the HESS performance is investigated via a case study. As shown in Eq. (5.9), the battery price dramatically influences the optimization result. In specific, Analysts predicted that in 2020 and 2025, the battery price could decrease to \$200/kW h and \$160/kW h [26], respectively, while the current price of a lithium-ion battery pack is about \$257/kW h. Therefore the battery prices of \$200 and \$143/kW h, assumed to be realized after 5 and 10 years, are considered in this section. It has been verified that the optimal sizing results and the structure of the rule-based EMS are robust to different batter prices and temperatures [29]. Only the threshold P_{min} in the EMS should be slightly changed under different conditions (i.e., different temperatures and battery prices) [29]. In consequence, 50 supercapacitor modules and the rule-based EMS shown in Fig. 5.9 are employed. And a case study in Beijing is conducted. The average temperatures in different months in Beijing vary from −4°C (i.e., January) to 26°C (i.e., July), as shown in Fig. 5.10. It is obvious that the battery will have the best performance (i.e., least degradation rate) in April and October given the

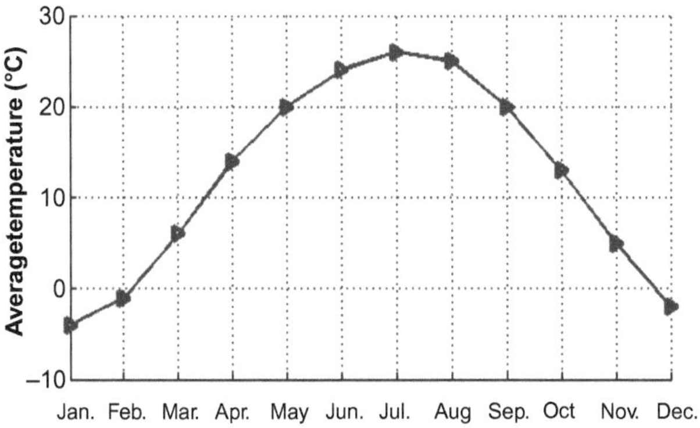

Fig. 5.10 The average temperatures of Beijing.

appropriate temperatures, as indicated in Eq. (5.8). It is assumed that the electric bus runs 365 days a year. The daily mileage is set to 200 km based on the exiting analysis in [31], meaning that the daily operation includes approximately 34 CBDCs.

The simulation results of HESS and BESS corresponding to the current battery price (i.e., \$257/kW h) are compared in Fig. 5.11. It shows that the electricity cost is around \$0.08/km and slightly changes with the temperature, while it is not influenced by the battery price. The battery degradation cost is significantly influenced by the temperature and for the HESS it varies from \$0.029 to \$0.047/km, which are much less than the electricity cost, as shown in Fig. 5.11A. In comparison, for the BESS, as shown in Fig. 5.11B, the battery degradation cost is higher than the electricity cost for most months except April and October when the battery has the least degradation rate. Note that the electricity cost of BESS is slightly lower than that of HESS because no DC/DC converter is used in BESS. It can be seen that the variation of the operation cost is dominated by the battery degradation cost and therefore they have the same trend. Since the battery degradation cost of HESS is much lower than the one of BESS, the effectiveness of HESS is validated.

HESS has been widely developed to extend the battery lifetime and avoid the battery replacement. It is clear that the benefit of HESS in EV applications will be weakened if the battery price continuously decreases. As a result, the economy analysis of HESS considering different battery prices is necessary and a 10-year scenario is selected. There is common sense that the battery should be replaced when its capacity decreases to 80% of the initial capacity. The simulation results of HESS and BESS under different battery prices are summarized in Table 5.1. We point out that the capital

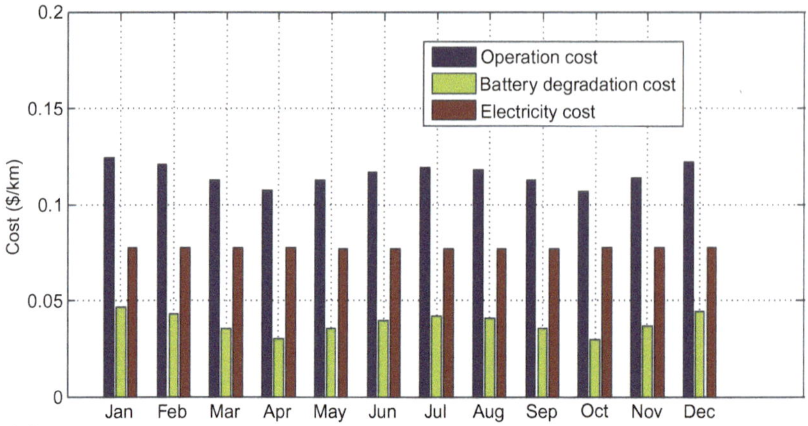

(A)

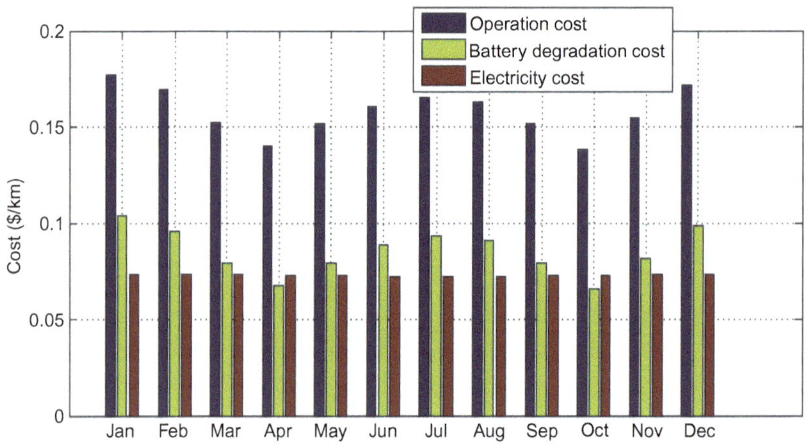

(B)

Fig. 5.11 The cost of HESS and BESS during one whole year. (A) HESS and (B) BESS.

Table 5.1 HESS and BESS cost over 10-year operation in Beijing.

Type	HESS			BESS		
Battery price ($/kW h)	143	200	257	143	200	257
Capital cost (k$)	34.1	41.7	49.2	16.5	23.2	29.8
Electricity cost (k$)	56.4	56.4	56.4	53.2	53.2	53.2
Battery replacement cost (k$)	0	0	0	33.1	46.3	59.5
Total cost (k$)	90.5	98.1	105.6	102.8	122.7	142.5

cost of supercapacitor and DC/DC converter is conservatively assumed to decline by 1% annually. First, due to the adoption of DC/DC converter, there is 6% more electricity cost for HESS when compared to BESS. Second, due to the capital cost of supercapacitor, the initial cost of HESS is 39.4% higher than that of BESS. Third, the replacement of battery pack can be avoided when HESS is used, and for BESS, the battery pack needs to be replaced twice during the 10-year running. Based on the current battery price, the operation cost of HESS, including capital cost, electricity cost, and battery replacement cost, can be reduced by up to 25.9% when compared to BESS. Even though the effectiveness of HESS is impaired when the battery price decreases, the operation cost of HESS is still 12% less than that of BESS even if the battery price can reduce to \$143/kW h after 10 years. Hence, HESS will still be beneficial in EV applications in the future.

5.4 The optimization of HESS in PHEVs

The satisfactory performance of HESS in electric buses has been verified, and in this section, the effectiveness of HESS in the plug-in hybrid electric bus is examined. The powertrain of the studied series-parallel PHEV, which consists of a diesel ICE, an integrated starter generator (ISG), a traction motor, a clutch, a supercapacitor pack, a battery pack, a bidirectional DC/DC converter, and the control system, is depicted in Fig. 5.12.

Among all configurations of the PHEV powertrain, the series-parallel one needs the most complex EMS because there are so many operation modes. For example, the ICE and the motor can deliver power to the traction wheels independently or simultaneously. The ISG may absorb

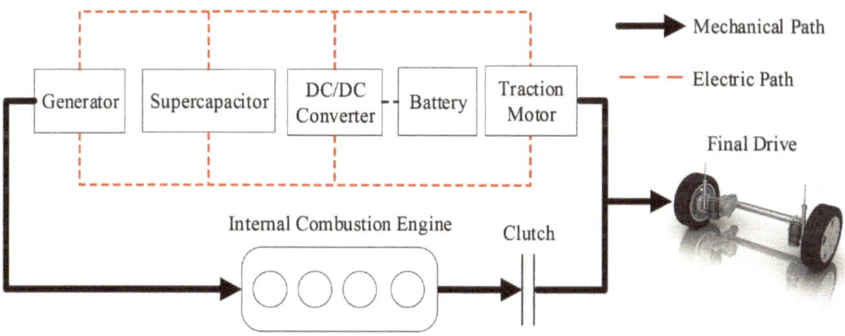

Fig. 5.12 The powertrain of the series-parallel PHEV.

the excess power from the ICE and charge the battery or the supercapacitor. Specifically, the modes of the studied PHEV can be summarized as follows:

(1) There is a predefined rule that the clutch is disengaged and the motor independently drives the vehicle when the vehicle speed is low (i.e., below 20 km/h) and the battery SoC is high. This is denoted as the pure electric mode.

(2) When both the vehicle speed and the battery SoC are low, the engine is turned on to drive the ISG for charging the HESS and supply power to the traction wheels. This is denoted as the series mode.

(3) When the vehicle speed is high (i.e., above 20 km/h) and the battery SoC is high, the engine and the motor directly supply power to the traction wheel together or independently. This is denoted as the parallel mode.

(4) When the vehicle speed is high and the battery SoC is extremely low, the clutch is engaged and the engine can drive the traction wheels and the ISG. Therefore the HESS can be charged using the excess power from engine. This is denoted as the pure engine mode.

(5) All regenerative braking energy is absorbed by the HESS through the clutch and the traction motor, which acts as a generator. This is denoted as the regenerative braking mode.

Similar to Eqs. (5.1) and (5.2), the following equations must be satisfied for the series-parallel PHEV shown in Fig. 5.12.

(1) Traction mode:

$$\begin{cases} P_{\mathrm{demand}} = \left[P_{\mathrm{eng}} + P_{\mathrm{bat}}\left(\eta_{\mathrm{DC/DC}}\right)^{z} + P_{\mathrm{SC}} - P_{\mathrm{ISG,loss}} - P_{\mathrm{motor,loss}}\right]\eta_{\mathrm{fd}} \\ T_{\mathrm{demand}} = \left(T_{\mathrm{eng}} + T_{\mathrm{mot}}\right)\delta_{\mathrm{fd}} \end{cases}. \tag{5.10}$$

(2) Regenerative braking mode:

$$\begin{cases} P_{\mathrm{demand}}\eta_{\mathrm{fd}} = \left[P_{\mathrm{eng}} + P_{\mathrm{bat}}\left(\eta_{\mathrm{DC/DC}}\right)^{z} + P_{\mathrm{SC}} - P_{\mathrm{ISG,loss}} - P_{\mathrm{motor,loss}}\right] \\ T_{\mathrm{demand}} = T_{\mathrm{mot}}\delta_{\mathrm{fd}} + T_{\mathrm{mb}} \end{cases}. \tag{5.11}$$

where T_{demand} is the torque demand, P_{eng} is engine power, η_{fd} is the average efficiency of the final drive (i.e., 97%), δ_{fd} is the reduction ratio of the final drive (i.e., 6.17), $\eta_{\mathrm{DC/DC}}$ is the average efficiency of the DC/DC converter (i.e., 90%), T_{eng}, T_{mot}, and T_{mb} are the torque of engine, motor, and mechanical braking system, respectively, z is the mode factor (i.e., 1 for discharging and −1 for charging), $P_{motor,loss}$ is the motor power loss, and $P_{ISG,loss}$ is the ISG power loss. The quasi-static models for the ICE and the traction motor are adopted and their dynamics are neglected since this work only focuses on the system efficiency optimization. The fuel consumption rate of ICE is determined by the torque and the rotational speed and the motor efficiency can be determined similarly, as shown in Fig. 5.13 [29].

The optimization for the PHEV is also conducted under CBDC, and the demanded power and torque profiles are shown in Fig. 5.14. The total power should be optimally split between battery, supercapacitor, and ICE to minimize the operation cost including the fuel cost, the electricity cost, and the battery degradation cost. Note that the optimization of PHEV with the HESS is much more complex than the optimization of EVs, as the ICE is involved and many operation modes should be considered.

It is intuitive that the HESS should be used as much as possible given the electricity is much cheaper than the fuel. However, the energy density of supercapacitor is not high enough to last for a long period when the power demand is high. The battery can also be used to substitute electricity for fuel, but the frequent operations will induce a high degradation cost. Therefore the engine should be involved frequently to supply the required power and to reduce the fuel cost the engine should operate in its high-efficiency region. Thus different energy sources should be carefully coordinated and a two-dimensional optimization problem should be solved. The operation cost of the studied PHEV, consisting of the fuel cost, the electricity cost, and the battery degradation cost, can be expressed as

$$J(x(t),u(t),t) = p_{fuel}\int_{t_0}^{t_f} \dot{m}(u(t),t)dt + p_{bat}\frac{\int_{t_0}^{t_f}\dot{Q}_{loss}(u(t),t)dt}{2Q_0} + p_{ele}E_{bat}(SOC_{bat}(t_0) - SOC_{bat}(t_f)), \tag{5.12}$$

where p_{fuel} is the fuel price and set to 1\$/L. The control variables are $u(t) = \begin{bmatrix} P_{bat}(t) \\ P_{SC}(t) \end{bmatrix}$, the system states are $x(t) = \begin{bmatrix} SOC_{bat}(t) \\ SOE_{SC}(t) \end{bmatrix}$, Q_0 is the total degradation capacity of battery (i.e., 0.2), and t_0 and t_f are the start and finish time. Note that the state of energy (SOE) of the supercapacitor (i.e., SOE_{SC}) indicates how much percentage of energy is reserved. The state transfer equations are as follows:

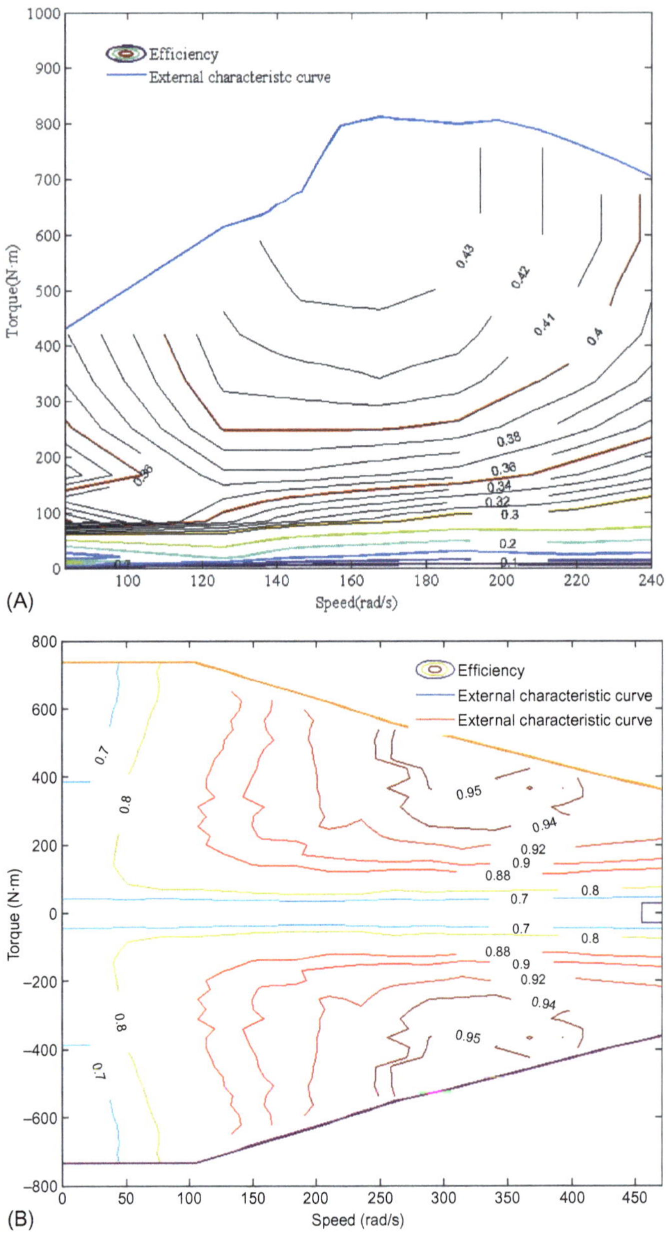

Fig. 5.13 Efficiency maps of the ICE and the traction motor [29]. (A) Efficiency map of diesel ICE and (B) Efficiency map of traction motor.

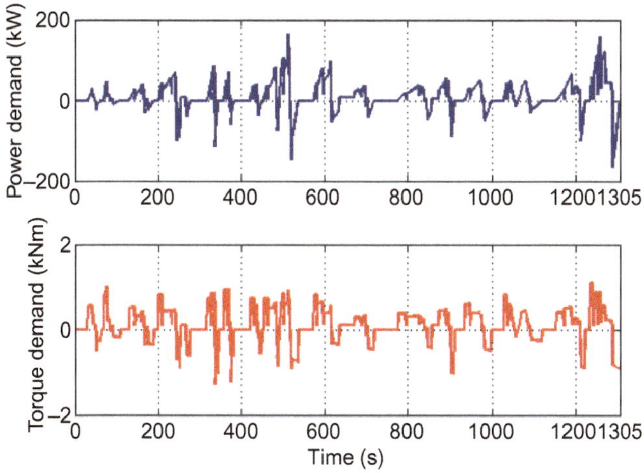

Fig. 5.14 The power and torque demand profiles.

$$S\dot{O}C_{\text{bat}} = -\frac{i_{\text{bat}}}{C_{\text{bat}}} = -\frac{V_{\text{bat}} - \sqrt{V_{\text{bat}}{}^2 - 4P_{\text{bat}}R_{\text{bat}}}}{2R_{\text{bat}}C_{\text{bat}}}, \tag{5.13}$$

$$SOE\cdot_{\text{SC}} = -\frac{P_{\text{SC}}}{E_{\text{SC}}} = \frac{-2P_{\text{SC}}}{C_{\text{SC}}V_{\text{SC, max}}^2}, \tag{5.14}$$

where E_{SC} is the maximum energy that the supercapacitor can store. The optimization is conducted over six CBDCs to minimize the operation cost. Specifically, the bus line is generally shorter than 50 km and therefore a 35-km trip (i.e., six consecutive CBDCs) is chosen. We assume that the daily range of the PHEV is 210 km (i.e., 6 trips) and the PHEV can be fully charged at noon and midnight (i.e., after every three trips). Therefore the battery SoC range is set to be from [0.67, 0.9] in the optimization, since the recommended SoC range is [0.2, 0.9]. As aforementioned, the supercapacitor voltage is above half of its maximum voltage, so the SOE_{SC} range is set to be [0.25, 0.81] accordingly.

The PMP, which is a global optimization algorithm, is utilized for the optimization problem defined above. Similar to the DP approach, the PMP can also help us understand how to use the HESS efficiently. Based on the cost function shown in Eq. (5.12), the Hamiltonian H can be defined as:

$$H(x(t), u(t), \lambda(t), t) = p_{\text{fuel}}\dot{m}(u(t), t) + p_{\text{bat}}\frac{\dot{Q}_{\text{loss}}(u(t), t)}{2Q_0} + \lambda_1(t)f_1(x(t), u(t), t) + \lambda_2(t)f_2(x(t), u(t), t), \tag{5.15}$$

where $\lambda_1(t)$ and $\lambda_2(t)$ are costates corresponding to the battery SoC and the SOE_{SC} differentials. The necessary conditions for achieving the global optimal solution can be given as:

$$u^*(t) = \arg\min\left\{p_{fuel}\dot{m}(u(t),t) + p_{bat}\frac{Q\cdot_{loss}(u(t),t)}{2Q_0} + \lambda_1(t)S\dot{O}C_{bat}(t) + \lambda_2(t)SOE\cdot_{SC}(t)\right\}, \tag{5.16}$$

$$\begin{cases} \dot{\lambda}_1(t) = -\dfrac{\partial H}{\partial SOC_{bat}}, \\ \dot{\lambda}_2(t) = -\dfrac{\partial H}{\partial SOE_{SC}} \approx 0. \end{cases} \tag{5.17}$$

$$\begin{cases} S\dot{O}C_{bat} = \dfrac{\partial H}{\partial \lambda_1} = f_1(x(t), u(t), t), \\ S\dot{O}E_{SC} = \dfrac{\partial H}{\partial \lambda_2} = f_2(x(t), u(t), t). \end{cases} \tag{5.18}$$

To obtain the optimal sizes of battery and supercapacitor, the proposed PMP is used to optimize all systems with various sizes. We point out that the initial values of the costates are hard to obtain [32] and manually tuned for different HESS candidates. The sizing results are shown in Fig. 5.15. As for the battery, the operation cost can be reduced from 5.52$ (i.e., 38.64¥) to 4.8$ (33.59¥) for one trip when the battery size increases from 30.4 to 60.8 kW h, and the supercapacitor capacitance is fixed at 165 F and its maximum voltage is 48 V. From the viewpoint of the PHEV market, the manufacturers generally would like to adopt the minimum battery pack to satisfy the demand of pure-electric mileage. In consequence, the 30.4 kW h is chosen as the optimal battery size. As for the supercapacitor size optimization, as shown in Fig. 5.15B, the operation cost strictly decreases when the supercapacitor size increases. As aforementioned, the "knee point" (i.e., 165 F, 48 V) can be regarded as the optimal solution.

The power split results with the optimal sizes of battery and supercapacitor are shown in Fig. 5.16. The supercapacitor SoE fluctuates similarly under different driving cycles in a wide range, meaning that the supercapacitor is effectively used, as shown in Fig. 5.16A. As shown in Fig. 5.16B, the PMP result along one CBDC shows that the supercapacitor is used as the primary power source to provide peak power for short periods and absorb all regenerative power. The battery is used as the secondary power source to follow the small power demand. The engine only works when the power demand is continuously high, and under this condition, the supercapacitor is charged sometimes to keep the engine operating efficiently. To verify the HESS effectiveness in PHEV applications, the PHEV with only battery

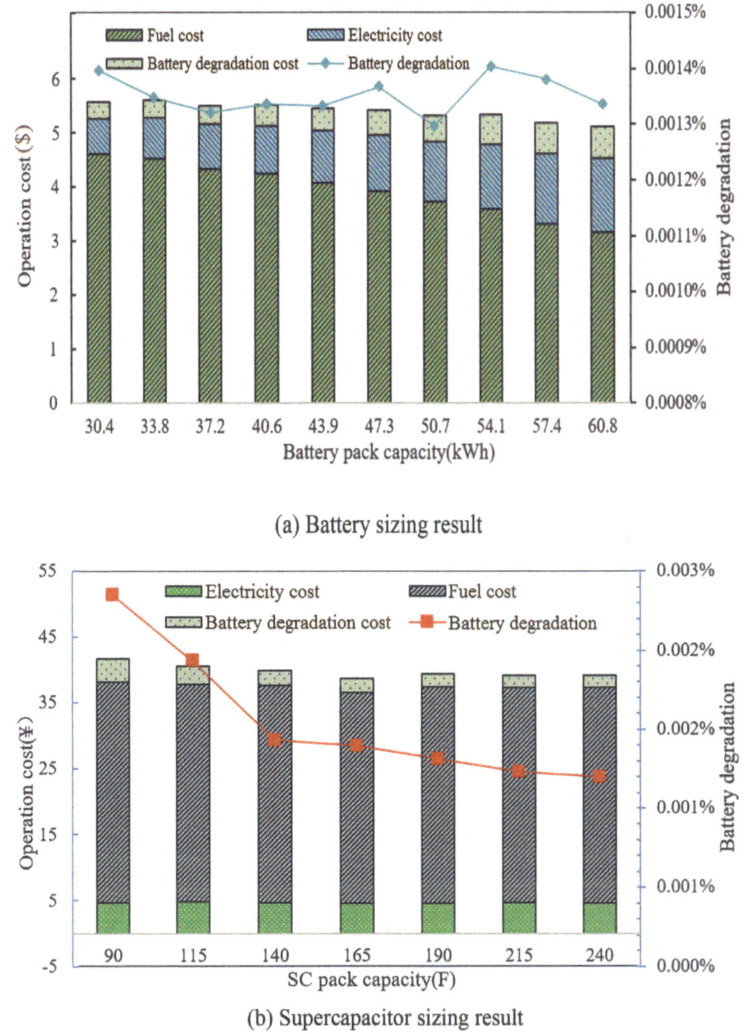

Fig. 5.15 The HESS sizing results [29]. (A) Battery sizing result and (B) supercapacitor sizing result. *(With permission from Elsevier).*

(i.e., the traditional PHEV) is also investigated. For the fair comparison, the same battery capacity (i.e., 30.4 kW h) is considered and the proposed PMP is also used to optimally split power between the engine and the battery. In addition, all the simulation conditions remain the same. The typical PHEV has been widely studied, so this content is not extended in this chapter.

The HESS operates timely and effectively, thus the fuel cost and the battery degradation cost are dramatically reduced. Although the proposed PMP

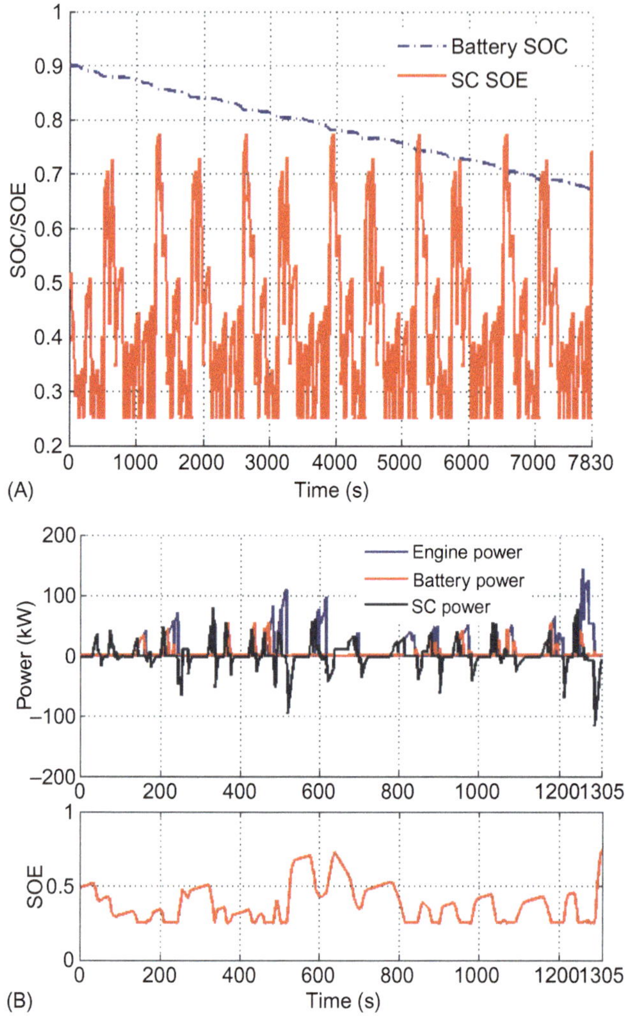

Fig. 5.16 PMP result of the PHEV [23]. (A) Supercapacitor SOE and battery SoC curves and (B) PMP result along one CBDC. *(With permission from Elsevier).*

is hard to be directly implemented, the power split result shown in this paper provides a helpful guideline when designing an online EMS. Due to the space limitation, this topic is not further extended. Simulation results show that for the PHEV with the HESS, the battery degradation can be reduced by up to 83.5% during one drive cycle when compared to the one with battery only. When the studied PHEV works for 300 days per year, the battery in the HESS can last for 8 years, while the battery in the typical PHEV can only last for 1.3 years. In addition, the operation cost can be reduced by up to 28.6% when the HESS is employed. Therefore the effectiveness of HESS in PHEV applications is verified.

5.5 A brief examination of microsupercapacitor in vehicle applications

The HESS has been implemented in the plug-in hybrid electric buses in China, given its good performance. However, for the current volumetric power density of the supercapacitor, there is still a barrier to adopt supercapacitor in the passenger vehicle due to the limited onboard space. The development of the microsupercapacitor, which has a much high power density [33], draws more and more attention and may be able to expedite the utilization of HESS in passenger vehicles. To be specific, the volumetric power density of the microsupercapacitor can reach 495 W/cm^3, which is 35 times higher than the one of commercial supercapacitor (i.e., 14 W/cm^3) [34]. For example, based on the analysis provided in this chapter, the supercapacitor volumes for the electric bus and the plug-in hybrid electric bus are around 200 and 22 L, which account for the large space even for the buses. In comparison, when the microsupercapacitor can be used, the volumes can be reduced to 5.7 and 0.6 L accordingly, thus the significant space can be saved. It is always favorable for the vehicle manufacturer to save space while optimizing the layout of the whole system. Therefore as long as the price of microsupercapacitor is acceptable, there will be a huge chance for it in the vehicle market.

5.6 Conclusions

The battery/supercapacitor HESS is introduced in this chapter and adopted in EVs and PHEVs. The integrated optimization, indicating that the component size and the EMS are optimized simultaneously, is conducted to derive the optimal design and therefore reduce the operation cost, which consists of the electricity cost, the battery degradation cost, and the fuel cost. Considering various battery prices and temperatures, a case study is conducted to verify the effectiveness of the HESS in the near future. Finally, the feasibility of adopting microsupercapacitor in vehicle applications is briefly examined and there will be a huge chance in the vehicle market for the microsupercapacitor, as long as its price is acceptable.

Glossary

All-electric mileage The distance that vehicle can drive using only electricity
Battery degradation model A model which can characterize the battery capacity reduction in dynamic processes
Bidirectional DC/DC converter The DC/DC converter allowing bidirectional energy flows
Capital cost One-time expenses incurred on the purchase
Component sizing A process optimizing the sizes of components
DC/DC converters An electronic device converting a source of direct current from one voltage level to another
Driving cycle A series of data points representing the speed of a vehicle versus time
Dynamic programming A global optimization method over plain recursion
Energy management strategy A strategy determining the energy flow between different energy sources and loads
Fully active configuration A configuration with two degrees of freedom
Hybrid energy storage system The system including multiple energy storage components
Normalized capacity The capacity normalized from 0 to 1
Off-line strategy The strategy which can only be used off-line
Online strategy The strategy which can be used online
Open-circuit voltage The voltage of a device when disconnected from any circuit
Passive configuration A configuration without the degree of freedom
Plug-in hybrid electric vehicle A hybrid electric vehicle whose battery can be recharged by an external source of electric power
Power demand profile A series of data points representing the power demand versus time
Quasi-static model The system model only considering the steady-state responses
Regenerative braking A braking process converting the kinetic energy of vehicles into electric energy
***Rint* model** The equivalent circuit model only considering open circuit voltage and internal resistance
Rule-based strategy The strategy constructed based on various rules
Semiactive configuration A configuration with one degree of freedom
State of charge The level of charge relative to its capacity
State of energy The level of stored energy

References

[1] S.F. Tie, C.W. Tan, A review of energy sources and energy management system in electric vehicles, Renew. Sustain. Energy Rev. 20 (2013) 82–102.
[2] S.M. Lukic, J. Cao, R.C. Bansal, F. Rodriguez, A. Emadi, Energy storage systems for automotive applications, IEEE Trans. Ind. Electron. 55 (6) (2008) 2258–2267.
[3] L. Lu, X. Han, J. Li, J. Hua, M. Ouyang, A review on the key issues for lithium-ion battery management in electric vehicles, J. Power Sources 226 (2013) 272–288.
[4] Z. Song, H. Hofmann, J. Li, J. Hou, X. Zhang, M. Ouyang, The optimization of a hybrid energy storage system at subzero temperatures: energy management strategy design and battery heating requirement analysis, Appl. Energy 159 (2015) 576–588.

[5] J. Li, Z. Song, Z. Shuai, L. Xu, M. Ouyang, Wheel slip control using sliding-mode technique and maximum transmissible torque estimation, J. Dyn. Syst. Measurem. Control 137 (11) (2015), 111010.
[6] S.B. Peterson, J. Apt, J.F. Whitacre, Lithium-ion battery cell degradation resulting from realistic vehicle and vehicle-to-grid utilization, J. Power Sources 195 (8) (2010) 2385–2392.
[7] A. Khaligh, Z. Li, Battery, ultracapacitor, fuel cell, and hybrid energy storage systems for electric, hybrid electric, fuel cell, and plug-in hybrid electric vehicles: state of the art, IEEE Trans. Vehicular Technol. 59 (6) (2010) 2806–2814.
[8] Z. Song, J. Hou, S. Xu, M. Ouyang, J. Li, The influence of driving cycle characteristics on the integrated optimization of hybrid energy storage system for electric city buses, Energy 135 (2017) 91–100.
[9] Z. Song, J. Hou, H. Hofmann, J. Li, M. Ouyang, Sliding-mode and Lyapunov function-based control for battery/supercapacitor hybrid energy storage system used in electric vehicles, Energy 122 (2017) 601–612.
[10] Z. Song, H. Hofmann, J. Li, X. Han, X. Zhang, M. Ouyang, A comparison study of different semi-active hybrid energy storage system topologies for electric vehicles, J. Power Sources 274 (2015) 400–411.
[11] Z. Song, J. Li, X. Han, L. Xu, L. Lu, M. Ouyang, H. Hofmann, Multi-objective optimization of a semi-active battery/supercapacitor energy storage system for electric vehicles, Appl. Energy 135 (2014) 212–224.
[12] Z. Song, H. Hofmann, J. Li, J. Hou, X. Han, M. Ouyang, Energy management strategies comparison for electric vehicles with hybrid energy storage system, Appl. Energy 134 (2014) 321–331.
[13] H. Liu, Z. Wang, J. Cheng, D. Maly, Improvement on the cold cranking capacity of commercial vehicle by using supercapacitor and lead-acid battery hybrid, IEEE Trans. Vehicular Technol. 58 (3) (2008) 1097–1105.
[14] J.P. Trovão, P.G. Pereirinha, H.M. Jorge, C.H. Antunes, A multi-level energy management system for multi-source electric vehicles—an integrated rule-based meta-heuristic approach, Appl. Energy 105 (2013) 304–318.
[15] Y.H. Hung, C.H. Wu, An integrated optimization approach for a hybrid energy system in electric vehicles, Appl. Energy 98 (2012) 479–490.
[16] J. Du, J. Chen, Z. Song, M. Gao, M. Ouyang, Design method of a power management strategy for variable battery capacities range-extended electric vehicles to improve energy efficiency and cost-effectiveness, Energy 121 (2017) 32–42.
[17] Z. Song, H. Hofmann, J. Li, X. Han, M. Ouyang, Optimization for a hybrid energy storage system in electric vehicles using dynamic programing approach, Appl. Energy 139 (2015) 151–162.
[18] A.L. Allegre, A. Bouscayrol, R. Trigui, Influence of control strategies on battery/supercapacitor hybrid energy storage systems for traction applications, in: 2009 IEEE Vehicle Power and Propulsion Conference, IEEE, 2009, September, pp. 213–220.
[19] M. Hadartz, M. Julander, Battery-supercapacitor energy storage, Chalmers University of Technology, 2008.
[20] R. Carter, A. Cruden, P.J. Hall, Optimizing for efficiency or battery life in a battery/supercapacitor electric vehicle, IEEE Trans. Vehicular Technol. 61 (4) (2012) 1526–1533.
[21] J. Hou, Z. Song, H. Park, H. Hofmann, J. Sun, Implementation and evaluation of real-time model predictive control for load fluctuations mitigation in all-electric ship propulsion systems, Appl. Energy 230 (2018) 62–77.
[22] H. He, R. Xiong, K. Zhao, Z. Liu, Energy management strategy research on a hybrid power system by hardware-in-loop experiments, Appl. Energy 112 (2013) 1311–1317.

[23] Z. Song, X. Zhang, J. Li, H. Hofmann, M. Ouyang, J. Du, Component sizing optimization of plug-in hybrid electric vehicles with the hybrid energy storage system, Energy 144 (2018) 393–403.
[24] J. Du, X. Zhang, T. Wang, Z. Song, X. Yang, H. Wang, X. Wu, Battery degradation minimization oriented energy management strategy for plug-in hybrid electric bus with multi-energy storage system, Energy 165 (2018) 153–163.
[25] R. Xiong, J. Cao, Q. Yu, Reinforcement learning-based real-time power management for hybrid energy storage system in the plug-in hybrid electric vehicle, Appl. Energy 211 (2018) 538–548.
[26] R. Hensley, J. Newman, M. Rogers, M. Shahinian, Battery technology charges ahead, McKinsey Quart. 3 (2012) 5–50.
[27] T. Wang, X. Wu, S. Xu, H. Hofmann, J. Du, J. Li, Z. Song, Performance of plug-in hybrid electric vehicle under low temperature condition and economy analysis of battery pre-heating, J. Power Sources 401 (2018) 245–254.
[28] J. Wang, P. Liu, J. Hicks-Garner, E. Sherman, S. Soukiazian, M. Verbrugge, P. Finamore, Cycle-life model for graphite-LiFePO4 cells, J. Power Sources 196 (8) (2011) 3942–3948.
[29] Z. Song, J. Li, J. Hou, H. Hofmann, M. Ouyang, J. Du, The battery-supercapacitor hybrid energy storage system in electric vehicle applications: a case study, Energy 154 (2018) 433–441.
[30] M. Masih-Tehrani, M.R. Ha'iri-Yazdi, V. Esfahanian, A. Safaei, Optimum sizing and optimum energy management of a hybrid energy storage system for lithium battery life improvement, J. Power Sources 244 (2013) 2–10.
[31] Z. Luo, Y. Song, Z. Hu, Z. Xu, X. Yang, K. Zhan, Forecasting charging load of plug-in electric vehicles in China, in: 2011 IEEE Power and Energy Society General Meeting, IEEE, 2011, July, pp. 1–8.
[32] C. Hou, M. Ouyang, L. Xu, H. Wang, Approximate Pontryagin's minimum principle applied to the energy management of plug-in hybrid electric vehicles, Appl. Energy 115 (2014) 174–189.
[33] P. Huang, C. Lethien, S. Pinaud, K. Brousse, R. Laloo, V. Turq, B. Chaudret, On-chip and freestanding elastic carbon films for micro-supercapacitors, Science 351 (6274) (2016) 691–695.
[34] K.U. Laszczyk, K. Kobashi, S. Sakurai, A. Sekiguchi, D.N. Futaba, T. Yamada, K. Hata, Lithographically integrated microsupercapacitors for compact, high performance, and designable energy circuits, Adv. Energy Mater. 5 (18) (2015), 1500741.

CHAPTER 6

Microbatteries

Karolina Urszula Laszczyk[a] **and Kazufumi Kobashi**[b]
[a]Faculty of Microsystem Electronics and Photonics, Wroclaw University of Science and Technology, Wrocław, Poland
[b]CNT-Application Research Center, National Institute of Advanced Industrial Science and Technology, Ibaraki, Japan

Microbatteries (MBs) also named microscale batteries (μBs) are, aside microsupercapacitors (MSCs), one of the miniaturized electrochemical energy storage devices that could be directly placed with microelectronics. Although the MBs have achieved substantial performance improvements (cyclability, energy density) over the past two decades, they do not fulfill the actual demand to power microdevices in the milliwatt range. The MBs, similarly to the MSCs, consist of an electrolyte, electrodes (anode and cathode), and a current collector. In the case of sandwiched or fiber structures, the insulating separator must be added. The main difference between the MSCs and the MBs is the mechanism by which the energy is stored. In the MBs, the energy storage takes place through chemical reactions (faradaic oxidization and reduction of the electrochemically active reagents) and the energy is slowly released on demand. Meanwhile, the MSCs store the charges electrostatically, i.e., reversibly absorb and desorb electrolyte ions at a faster rate (the nonfaradaic process in electric-double layer capacitors) or undergo a rapid faradaic reaction at the interface between the electrolyte and electrodes (pseudocapacitors) [1]. In general, MBs deliver a low self-discharge rate and high energy density (20–200 mWh cm^{-3}, 0.01–1 W cm^{-3}) [2]. MBs are dedicated for:

- Internet of Things, IoT (sensors, tracking devices, smart meters, wearable electronics, or smart textiles),
- automotive (tire pressure monitoring system, TPMS, keyless entry or smart key, electronic toll collection devices), and
- medicine (insulin pump, continuous glucose monitoring, CGM, capsule endoscopy).

From the above-pointed applications of the MBs, the IoT devices and implants must be robust, maintenance-free, and corrosion-resistant.

https://doi.org/10.1016/B978-0-08-102888-9.00011-2

Additionally, for the implants, the MBs must resist sterilizing temperature and be nontoxic, in case of unsealing. Unfortunately, the MB reliability is greatly affected by a restricted lifetime (<1000 cycles) that occurs due to power consumption and load peak current which depends on the ambient temperature, as the temperature increases self-discharge tends to increase.

In the market, we could find few products named as MBs being indeed in most cases coin-type batteries (Table 6.1), which average lifetime is expected to endure up to 5 years at 45 °C. Recently, AAA-type batteries are sold by Panasonic offering up to 10 years of the shelf-life[a] thanks to an increase of the compact fine particle density in the electrode. Another company, SAFT, offers the bobbin-type construction instead of a spiral one as ensuring long-term operational stability, even up to 20 years.[b] However, to make self-sustainable systems really small and ultra-high densely packed, the MBs should be flat and thin to not enlarge the whole system as well as flexible and stretchable to be applicable for wearable electronics.

The MBs, which are reported in the scientific literature, are distinguished by their chemical content as microscale lithium ion, lithium air, lithium sulfur, sodium ion batteries, and many others. Lithium batteries (LiBs) power cell phones, computers, power tools, and EVs. The LIBs have a theoretical capacity of 3861 mAh g^{-1} [3] or 2061 mAh cm^{-3} [4] resulting in a cell cycling for thousands of hours with small increases of voltage polarization [5]. On the other hand, the LIBs suffer from low power density and poor rate capability due to low electrical conductivity of electrode films and inferior ionic conductivity of solid-state or liquid electrolyte at room temperature. The alternative to the solid-state or liquid electrolytes might be an inorganic ceramic electrolyte. It can provide superior ionic conductions but first of all is complex to synthesize and possesses high interface resistance [6]. Additionally, although lithium is regarded as an abundant element, the cost of lithium-containing materials has also rapidly increased in the past years, resulting in increased prices for LIBs [7]. One of the proposed alternatives to lithium is sodium as attainable at low cost, and geographically distributed [8]. Hence, we could find it in sodium-ion batteries (SIBs) but also as a building material for the anodes in sodium-metal batteries. The theoretical capacity could be as high as 2500 mAh g^{-1} [9]. Meanwhile, the

[a] https://www.panasonic-batteries.com/en/alkaline/evolta-neo/evolta-neo-aaa, accessed September 24, 2020.

[b] https://www.saftbatteries.com/products-solutions/products/mp-small-vl?text=&tech=&market=327&sort=newest&submit=Search, accessed September 24, 2020.

Table 6.1 Properties of selected microbatteries available on the market.

Company	Material: cathode-anode	Nominal capacity (mAh)	Operational temp. (°C)	Nominal voltage (V)	Dia. (mm)	Area (mm^2)	Height (mm)	Mass (g)
Murata[a]	Silver oxide-zinc	20–160	−10–60	1.55	6.8–11.6	–	2.05–5.4	0.32–2.2
	Manganese dioxide-zinc	45–120	−10–60	1.5	7.9–11.6	–	3.05–5.4	0.57–2
	Manganese dioxide-lithium	30–2000	−40–125	3.0	20–36.5	–	3.2–7.7	0.67–20
I-Ten[b]	–	$(50–500) \times 10^{-3}$	−40–85	2.5	–	8–14.4	0.6–2.3	–
Varta[c]	Nickel-metal hydride	65–500	−20–85	1.2–7.2	–	$(1.1–2.54) \times 10^3$	6.6–8	4.5–90
		6–19	−20–85	1.2–3.6	6.8–12	–	2.3–7.3	0.29–3
Saft[2]	Manganese dioxide-lithium	500	−40–70	3.0	16.7	–	16.33	8

[a]https://www.murata.com/en-eu/products/batteries/micro, accessed September 2020.
[b]https://www.iten.com/en/micro-batteries, accessed September 24, 2020.
[c]https://www.varta-microbattery.com/en/products/nickel-metal-hydride/rechargeable-nimh-button-cells, accessed September 24, 2020.

practical samples reached up to 800 mAh g^{-1} [10]. But it has a common drawback with lithium anodes, i.e., dendrite formation that in consequence results in short circuits giving 10 times lower cycling compared to LIBs (~100 h) [11]. Another issue related to the cost reduction is the problem to exchange copper for aluminum in the current collector because sodium metal cannot form the binary alloy with aluminum [3]. Additionally, the ionic radius of sodium cation is three times larger than lithium; thus the successful reversible intercalation hosts must possess large-enough channels and interstitial sites to accept sodium cations. This demands exist searching for alternatives to metallic sodium, such as intercalation compounds or alloys, as anodes, but it should be considered that for now the alloys are impeded by huge volume expansion (126%–520%) and derive limited cycle life [12].

Silicon has been also widely studied as a promising anode material due to the high theoretical gravimetric capacity (4200 mAh g^{-1}) though the laboratory prototypes reached the capacity up to 1500 mAh g^{-1} [13]. Due to the large volume change of Si during the lithiation and delithiation process (volume change up to 400%) as well as inferior electric conductivity, direct use of Si for the anode is very challenging to be industrially cost-efficient. To get larger specific capacity, a larger volume following a more complex fabrication process and ultimately a higher cost and lower yield to form a larger buffering zone reserved in the composite [14].

Beyond the mentioned list, there are yet so-called postlithium batteries (PLIBs). Theoretically, they could offer even at least five times the higher volumetric capacity (up to ~11,000 mAh mL^{-1} vs ~2000 mAh mL^{-1} for the LIBs) [15]. Such PLIBs apply magnesium, calcium, or aluminum for the anode and the electrolyte. Especially, Mg is pointed to be perceived for high volumetric capacity (3833 mAh cm^{-3}) [16] and the lack of dendrite formation during charging, hence becoming much safer when compared to LIBs. However, the research on these elements for battery application needs to be yet developed, and it is considered as substantially more complex than that of the alkali metals. It encompasses the issues related to the compatibility of the electrolyte with both the anode and the cathode and the discovery of high-voltage cathodes. For example, there is a difficulty in hosting Mg^{2+} ion due to its large charge density, and compared to monovalent cations (e.g., Li^{+}, Na^{+}, and K^{+}), Mg^{2+} is very "sticky" when migrating through the crystal structures of ionic compounds due to the strong electrostatic interactions between Mg^{2+} and the packing anions [17]. Usually, the as-made magnesium-based batteries have the capacity in the range of 50–900 $mAh\ g^{-1}$ [18], but recently Ikhe et al. [19] reported the capacity of 1243 $mAh\ g^{-1}$.

In general, metal-air batteries generate electricity through a redox reaction between metal and oxygen that is drawn from the air into a cathode being open structure. Especially, the Li-air system releases an extremely high theoretical specific energy of nearly 11,700 $Wh\ kg^{-1}$, which is close to the value for one of the most energy-dense liquids—gasoline [20]. Zinc-air batteries have a high theoretical energy density of 1086 $Wh\ kg^{-1}$ (including oxygen), about five times higher than the current lithium-ion technology, and have potentially very low cost to be fabricated ($<10\$\ kW^{-1}\ h^{-1}$) [21], in practice up to $\sim$987 $mAh\ g^{-1}$ [22]. It is proposed that the discharge products can be regenerated off-site using renewable energy such as the heat from concentrated solar radiation. Traditional zinc-air batteries are present commercially supplying the power to hearing aids and buoys. Though they are nonrechargeable, interesting is that zinc-air batteries can be mechanically recharged via physically replacing the zinc electrode and electrolyte. At the present, the estimated zinc reserves per person per year will reach a maximum in 2100 and decline after 2100 [23].

In general, electrodes of the MBs are either sandwiched films, interdigitated or comb-type structures, and fibers. The sandwiched film thickness of the sequentially stacked layers differs from few micrometers, for each electrode, up to about 20 μm for a separator if a liquid electrolyte is used [24]. Such a stack, yet with the packaging, greatly limits effective volumetric capacity due to quite a fraction of the total stack thickness. This architecture is the simplest to test newly developed materials and the chemical reactions related to the energy storage mechanism and also to develop new and cost-efficient flexible and stretchable battery materials. As an example, carbon cloth (CC)-based electrodes (Fig. 6.1) demonstrated by Jeong et al. [25] having the capacity up to 2200 $mAh\ g^{-1}$. Most presented CC possesses electrochemically active materials only on the surface [26, 27], such as the nitrogen-doped porous CC suitable for storing more Li ions developed by Balogun et al. [28]. Though the obtained battery achieved a lower capacity of 168 $mAh\ g^{-1}$ (1.65 $mAh\ cm^{-2}$), the work revealed the technique to explore the interior of the fibers.

The planar design is often fabricated by thin-film techniques, which mostly used deposition techniques such as atomic deposition layer (ADL), chemical vapor deposition (CVD), or sputtering. Another method to produce planar MB electrodes is ink-jet printing of a cathode [29, 30] and anode [31, 32]. This printing technique enables controlling precisely the thickness

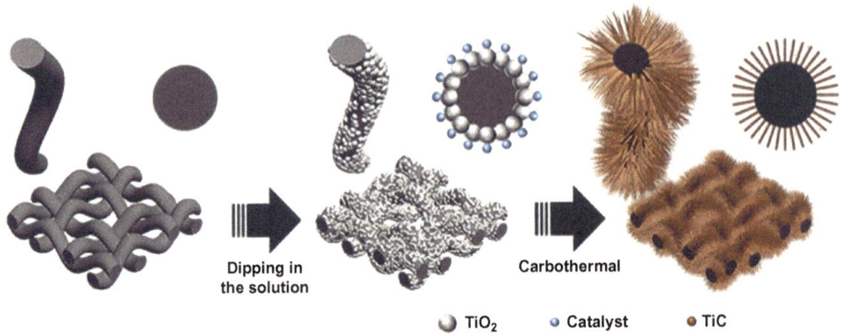

Fig. 6.1 Schematic representation of TiC-coated carbon cloth fabrication from a piece of cotton T-shirt. *Reprinted with permission from M. Gi Jeong, W.-J. Kwak, H.-Ji Shin, Y.-K. Sun, H.-Gi Jung, Perpendicularly aligned TiC-coated carbon cloth cathode for high-performance Li-O2 batteries, Chem. Eng. J. 399 (2020), 125699.*

of the electrodes, which can be tuned by the number of layers printed on top of one another. Comparing to thin-film fabrication techniques, ink-jet printing offers minimal wasted material and scalability, and the deposition takes place at room temperature and ambient conditions. From the photo-patterning techniques, Cao et al. [33] proposed the photolithography of commonly known thick photosensitive polymer (SU-8) modified with a portion of the radical polymer and ionic conductor $LiClO_4$ to form electrochemical energy storage material with circular features 30 μm in diameter. Moreover, the modified SU-8 was used to pattern an array 67 × 67 of MBs on a 4 × 4 mm^2 silicon substrate. Another example was reported by Ning et al. [34], who combined 3D holographic lithography with conventional photolithography followed by template-assisted electrodepositions to produce thin layers of active materials. The process allowed obtaining well-defined, periodically mesostructured porous electrodes. However, to generate high capacity, 2D electrodes must cover a large footprint area, which hampers the integration into portable microelectronics. On the other hand, assembling the 2D electrodes into 3D architecture became an obstacle. With the development of new additive manufacturing techniques, the next-generation 3D MBs offer freedom in the choice of arbitrary shapes and sizes, hence could be used to improve the areal energy density of the investigated materials. Besides, 3D printing offers great chances for efficiently integrating printed batteries and electronic devices. It has been already demonstrated MBs printed using various methods, e.g., fused deposition modeling (FDM), stereolithography (SLA), or ink-jet printing, and many others [35, 36].

Moreover, for MBs a wide spectrum of materials is used, including graphite [37], graphene [38], carbon nanotubes [39], cellulose nanofiber [40], lithium titanate (LTO), and lithium iron phosphate (LFP) [41], silver [42], and many others. There already reported materials and techniques not only to print cathodes or anodes but also working reference electrodes that are used for testing electrochemical cells [43] and electrolytes [44]. Another extraordinary example of a 3D printing method was reported by the formation of high structural integrity graphene aerogel (GA) by integrating 3D printing of ice and freeze casting [45]. Instead of heating up or extruding out the material at room temperature, the water-based GO suspension was rapidly frozen and selectively solidified to the aqueous droplets into ice crystal on a cold sink (−25 °C), well below water's freezing point. Contrary to the continuous printing that results in insufficient bonding at the interface of the layers, the deposited materials freeze, and firmly bond together with the previous layer via hydrogen bond. Additionally, the large surface area of the graphene benefits the loading of active materials and ensures structural integrity under physical deformation including stretching [46]. By analogy Sun et al. [47] reported the formation of thick (up to 1.5 mm) and robust mesh electrodes with the area of $7 \times 7\ mm^2$ by gelation of the ternary nanocomponents. The amalgamation of a highly conductive AgNW network, interconnected 3D graphene scaffold, and the hierarchical porous structure can allow the 3D-printed electrode to demonstrate much-improved charge and ion transport, and to alleviate internal stresses during the charging and discharging process even under ultrahigh thickness state. The ink was dispensed via syringe in a layer-by-layer arrangement on a copper disc by adjusting the compressed air pressure and printing rate. After the desired structure is printed, an overnight freeze-drying process is employed to dry the electrodes (Fig. 6.2). As a result, the LIBs utilizing these thick electrodes have a rate capability of $121\ mAh\ g^{-1}$ and high areal capacities up to $4.74\ mAh\ cm^{-2}$ with a capacity retention of ~95.5% after 100 cycles.

Highly porous metal foams have been also demonstrated as 3D current collectors to create high-performance MBs. Wei et al. [48] presented electrodes with 540 μm thickness made by the use of direct ink writing. The electrodes consist of dispersed LFP or LTO particles mixed with attractive conductive carbon particles that form a percolative network within a lithium-based electrolyte solution. Fully 3D printed and packaged LIBs composed of thick LFP/LTO electrodes with glassy carbon current collectors, customized separator, and packaging materials deliver an areal capacity of $4.45\ mAh\ cm^{-2}$. On the other hand, having thick electrodes leads

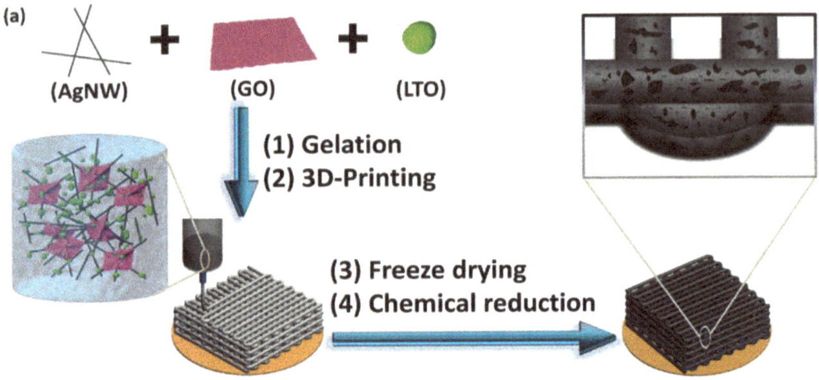

Fig. 6.2 Schematic illustration of the fabrication process of the proposed 3D printed architectures by using gelation [47].

mainly to cracking and delamination during drying and incomplete electrolyte infiltration as well as poor mechanical properties, and slow charge and ion transport [49, 50]. Despite the efforts put into the development of techniques to fabricated MBs, the technology compatibility and safety of the materials used for each component as well as ensuring mechanical toughness to withstand internal stress fluctuations during charging and discharging remains a big challenge.

References

[1] J. Liu, J. Wang, C. Xu, H. Jiang, C. Li, L. Zhang, J. Lin, Z.X. Shen, Advanced energy storage devices: basic principles, analytical methods, and rational materials design, Adv. Sci. 5 (2018), 1700322.

[2] S. Zheng, X. Shi, P. Das, Z.-S. Wu, X. Bao, The road towards planar microbatteries and micro-supercapacitors: from 2D to 3D device geometries, Adv. Mater. 31 (2019), 1900583.

[3] N. Yabuuchi, K. Kubota, M. Dahbi, S. Komaba, Research development on sodium-ion batteries, Chem. Rev. 114 (2014) 11636–11682.

[4] W. Lu, B. Jiang, P.E. Vullum, A.M. Svensson, A. Erbe, S.M. Selbach, H. Xu, F. Vullum-Bruer, High interfacial charge storage capability of carbonaceous cathodes for Mg batteries, ACS Nano 12 (2018) 2998–3009.

[5] X.B. Cheng, R. Zhang, C.Z. Zhao, Q. Zhang, Toward safe lithium metal anode in rechargeable batteries: a review, Chem. Rev. 117 (2017) 10403–10473.

[6] S.J. Tan, X.-X. Zeng, Q. Ma, X.-W. Wu, Y.-G. Guo, Recent advancements in polymer-based composite electrolytes for rechargeable lithium batteries, Electrochem. Energy Rev. 1 (2018) 113–138.

[7] A. El Kharbachi, O. Zavorotynska, M. Latroche, F. Cuevas, V. Yartys, M. Fichtner, Exploits, advances and challenges benefiting beyond Li-ion battery technologies, J. Alloys Compd. 817 (2020), 1532612.

[8] A. Ponrouch, D. Monti, A. Boschin, B. Steen, P. Johansson, M.R. Palacín, Non-aqueous electrolytes for sodium-ion batteries, J. Mater. Chem. A 3 (2015) 22–42.
[9] N. Yabuuchi, Y. Matsuura, T. Ishikawa, S. Kuze, J.-Y. Son, Y.-T. Cui, H. Oji, S. Komaba, Phosphorus electrodes in sodium cells: small volume expansion by sodiation and the surface-stabilization mechanism in aprotic solvent, ChemElectroChem (2014) 580–589.
[10] Q. Sun, Q.-Q. Ren, H. Li, Z.-W. Fu, High capacity Sb2O4 thin film electrodes for rechargeable sodium battery, Electrochem. Commun. 13 (2011) 1462–1464.
[11] Y. Zhao, K.R. Adaira, X. Sun, Recent developments and insights into the understanding of Na metal anodes for Na-metal batteries, Energ. Environ. Sci. 11 (2018) 2673–2695.
[12] W. Chen, X. Zhang, L. Mi, C. Liu, J. Zhang, S. Cui, X. Feng, Y. Cao, C. Shen, Flexible freestanding anode with hierarchical 3D carbon-networks/Fe7S8/graphene for applicable sodium-ion batteries, Adv. Mater. 31 (2019), 1806664.
[13] Y. Qi, G. Wang, S. Li, T. Liu, J. Qiu, H. Li, Recent progress of structural designs of silicon for performance-enhanced lithium-ion batteries, Chem. Eng. J. 397 (2020), 125380.
[14] M.-H. Park, M.G. Kim, J. Joo, K. Kim, J. Kim, S. Ahn, Y. Cui, J. Cho, Silicon nanotube battery anodes, Nano Lett. 9 (2009) 3844–3847.
[15] J. Muldoon, C.B. Bucur, T. Gregory, Quest for nonaqueous multivalent secondary batteries: magnesium and beyond, Chem. Rev. 114 (2014) 11683–11720.
[16] Y. Wang, Z. Liu, C. Wang, X. Yi, R. Chen, L. Ma, Y. Hu, G. Zhu, T. Chen, Z. Tie, J. Ma, J. Liu, Z. Jin, Highly branched VS4 nanodendrites with 1D atomic-chain structure as a promising cathode material for long-cycling magnesium batteries, Adv. Mater. 30 (2018), 1802563.
[17] I.A. Rodríguez-Pérez, Y. Yuan, C. Bommier, X. Wang, L. Ma, D.P. Leonard, M.M. Lerner, R.G. Carter, T. Wu, P.A. Greaney, J. Lu, X. Ji, Mg-ion battery electrode: an organic solid's herringbone structure squeezed upon Mg-ion insertion, J. Am. Chem. Soc. 139 (2017) 13031–13037.
[18] R. Attias, M. Salama, B. Hirsch, Y. Goffer, D. Aurbach, Anode-electrolyte interfaces in secondary magnesium batteries, Joule 3 (2019) 27–52.
[19] A.B. Ikhe, S.C. Han, S.J.R. Prabakar, W.B. Park, K.-S. Sohn, M. Pyo, 3Mg/Mg2Sn anodes with unprecedented electrochemical performance towards viable magnesium-ion batteries, J. Mater. Chem. A 8 (2020) 14277–14286.
[20] F. Cheng, J. Chen, Metal–air batteries: from oxygen reduction electrochemistry to cathode catalysts, Chem. Soc. Rev. 41 (2012) 2172–2192.
[21] Y. Li, H. Dai, Recent advances in zinc–air batteries, Chem. Soc. Rev. 43 (2014) 5257.
[22] X. Han, X. Ling, Y. Wang, T. Ma, C. Zhong, W. Hu, Y. Deng, Generation of nanoparticle, atomic-cluster, and single-atom cobalt catalysts from zeolitic imidazole frameworks by spatial isolation and their use in zinc–air batteries, Angew. Chem. 58 (2019) 5359–5364.
[23] H.U. Sverdrupa, A.H. Olafsdottira, K.V. Ragnarsdottir, On the long-term sustainability of copper, zinc and lead supply, using a system dynamics model, Resour. Conserv. Recycl. X4 (2019), 100007.
[24] J.F.M. Oudenhoven, L. Baggetto, P.H.L. Notten, All-solid-state lithium-ion microbatteries: a review of various three-dimensional concepts, Adv. Energy Mater. 1 (2011) 10–33.
[25] M. Gi Jeong, W.-J. Kwak, H.-J. Shin, Y.-K. Sun, H.-G. Jung, Perpendicularly aligned TiC-coated carbon cloth cathode for high-performance Li-O2 batteries, Chem. Eng. J. 399 (2020), 125699.
[26] R. Zou, Z. Zhang, M.F. Yuen, M. Sun, J. Hu, C.-S. Lee, W. Zhang, Three-dimensional-networked NiCo2S4 nanosheet array/carbon cloth anodes for high-performance lithium-ion batteries, NPG Asia Mater. 7 (2015), e195.

[27] B. Liu, J. Zhang, X. Wang, G. Chen, D. Chen, C. Zhou, G. Shen, Hierarchical three-dimensional ZnCo2O4 nanowire arrays/carbon cloth anodes for a novel class of high-performance flexible lithium-ion batteries, Nano Lett. 12 (2012) 3005–3011.
[28] M.-S. Balogun, W. Qiu, F. Lyu, Y. Luo, H. MengJiantao, L. Wenjie, M. Liqiang, M.Y. Tong, All-flexible lithium ion battery based on thermally-etched porous carbon cloth anode and cathode, Nano Energy 26 (2016) 446–455.
[29] Y. Gu, A. Wu, H. Sohn, C. Nicoletti, Z. Iqbal, J.F. Federici, Fabrication of rechargeable lithium ion batteries using water-based inkjet printed cathodes, J. Manuf. Process. 20 (2015) 198–205.
[30] F.M. Courtel, S. Niketic, D. Duguay, Y. Abu-Lebdeh, I.J. Davidson, Water-soluble binders for MCMB carbon anodes for lithium-ion batteries, J. Power Sources 196 (2011) 2128–2134.
[31] Y. Zhao, Q. Zhou, L. Liu, J. Xu, M. Yan, Z. Jiang, A novel and facile route of ink-jet printing to thin film SnO2 anode for rechargeable lithium ion batteries, Electrochim. Acta 51 (2006) 2639–2645.
[32] S. Lawes, Q. Sun, A. Lushington, B. Xiao, Y. Liu, X. Sun, Inkjet-printed silicon as high performance anodes for Li-ion batteries, Nano Energy 36 (2017) 313–321.
[33] L. Cao, G. Fang, H. Cao, X. Duan, Photopatterning and electrochemical energy storage properties of an on-chip organic radical microbattery, Langmuir 35 (2019) 16079–16086.
[34] H. Ning, J.H. Pikul, R. Zhang, X. Li, S. Xu, J. Wang, J.A. Rogers, W.P. King, P.V. Braun, Holographic patterning of high-performance on-chip 3D lithium-ion microbatteries, PNAS 112 (2015) 6573–6578.
[35] V. Egorov, U. Gulzar, Y. Zhang, S. Breen, C. O'Dwyer, Evolution of 3D printing methods and materials for electrochemical energy storage, Adv. Mater. 32 (2020), 2000556.
[36] F. Zhang, M. Wei, V.V. Viswanathan, B. Swart, Y. Shao, G. Wu, C. Zhou, 3D printing technologies for electrochemical energy storage, Nano Energy 40 (2017) 418–431.
[37] A. Maurel, M. Courty, B. Fleutot, H. Tortajada, K. Prashantha, M. Armand, S. Grugeon, S. Panier, L. Dupont, Highly loaded graphite–polylactic acid composite-based filaments for lithium-ion battery three-dimensional printing, Chem. Mater. 30 (2018) 7484.
[38] D. Vernardou, K.C. Vasilopoulos, G. Kenanakis, 3D printed graphene-based electrodes with high electrochemical performance, Appl. Phys. A Mater. Sci. Process. 123 (2017) 623.
[39] J. Wang, Q. Sun, X. Gao, C. Wang, W. Li, F.B. Holness, M. Zheng, R. Li, A.D. Price, X. Sun, T.K. Sham, X. Sun, Toward high areal energy and power density electrode for Li-ion batteries via optimized 3D printing approach, ACS Appl. Mater. Interfaces 10 (2018) 39794.
[40] D. Cao, Y. Xing, K. Tantratian, X. Wang, Y. Ma, A. Mukhopadhyay, Z. Cheng, Q. Zhang, Y. Jiao, L. Chen, H. Zhu, 3D printed high-performance lithium metal microbatteries enabled by nanocellulose, Adv. Mater. 31 (2019) 1807313.
[41] C. Reyes, R. Somogyi, S. Niu, M.A. Cruz, F. Yang, M.J. Catenacci, C.P. Rhodes, B.J. Wiley, Three-dimensional printing of a complete lithium ion battery with fused filament fabrication, ACS Appl. Energy Mater. 1 (2018) 5268.
[42] C.C. Ho, K. Murata, D.A. Steingart, J.W. Evans, P.K. Wright, A super ink jet printed zinc–silver 3D microbattery, J. Micromech. Microeng. 19 (2009), 094013.
[43] N. Rohaizad, C.C. Mayorga-Martinez, F. Novotný, R.D. Webster, M. Pumera, 3D-printed Ag/AgCl pseudo-reference electrodes, Electrochem. Commun. 103 (2019) 104–108105.
[44] D.W. McOwen, S. Xu, Y. Gong, Y. Wen, G.L. Godbey, J.E. Gritton, T.R. Hamann, J. Dai, G.T. Hitz, L. Hu, E.D. Wachsman, 3D-printing electrolytes for solid-state batteries, Adv. Mater. 30 (2018) 1707132.

[45] Q. Zhang, F. Zhang, S.P. Medarametla, H. Li, C. Zhou, D. Lin, 3D printing of graphene aerogels, Small 12 (2016) 1702–1708.
[46] S. Kang, S.Y. Hong, N. Kim, J. Oh, M. Park, K.Y. Chung, S.-S. Lee, J. Lee, J.G. Son, Stretchable lithium-ion battery based on re-entrant micro-honeycomb electrodes and cross-linked gel electrolyte, ACS Nano 14 (2020) 3660–3668.
[47] C. Sun, S. Liu, X. Shi, C. Lai, J. Liang, Y. Chen, 3D printing nanocomposite gel-based thick electrode enabling both high areal capacity and rate performance for lithium-ion battery, Chem. Eng. J. 381 (2020) 122641.
[48] T.-S. Wei, B.Y. Ahn, J. Grotto, J.A. Lewis, 3D printing of customized Li-ion batteries with thick electrodes, Adv. Mater. 30 (2018) 1703027.
[49] D. Lv, J. Zheng, Q. Li, X. Xie, S. Ferrara, Z. Nie, L.B. Mehdi, N.D. Browning, J.-G. Zhang, G.L. Graff, J. Liu, J. Xiao, High energy density lithium–sulfur batteries: challenges of thick sulfur cathodes, Adv. Energy Mater. 5 (2015), 1402290.
[50] J.H. Prosser, T. Brugarolas, S. Lee, A.J. Nolte, D. Lee, Avoiding cracks in nanoparticle films, Nano Lett. 12 (2012) 5287.

CHAPTER 7

Energy harvesting

Karolina Urszula Laszczyk[a], Piotr Śliwiński[a], and Kazufumi Kobashi[b]
[a]Faculty of Microsystem Electronics and Photonics, Wroclaw University of Science and Technology, Wrocław, Poland
[b]CNT-Application Research Center, National Institute of Advanced Industrial Science and Technology, Ibaraki, Japan

Energy harvesting is based on obtaining energy from the environment and is strongly associated with the trend of miniaturization of consumer electronics and the dissemination of the concept of the Internet of Things (IoT). The latest is defined as a network of communicating electronic devices and often requires wireless operation without the need to frequently replace or recharge batteries [1]. Sensors and sensing systems to monitor the condition of machines, portable and wearable electronics are used in communication exchange to set and manage automatically, e.g., appropriate in-door conditions for an individual flat room. Some of these sensors are intended to be mounted in hard-to-reach places or exposed to unstable outdoor conditions (weather stations). Because the IoT needs small and lightweight devices, the microelectromechanical system (MEMS) technology is often in usage as the model to prototype, novel, tiny energy harvesters. On the other hand, Li-ion batteries or other energy storage devices are commonly used as a stable power supply for driving microelectronic devices. The limited capacity of energy stored by microbatteries (MBs) or microsupercapacitor (MSCs) is in demand for maintaining energy supply to the system by the integration of the energy storage unit with the energy harvester. Nevertheless, the main drawback of the available MBs is their limited lifetime which requires charging or replacement. This could become a critical issue by surgery, e.g., the artificial pacemaker implanted beneath the chest skin. Similarly, the replacement is nearly impossible or very tricky and costly in hazardous, harsh, and large terrain deployment. In this regard, coupling the harvester with the energy storage device is highly important for developing a sustainable self-charging power unit (SCPU) [2]. It is envisioned that the energy harvesters go forward to systems without implementing energy storage units (batteries or supercapacitors) for harsh environmental conditions like high temperature, fear of poisoning the natural system.

Microsupercapacitors
https://doi.org/10.1016/B978-0-08-102888-9.00006-9

The energy can be harvested from the human/animal or the surrounding environment. They are classified into two physical and chemical sources: physical one for the energy harvested from the heat emitted by the body, blood pressure, breathing (airflow), foot/hand pressure, kinetic energy of body movements (Fig. 7.1) [3], and chemical one from chemical reactions, e.g., blood glucose [4], devices that can use the energy of living organisms, e.g., transmitters attached to animals [5], wearable electronics (watches, sports bands) [6] and epidermal electronics [7], autonomous prostheses, or implanted biomedical devices (IMD) [8].

Several types of energy harvesters have been so far proposed (Fig. 7.2) [9] by mechanical, thermal, radiant, and biochemical sources with several mechanisms including piezoelectricity [10], electrostatics [11], electromagnetism [12], and others. At the microscale and below, the piezoelectric mechanism becomes more attractive as compared to electromagnetics. Comparing the respective conversion mechanisms by taking into account constitutive equations, the piezoelectric output power is proportional to V^3, while the electromagnetic one is proportional to V^2 [13]. Additionally, piezoelectric generators are the most attractive for application in

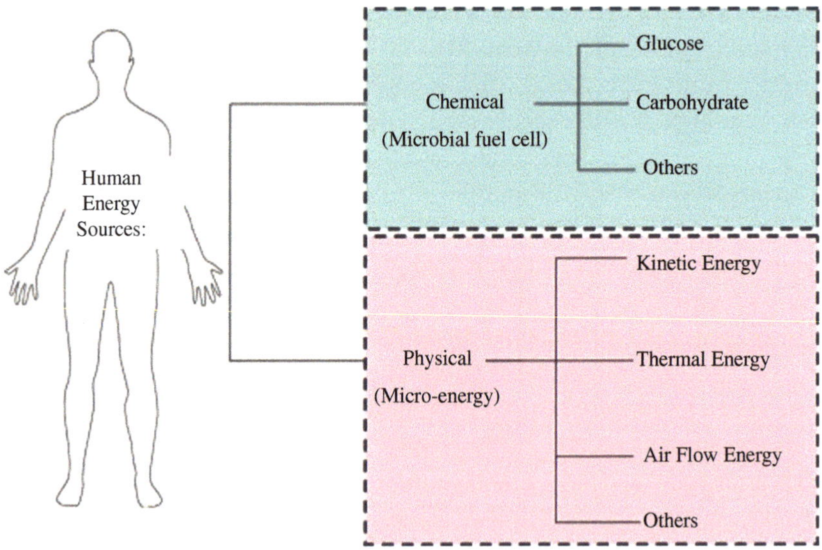

Fig. 7.1 Classification of human energies. *Reprinted with permission from C.-Y. Sue, N.-C. Tsai, Human powered MEMS-based energy harvest devices, Appl. Energy 93 (2012) 390–403.*

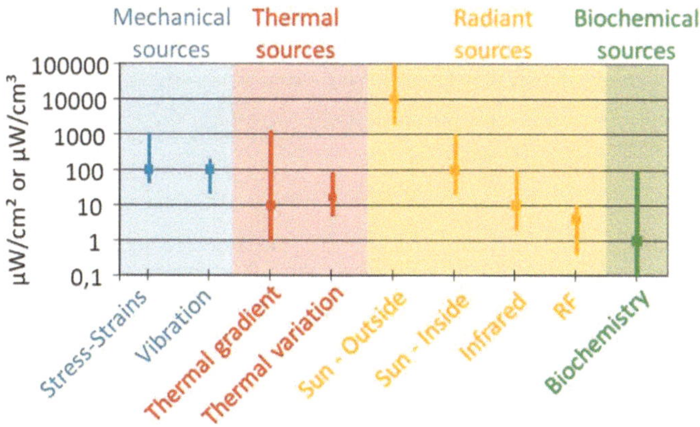

Fig. 7.2 Ambient source power densities before conversion. *Reprinted with permission from D. Zabek, F. Morini, Solid state generators and energy harvesters for waste heat recovery and thermal energy harvesting, Therm. Sci. Eng. Prog. 9 (2019) 235–247.*

self-powered biomedical electronics. They can scavenge inexhaustible biomechanical energies such as cardiac motion, muscle contraction/relaxation, and blood circulation and convert them into electrical ones [14]. For example, Hwang et al. [15] demonstrated the conversion through the periodic mechanical motions of bending and unbending of a flexible single crystalline piezoelectric thin film energy harvester placed on a plastic substrate to achieve a self-powered artificial pacemaker with significant 145 μA output current and voltage of 8.2 V. Moreover, the authors implemented the generator in the cardiac muscle of a live rat, enabling real-time functional electrical stimulation. The disadvantages of application for piezoelectric generators are the limited environments with frequent vibration and the variable output performance.

Thermoelectric generators (TEGs) are active devices to convert thermal energy into an electrical one. TEGs are composed of dissimilar thermocouples, based on the Seebeck effect, connected electrically in series and parallel (Fig. 7.3). The features are energy efficiency, free maintenance, and long lifetime, design simplicity, fast response, no noise, and the absence of moving parts [16]. Microscale TEGs work with low wasted heat and generate electrical power in the range of μW to a few mW. The microscale TEGs are typically less than 1 mm thick and suitable to recover waste heat below 230 °C to power wearable electronics, wireless sensors, or medical devices [17]. The popular solutions among TEGs are based on either fully organic thermoelectrics or inorganic/organic hybrids forming so-called flexible

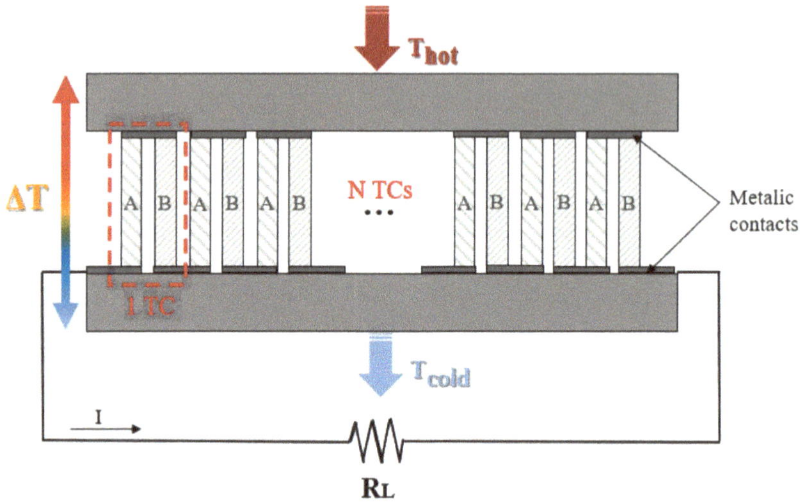

Fig. 7.3 A schematic presenting a basic principle of thermoelectric generator. *Reprinted with permission from N. Jaziri, A. Boughamour, J. Müller, B. Mezghani, F. Tounsi, M. Ismail, A comprehensive review of thermoelectric generators: technologies and common applications, Energy Rep. 6 (7) (2020) 264–287.*

thermoelectrics (TE) [18]. Most of the conducting polymers used for TE composites are p-type materials (e.g., PEDOT, PANI, PPY). N-type conducting polymers possess poor stability in the air hence are not developed for the TE. On the other hand, nonconducting polymers work as suitable matrices for inorganic-nanostructure/polymer composites, which include poly(vinyl acetate) (PVAc) polyvinylidene fluoride (PVDF), and polylactic acid (PLA). For microcombustion of fuels, Yadav et al. [19] presented a prototype microthermoelectric power generator (micro-TEG) system with a multistep microcombustor to convert the chemical energy of the fuel (propane) into heat. The system delivers a maximum electric power of 2.35 W with a voltage of 4.34 V and a current of 0.54 A at 8 Ω load resistance across the system.

From different radio frequencies (RF) of the environment, the energy can be also converted [20]. Zhang et al. [21] illustrated the power consumption of typical IoT applications and the measured ambient RF power across different locations. They predict that the convergence of the increasing reach of RF signals and the development of ultralow-power circuits represent an untapped opportunity as a ubiquitous energy solution for the era of IoT. The RF energy is typically achieved by rectennas, i.e., the combination of an antenna for collecting electromagnetic radiation from the environment and an ultrafast diode with nonlinear current-voltage characteristics that rectifies the RF signals generating the DC power. Such a solution covers most of the unlicensed industrial, scientific, and medical radio bands, including the WiFi channels, yet battery-free, as reported by Zhang et al. [22]. They demonstrated an atomically thin passive electronic device that directly converts an incident electromagnetic wave into DC. The device can provide power to a load connected to its output with a cutoff frequency of 10 GHz, moreover, be integrated with various flexible electronic systems. The receiver of the waves is an antenna and the rectifier is performed by a two-dimensional structure based on MoS_2, which creates a Schottky diode.

The other ways to gain energy are from the surrounding area temperature differences [23], electroacoustic transduction [24], conversion of the kinetic energy of vibrations as well as rotational and aeroelastic mechanisms [25]. It is worth remembering that the intensity and nature of each source are not linear but individual (Fig. 7.4).

Due to common availability in our daily life, mechanical energy harvesting such as piezoelectric and triboelectric generator (TENG) has become the most significant for building self-powered systems [26]. TENG relies on the coupling effect of contact electrification and electrostatic induction and

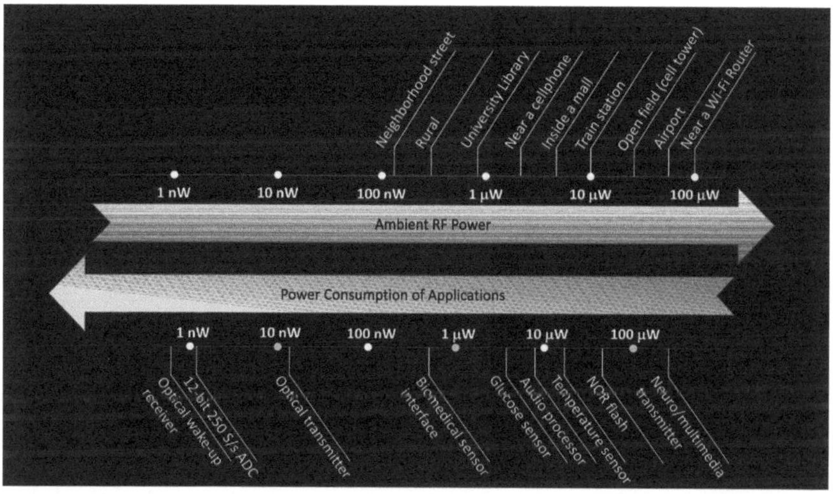

Fig. 7.4 Illustration of measured ambient RF power across different locations (*upper arrow*) and the power consumption of typical IoT applications (*lower arrow*). *Reprinted with permission from Xu Zhang, J. Graja, M. Lopez-Vallejo, E. McVay, T. Palacios, Opportunities and challenges of ambient radio-frequency energy harvesting, Joule 4 (2020) 1148–1152.*

offers high output performance, high energy conversion, low cost, and lightweight features. In general, two different triboelectric materials (having opposite polarities) come in contact and then separate (vertical contact-separation motion) or slid against each other (sliding motion). The electron transfer takes place between the materials and electrostatic charges of opposite polarities are developed on the surfaces of the triboelectric materials. In sliding-type TENG, rotating and freestanding designs are generally used, whereas, in vertical contact separation, pressures and vibrations are converted into electricity [27]. TENGs are workable by coupling supercapacitors than batteries, which ideally need to be charged with constant voltage. Various examples of coupling TENGs and supercapacitors have been already proposed. However, direct integration of a TENG with an energy storage device is challenging due to extremely low charging efficiency and large power loss by a large impedance mismatch between TENGs (high impedance) and energy storage units (low impedance). This issue needs to be solved by an appropriately designed power management circuit (90% of efficiency with a continuous DC electricity of 1.044 mW) [28]. Guo et al. [29] developed an ultralight cut-paper-based self-charging power unit (PC-SCPU) capable of simultaneously harvesting and storing energy by combining the TENG and the supercapacitor. The use of perforated sandpaper coated with Al film as a substrate of supercapacitor enabled to fill the active material-dispersed graphene in the small cavities. Meanwhile, a PVA-wetted separator working at once as an electrolyte was sandwiched between two pieces of the as-prepared graphite electrodes. In the case of the TENG, the active fluorinated ethylene propylene (FEP) thin film was formed on several parallel crevices and enclosed between Au-film-coated paper pieces for the next assembling process.

References

[1] I.C.L. Ng, S.Y.L. Wakenshaw, The internet-of-things: review and research directions, Int. J. Res. Mark. 34 (2017) 3–21.
[2] X. Pu, L. Li, H. Song, C. Du, Z. Zhao, C. Jiang, G. Cao, W. Hu, Z.L.A. Wang, Self-charging power unit by integration of a textile triboelectric nanogenerator and a flexible lithium-ion battery for wearable electronics, Adv. Mater. 27 (2015) 2472–2478.
[3] D.F. Berdy, D.J. Valentino, D. Peroulis, Kinetic energy harvesting from human walking and running using a magnetic levitation energy harvester, Sens. Actuators A 222 (2015) 262–271.
[4] S. Kerzenmacher, J. Ducree, R. Zengerle, F. von Stetten, Energy harvesting by implantable abiotically catalyzed glucose fuel cells, J. Power Sources 182 (2008) 1–17.
[5] H. Li, C. Tian, J. Lu, M.J. Myjak, J.J. Martinez, R.S. Brown, Z.D. Deng, An energy harvesting underwater acoustic transmitter for aquatic animals, Sci. Rep. 6 (2016) 33804.

[6] V. Leonov, R.J.M. Vullers, Wearable electronics self-powered by using humanbody heat: the state of the art and the perspective, J. Renew. Sustain. Energy 1 (2009), 062701.

[7] D.-H. Kim, N. Lu, R. Ma, Y.-S. Kim, R.-H. Kim, S. Wang, J. Wu, S.M. Won, H. Tao, A. Islam, K.J. Yu, T.-I. Kim, R. Chowdhury, M. Ying, L. Xu, M. Li, H.-J. Chung, H. Keum, M. McCormick, P. Liu, Y.-W. Zhang, F.G. Omenetto, Y. Huang, T. Coleman, J.A. Rogers, Epidermal electronics, Science 333 (2011) 838–843.

[8] C.-Y. Sue, N.-C. Tsai, Human powered MEMS-based energy harvest devices, Appl. Energy 93 (2012) 390–403.

[9] D. Zabek, F. Morini, Solid state generators and energy harvesters for waste heat recovery and thermal energy harvesting, Therm. Sci. Eng. Prog. 9 (2019) 235–247.

[10] S. Saadon, O. Sidek, A review of vibration-based MEMS piezoelectric energy harvesters, Energ. Conver. Manage. 52 (2011) 500–504.

[11] P. Basset, D. Galayko, A.M. Paracha, F. Marty, A. Dudka, T. Bourouina, A batch-fabricated and electret-free silicon electrostatic vibration energy harvester, J. Micromech. Microeng. 19 (2009) 115025.

[12] P. Wang, K. Tanaka, S. Sugiyama, X. Dai, X. Zhao, J. Liu, A micro electromagnetic low level vibration energy harvester based on MEMS technology, Microsyst. Technol. 15 (2009) 941–951.

[13] M.A.S. Bressers, S. Priya, Multiple cell configuration electromagnetic vibration energy harvester, J. Phys. D Appl. Phys. 44 (2011) 295501.

[14] C.K. Jeong, Toward bioimplantable and biocompatible flexible energy harvesters using piezoelectric ceramic materials, MRS Comm. 10 (2020) 365–378.

[15] G.-T. Hwang, H. Park, J.-H. Lee, S. Oh, K.-I. Park, M. Byun, H. Park, G. Ahn, C.K. Jeong, K. No, H. Kwon, S.-G. Lee, B. Joung, K.J. Lee, Self-powered cardiac pacemaker enabled by flexible single crystalline PMN-PT piezoelectric energy harvester, Adv. Mater. 26 (2014) 4880–4887.

[16] N. Jaziri, A. Boughamour, J. Müller, B. Mezghani, F. Tounsi, M. Ismail, A comprehensive review of thermoelectric generators: technologies and common applications, Energy Rep. 6 (7) (2020) 264–287.

[17] S. Liu, B. Hu, D. Liu, F. Li, J.-F. Li, B. Li, L. Li, Y.-H. Lin, C.W. Nan, Micro-thermoelectric generators based on through glass pillars with high output voltage enabled by large temperature difference, Appl. Energy 225 (2018) 600–610.

[18] Y. Du, J. Xu, B. Paul, P. Eklund, Flexible thermoelectric materials and devices, Appl. Mater. Today 12 (2018) 366–388.

[19] S. Yadav, P. Sharma, P. Yamasani, S. Minaev, S. Kumar, A prototype micro-thermoelectric power generator for micro-electromechanical systems, Appl. Phys. Lett. 104 (2014) 123903.

[20] S. Keyrouz, H.J. Visser, A.G. Tijhuis, Multi-band simultaneous radio frequency energy harvesting, in: Proceedings of 7th European Conference on Antennas and Propagation (EUCAP), 2013, pp. 3058–3061.

[21] Z. Xu, J. Graja, M. Lopez-Vallejo, E. McVay, T. Palacios, Opportunities and challenges of ambient radio-frequency energy harvesting, Joule 4 (2020) 1148–1152.

[22] Z. Xu, J. Graja, J.L. Vazquez-Roy, U. Radhakrishna, X. Wang, W. Chern, L. Zhou, Y. Lin, P.-C. Shen, X. Ji, X. Ling, A. Zubair, Y. Zhang, H. Wang, M. Dubey, J. Kong, M. Dresselhaus, T. Palacios, Two-dimensional MoS2-enabled flexible rectenna for Wi-Fi-band wireless energy harvesting, Nature 566 (2019) 368–372.

[23] S. Lee, K.-H. Shin, J.-S. Lee, T.-J. Lee, D.-M. Sim, D. Jung, H. Jung, J.-H. Kim, Heat energy harvesting by utilizing waste heat with small temperature differences between heat source and sink, J. Mech. Sci. Technol. 34 (2020) 443–455.

[24] S.B. Horowitz, M. Sheplak, L.N. Cattafesta, T. Nishida, MEMS acoustic energy harvester, J. Micromech. Microeng. 16 (2006) S174–S181.

[25] L. Mateu, F. Moll, Review of energy harvesting techniques and applications for microelectronics (Keynote Address), SPIE Int. Soc. Opt. Eng. (2005).
[26] Z.L. Wang, L. Zhong, Triboelectric nanogenerators as new energy technology for self-powered systems and as active mechanical and chemical sensors, ACS Nano 7 (2013) 9533–9557.
[27] S. Rathore, S. Sharma, B. P. Swain, R. Kr Ghadai, A critical review on triboelectric nanogenerator. IOP Conf. Series: Materials Science and Engineering vol. 377 (2018), 012186.
[28] S. Niu, X. Wang, F. Yi, Y.S. Zhou, Z.L. Wang, A universal self-charging system driven by random biomechanical energy for sustainable operation of mobile electronics, Nat. Commun. 6 (2015) 8975.
[29] H. Guo, M.-H. Yeh, Y. Zi, Z. Wen, J. Chen, G. Liu, C. Hu, Z.L. Wang, Ultralight cut-paper-based self-charging power unit for self-powered portable electronic and medical systems, ACS Nano 11 (2017) 4475–4482.

PART III

Advances in Microsupercapacitor Development

CHAPTER 8

Design and technology processes used for microsupercapacitors

Karolina Urszula Laszczyk
Faculty of Microsystem Electronics and Photonics, Wroclaw University of Science and Technology, Wrocław, Poland

8.1 Introduction

With the appearance of the microsupercapacitor (MSCs) [1], the possible routes to fabricate them have been rapidly evolving and expanding for the last 10 years [2–4]. Various techniques have been adapted and combined, and not necessarily from the domain of the energy storage devices, opening routes to get innovative solutions. The MSC technology is a synergy of the microelectronics fabrication processes and traditional paper and fabric printing techniques. In general, the technology development might be split into six characteristic periods as follows: (1) exploring the potential of a planar architecture, such as fast charge-discharge rate and high specific capacitance, (2) ability to scale-up the foot-print area, i.e., to increase the number of simultaneously fabricated MSCs and tune the performance via arbitrary connections, (3) flexible and stretchable MSCs, (4) expansion along the third dimension, i.e., the capacity enhancement, hence energy, by increasing the thickness of the electrodes, (5) novel functionality of the MSCs by the synergy of various developed materials and techniques, e.g., photoswitchable or electrochromic MSCs [5, 6], and (6) the MSC as a part of the integrated circuits (ICs). Interestingly, the MSCs can be tailored towards specific services. On the other hand, there are still many technical challenges to overcome to make the MSCs attractive for commercial use. There is a lack of a universal approach that applies various types of materials to the MSC. Indeed, it is extraordinarily difficult because there are certain dependencies between materials for the electrodes, the electrolyte, as well as packaging or sealing. This causes limitations in available combinations of materials and requires to perform strict compatibility research. For example, there is a need to confirm the relationship between the affinity of the electrode material and the type of the electrolyte solvent, as well as the pore network of the electrode and the size of the ions or solvated ions.

Microsupercapacitors
https://doi.org/10.1016/B978-0-08-102888-9.00004-5

Additionally, the basic extraordinary performance features of active materials used to build MSC electrodes are lowered due to processing the raw materials into the adaptable formulas making the technology even more restricted. Nonetheless, the number of ideas to manufacture the MSCs is continuously increasing along with new methods such as stamping, self-healable sponge electrodes offering compressibility, and 3D printing (additive manufacturing) to project complex architectures. With the diverse MSCs in mind, the main purpose of the chapter is to present various approaches to build MSCs. The methods are grouped according to the two main developed architectures, i.e., planar or two- and three-dimensional. The special emphasis is put on the technology of the three-dimension electrodes, while the planar MSC technology description is complementary to the information from the other co-authors in the previous chapters. Some aspects for the electrode design and its influence on the performance of the MSCs are shown with the way to make packaging and sealing and the MSC integration with the other functional components.

8.2 Two-dimensional electrodes

8.2.1 Electrode geometry

Since Pech et al. [7] published their investigation on planar MSC made of carbon onion, the boom was created in search for the solutions to fabricate planar MSCs. The interdigital structure is considered to possess some salient features, such as ease of extension into the third dimension. On the other hand, the researchers started to ask the question on the influence and importance of the geometry and dimensions on the MSC performance. Commonly in use is the interdigitated structure which shows better capacitance performance compared to parallel plate capacitors (so-called sandwich structure). Because the electric field lines intensify at the electrode edges as compared to continuous or round-shape electrode designs. For example, it was shown that the edge planes can physically and chemically store approximately 10-times more charge than the basal planes [8–10]. The edge effect can be achieved by increasing the area of the edge planes. In the interdigitated architecture it is obtained via increasing the number of microelectrodes and the length of the interspace and reducing the width of the electrodes with the geometrical area maintained. Aside from the increase in the number of edge planes, the decrease in the width of electrodes leads to an increase in the capacitance because of efficient ion transport. Yoon et al. [11] utilized the edge effect of in-plane structured heat-treated graphene oxide. They showed the improved ion transport in electrodes, though the morphology of the active material was restacked, and the effective ion transport channels were removed. Although the geometrical area of the electrodes is reduced from 19.7 to 15.6 mm^2 keeping the interspace constant while increasing the number of electrodes from 2 to 92, reducing a single electrode width from 10 to 0.08 mm, and increasing a capacitance from 68 to 98 F cm^{-3}. The results are correlated with the electrode resistance and charge transfer resistance (Fig. 8.1). A similar trend was presented also by other researchers [12–14]. All the results suggest that the greater the number of electrodes, the higher the discharge current, and thus more power could be extracted because of the efficient decrease of the average ionic diffusion pathway between the adjacent electrodes. The fast ionic diffusion rate is important concerning the EDL mechanism, especially as the low diffusion rate causes low pseudo-capacitance and poor rate capability [15].

There were also explored other planar designs, e.g., basing on or excluding the edge effect. Hota et al. [16] applied Hilbert, Peano, and Moore fractal designs (Fig. 8.2) for the MSCs and confirmed experimentally their performance with the standard interdigitated design. To preserve the uniform of the film in thickness as well as its morphology for various MSCs, the sputtered RuO_2 film was used to build the electrodes. As the result, the deposited RuO_2 films are uniform for all devices possessing 100 μm wide electrodes and a 50 μm interelectrode spacing on a glass substrate. The Moore design (MD) shows an energy density of 23.2 mWh cm^{-3} vs 17.5 mWh cm^{-3} compared to conventional interdigital structures (IDE). The authors did detailed investigations of the design differences which might contribute to the various performance. They distinguish the geometrical sides of the electrodes which are under influence of the electric field and classify two types: the top side, which is indeed the foot-print area of the electrodes, and the wall side, which is indeed the walls having areas aligned adjacently between two opposite electrodes. The side areas vary only by a small amount less than 18%. For example, the top side areas are 0.29 cm^2 vs 0.34 cm^2 for the IDE and MD. Meanwhile, the wall sides are 0.0037 cm^2 vs 0.0045 cm^2, for the IDE and MD. On the other hand, the IDE design has 32% higher capacitance and the additional percentage is explained by the higher number of sharp edges per electrode foot-print area in the case of fractal electrodes. Esfahani et al. [17] confirmed that the trend and also noticed that the fractal-based MSC has the worst capacitance retention compared to the interdigitated design: 65% of its initial capacitance after 5000 cycles vs 85% after 25,000 cycles. To further explore

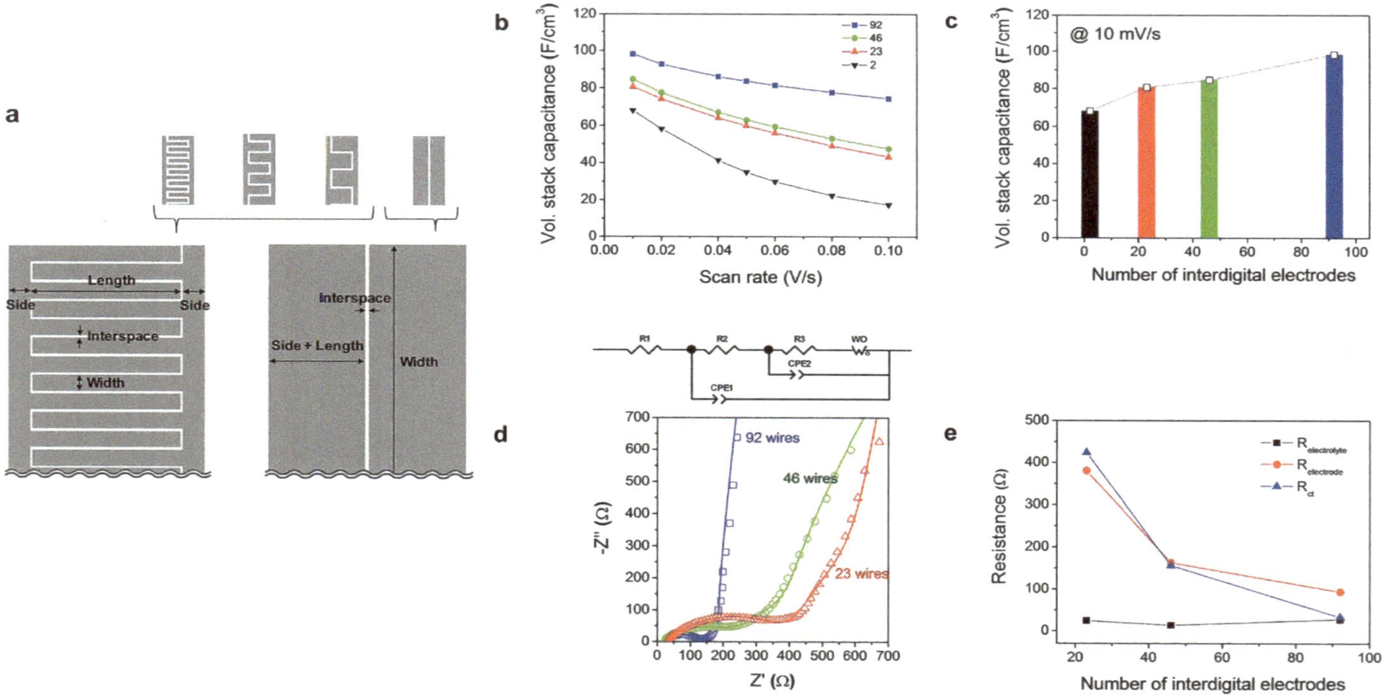

Fig. 8.1 The interdigitated architecture with the modified electrodes and related performance. *(Reprinted (adapted) with permission from J. Yoo, S. Byun, C.-W. Lee, C.-Y. Yoo, J. Yu, Precisely geometry controlled microsupercapacitors for ultrahigh areal capacitance, volumetric capacitance, and energy density, Chem. Mater. 30 (2018) 3979–3990. Copyright (2020) American Chemical Society).*

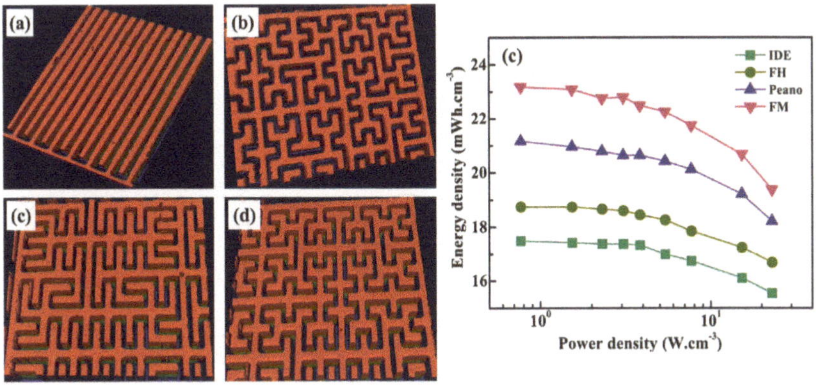

Fig. 8.2 Various electrode designs (*on the right*) including the interdigitated (IDE) (A) and fractal (FH, Peano, FM) (B–D) and related energy densities (*on left*). *(Reprinted with permission from M.K. Hota, Q. Jiang, Y. Mashraei, K.N. Salama, H.N. Alshareef, Fractal electrochemical microsupercapacitors, Adv. Electron. Mater. (2017) 1700185).*

the edge effect limitation on the MSC capacitive enhancement, Kumar et al. [18] proposed the sharp-edge electrode geometry resembling a star, that achieves a 68% enhancement in the electric field as compared to interdigitated electrodes. When combined with the carbon foam with pseudocapacitive metal oxides as the active material, the capacitance was enhanced by 235% compared with the interdigitated electrode MSC. For comparison, Tian et al. [19] demonstrated spiral-shaped MSCs, in which one spiral is situated closely to another spiral in the same plane. The authors fabricated three types of spiral geometries based on a circle and rectangular area and designed according to the principle concept of the Archimedean spiral (Fig. 8.3). Although the designs resulted in a bare enhancement of the performance, it should be underlined that the unconventional series/parallel combinations are easily realized in spiral-shaped design saving the footprint area.

Zhuang et al. [20] demonstrated that the size reduction and the narrowing of the interspace between electrodes are the most crucial factors in the enhancement of capacitive performance. Additionally, the new charge-storage mechanisms were discovered due to the considerable inner electric field force at the nanoscale. The electrodes were arranged in the interdigitated patterns from the ultrathin film (20 nm) synthesized using chemical vapor deposition (CVD) and then patterned by using focused ion beam (FIB) nanolithography technology. The minimum interspace between the adjacent electrodes equaled 100 nm. As compared to centimeter-scale

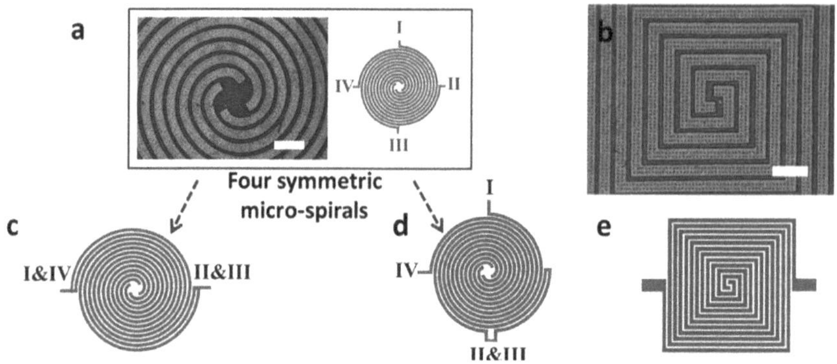

Fig. 8.3 Spiral-shaped electrodes of the MSCs. *(Reprinted with permission from X. Tian, M. Shi, X. Xu, M. Yan, L. Xu, A. Minhas-Khan, C. Han, L. He, L. Mai, Arbitrary shape engineerable spiral micropseudocapacitors with ultrahigh energy and power densities, Adv. Mater. 27 (2015) 7476–7482).*

MSCs, the obtained supercapacitor was smaller by a factor of 10^6 which is comparable to the size of a body cell, and exhibits remarkable nonfaradaic double-layer capacitance by the immense inner electric field force at the nanoscale. Since the electric field intensity is inversely proportional to the distance, the smaller interspace can induce much greater charge accumulation on or near the electrodes. As a result, the active material which is considered to be pseudocapacitive revealed the dominant mechanism in energy storage of the EDL comparing to the redox mechanism, thus enhancing the capacitive performances of the device.

8.2.2 Lithography

The lithography-based process flow is in general divided by two approaches. One is the local deposition of the active material onto lithographically patterned metallic current collectors on the insulating substrate, resulting in a continuous, uniform, ultrathin films. The deposition of the film is performed by CVD [7, 21, 22], electrochemical deposition [23, 24], spray or spin coating [25, 26], blade coating [27, 28], casting [29, 30].

The second is to first form a film of the active layer and to deposit the metal mask which functions as a metal current collector and the mask to preserve the material in selective removal to project the electrode patterns [31, 32]. This method allows for harsh processing of the film to modify its chemical and physical properties as presented by Wu et al. [33], who patterned heteroatom-doped carbon film. The active material (sulfur-decorated graphene) is established using the bottom-up approach, wherein the spin-coated film is covered by the sacrificial Au layer. Next, it went through the thermal treatment (at 400 and 800 °C) and patterned through the mask of the secondly deposited Au layer by oxidative etching in O_2.

The other techniques use pyrolysis of the patterned resist to make the porous carbon [34] or the laser beam as an alternative tool to UV or electron-beam projection to make fine patterning of the electrodes on the wafer-scale in polyimide [35]. Porous carbon electrodes were also laser machined using a CO_2-laser writer in which the channels were engraved on the film of active material and underlying metal current collector to form in-plan interdigitated electrodes [36]. Meanwhile, laser scribing was used to transform the insulating material such as graphene oxide (GO) into conductive electrodes of reduced graphene oxide (rGO) [37, 38].

Based on conventional approaches of photolithography, the planar or in-plane interdigital electrodes were readily constructed including all the MSC components on one single substrate. The wafer-scale processing (Fig. 8.4) was performed with a high resolution approaching submicrometers [39, 40]. On the other hand, it usually involves a complicated fabrication process, toxic chemical treatments, and high manufacturing costs. Also, the big challenge is to obtain the designed performance due to the laps and cracks (Fig. 8.5) which need better control at the nano and micro-scale. In the case of the lasers, the technique allows for in-situ patterning via the direct projection in the active film, and gains popularity to manufacture electrodes on a flexible and stretchable substrate [41, 42].

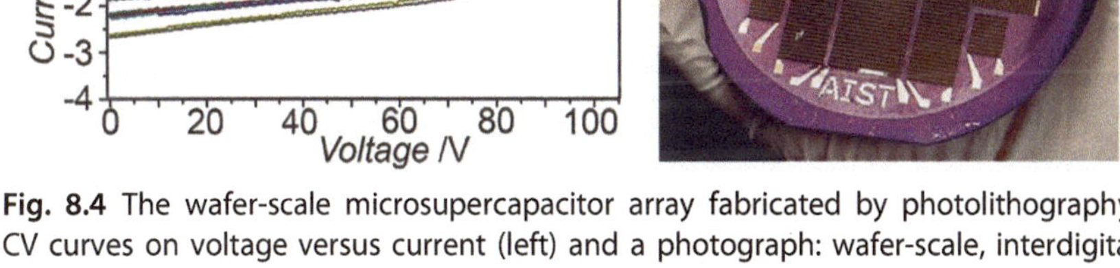

Fig. 8.4 The wafer-scale microsupercapacitor array fabricated by photolithography. CV curves on voltage versus current (left) and a photograph: wafer-scale, interdigital 4700 cells including MC pathways and cell barriers (right). *(Reprinted with permission from K.U. Laszczyk, K. Kobashi, S. Sakurai, A. Sekiguchi, D.N. Futaba, T. Yamada, K. Hata, Lithographically integrated microsupercapacitors for compact, high performance, and designable energy circuits, Adv. Energy Mater. (2015) 1500741).*

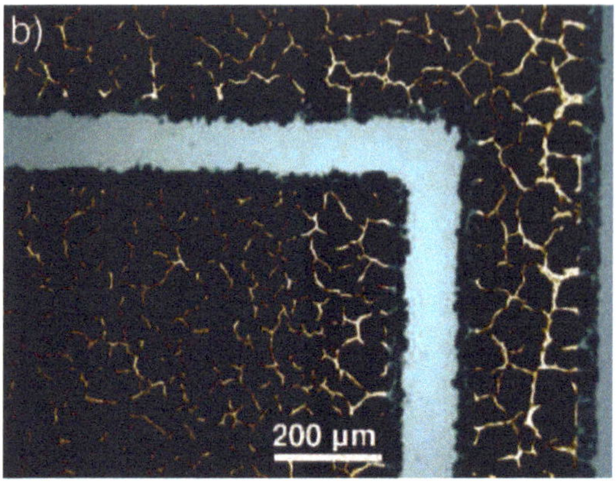

Fig. 8.5 Electrophoretically deposited onion-like carbon. *(Reprinted with permission from P. Hang, D. Pech, R. Lin, J.K. McDonough, M. Brunet, P.-L. Taberna, Y. Gogotsi, P. Simon, On-chip micro-supercapacitors for operation in a wide temperature range, Electrochem. Commun. 36 (2013) 53–56).*

8.2.3 Film printing

The continuous inkjet printing [43] and screen printing [44] became widely investigated as an efficient method for direct deposition of functional materials in predesigned patterns for the MSC electrodes. Both techniques are advantageous as compared to the spray-coating of which processing has much lower efficiency [26]. However, in the case of screen printing to achieve a resolution of a few micrometers, additional techniques must be used, such as, e.g., photolithography. The process is simple at low cost and applicable for large-scale fabrication of the transparent electronics for smart windows [45]. The main disadvantage of the technique is an average nanometer thickness of the electrodes. In the case of the ink-jet printing, the nonuniform morphology of the films is caused by a ubiquitous "coffee stain" effect, resulting in the disadvantage. It is characterized by the deposition of the majority of the solute only at the perimeter of the printed patterns [46, 47]. On the other hand, inkjet printing demands a printable ink with good knowledge of its fluid dynamic properties, including droplet formation and its spread on the substrate to achieve the designed patterns with the desired resolution. The obtained capacitance depends on the materials used and reaches up to 108 mF cm^{-2} [48]. So far, the inks for the MSCs contain such active material additives as activated carbon and nanocarbons [49, 50], metal oxide nanoparticles [51, 52], MXens [53, 54], and composites [55, 56]. The inks are used with the commercial inkjet office printers (Fig. 8.6) as well as a marker pen [57] also called direct ink writing (DIW) [58]. However, in the

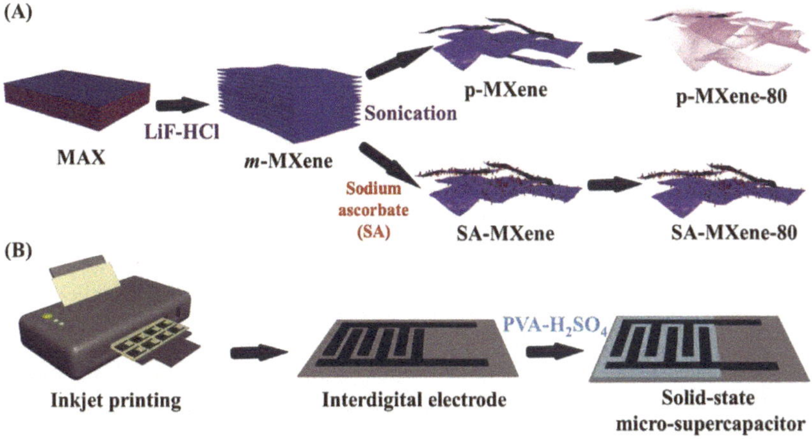

Fig. 8.6 Schematic representation of (A) the synthesis of stable SA-MXene nanocomposite and (B) solid-state micro-supercapacitor fabricated through inkjet of the SA-MXene nanocomposites. *(Reprinted with permission from C.-W. Wu, B. Unnikrishnan, I.-W.P. Chen, S.G. Harroun, H.-T. Chang, C.-C. Huang, Excellent oxidation resistive MXene aqueous ink for micro-supercapacitor application micro-supercapacitor application, Energy Storage Mater. 25 (2020) 563–571).*

latest, the inks need to be highly concentrated solutions or become rapidly cross-linked networks. The addition of viscous fillers or macromolecular cross-linking agents is necessary. This leads to unfavorable large agglomerates or inhomogeneous networks. On the other hand, the additional harsh removal procedures (e.g., high-temperature pyrolysis or chemical etching) are necessary to remove the additives altering intrinsic physicochemical properties [59].

8.3 Three-dimensional electrodes—Toward boosting the energy

The typical planar MSC electrodes are thin-film electrodes. Thus, the restrictions in the thickness of the electrodes provide insufficient energy for various applications [3]. On the other hand, the bottleneck of the typical planar MSCs is the limited number of solutions to boost the energy density. Among them so far there are the application of novel materials and their modifications for both electrodes and electrolytes [60, 61] as well as the connection of cells in parallel or in series [27], and the asymmetric configuration of electrodes [62]. However, none of these approaches can be submitted to the severe issue when considering the limited foot-print area of MSCs. For example, increasing the electrode area or utilizing parallel combinations of MSCs can readily increase the delivered capacitance and energy. But the approaches cause an increase in the footprint area, being on the contrary to the purpose of device miniaturization. Therefore, building three-dimensional (3D) electrodes is another approach to considerably enhance the ability of energy storage. A 3D architecture enlarges the surface areas and adds more active sites. Based on the concept the trend of electrodes has been created in optoelectronic devices, supercapacitors, and lithium-ion batteries [63–65] but considering the use of different materials. The 3D structures do not show the advantages for the on-board wireless devices for the IoT demand to be equipped with the power sources and energy storage together with the other specific units.

8.3.1 Electrodes with conducting buffers/spacers

The most common solution to increase the energy of 3D electrodes is to form the active material by mixing various conducting materials to make a "buffer" or a spacer between the nanosheets. Such a network becomes three-dimensional compared to the stacked sheets and prevents from a dense aggregation. It is self-restacking and aggregation of nanosheets due to the strong van der Waals interaction between adjacent nanosheets, which is inevitable during electrode fabrication processes. El-Kady et al. [60] combined the techniques of "top-down" LightScribe lithography with "bottom-up" selective electrodeposition to fabricate 3D hybrid electrodes for the MSC based on laser-scribed graphene (LSG) and pseudocapacitive MnO_2. First, the LSG was patterned into an interdigitated structure. Second, it was subsequently coated in situ with MnO_2 using a selective electrochemical deposition technique. In the latter, the mass loading was controlled by adjusting the deposition current and deposition time (at 0.25 mA/cm^2 an average deposition rate was estimated to be ~6 μg/min). The MnO_2 coating was made up of hundreds of ultrathin nanoflakes (10–20 nm) interconnected together to form a mesoporous film with a large accessible surface area, thus offering numerous electroactive sites available to the electrolyte which promotes fast surface Faradaic reactions. Moreover, the MnO_2 film was uniform and homogenous. The calculated area capacitance is approaching 400 mF/cm^{-2} at least four times higher than reported for the film electrodes, while the symmetric MSC resulted in the stack capacitance ~250 F cm^{-3}, in 1 M Na_2SO_4. The time constant was 23 ms.

The slow charge storage mechanism of pseudocapacitive additives, such as metal oxides (MnO_2, RuO_2, VS_2) or conducting polymers (PPy, PANI) hindered frequency response and rate handling capabilities. Therefore, the alternative is to use nonpseudocapacitive binders to the target electrode material. For example, Beidaghi et al. [26] made a solution of rGO and CNTs, with rGO making up the majority in weight ratio. The solutions were used for electrospray deposition through the mask on the metal interdigitated current collector. In the SEM images, several to hundreds of stacked graphene sheets were observed throughout the thickness of deposited films with almost no sign of stacked rGO sheets. The films had extended irregular porous structures, being potential diffusion channels. The time constant was 3.4 ms and the capacitance was 5.1 mF cm^{-2}, in 3 M KCl.

Another benefit of the use of mixed composition is to decrease the internal resistance between the nanosheets, especially as they are at least 1000 times shorter (hundreds nm) compared to the overall size of the electrodes (e.g., from several μm up to single mm). Ma et al. [66] sputtered Pt nanoparticles onto Co film, which was transformed into Co_3O_4/Pt used as the electrode material in so-called rapid thermal annealing process. The nanoparticles did not contribute to the capacitance but enhanced the transportation of electrons between the Co_3O_4 nanosheets by an order of magnitude. However, the capacitance increased about only 1.5 times from 22.3 to 35.7 F cm^{-3} in 1 M KOH. The time constant was 7.9 ms. Instead of the nanoparticles, Hu et al. [67] applied the continuous Au layers in change between MnO_2 films forming a layer stack of the interdigitated structures. The active and Au layers were 400 and 100 nm thick, respectively, and the total number of as-made layers was seven. The stack height is limited due to the difficulty of lift-off processing as the number of layers increased. The capacitance in 1 M Na_2SO_4 equaled 11.9 mF cm^{-2} and was comparable to similarly made MSC with a single layer of Au and MnO_2 (3.63 mF cm^{-2}). The configuration effectively increased the contact area between the active layer and the current collector improving the electrical conductivity, consequently lowering the charge transfer resistance. It should be noticed that in both cases, the exceeding number of the nanoparticles or metal layers blocked the storage mechanism, e.g., a thicker layer of Pt or Au, respectively, resulting in a lower or unchanged capacitance.

In the above examples, the expansion of the third dimension is not so spectacular and ranges from fractions of nanometers up to a few micrometers. Nonetheless, it makes a significant difference compared to the ultra-thin film, e.g., graphene, as a result, the obtained film forms a network of nanochannels enhancing permeabilities to ensure the efficient electrolyte penetration into the bulk electrodes thus fast ion diffusion. Recently, by the combination of various capacitive materials, the areal capacitance attained even ~665 mF cm^{-2} [68] by the integration of $Cu_{0.56}Co_{2.44}O_4$@MnO_2 core-shell nanoflowers (Fig. 8.7) and carbon nanotubes into a 3D hybrid asymmetric MSC.

8.3.2 CVD-based electrodes

The forest electrodes are as high as few hundreds of micrometers and made from silicon nanowires (SiNW, 50 μm high) [69] or silicon nanotrees (SiNTr, 50 μm high) [70], as well as CNTs [71] (200 μm high) and RuO_2 nanorods [72] (Fig. 8.8). Though the latest had an extremely lower height (1.9 μm) than the other forest-like electrodes in the MSCs, it showed a similar capacitance of ~21 mF cm^{-2} to the SiNTr electrodes ~14 mF cm^{-2}. On the other hand, the SiNW capacitance was at the level of 0.051 mF cm^{-2}, while the MSC with the electrodes of few-micrometer-thick SiNWs combined with PANI and graphene reached ~84 mF cm^{-2} [73]. In the case of Si as a building material, the advantage is integrated-circuit compatible. But due to entanglement morphology, the long Si forest is difficult to arrange in a planar structure with the short space between adjacent electrodes. Therefore, in most cases, we adopt the sandwich structure. Interdigitated structures are possible to be built up from cubic-like crystals as presented by Li et al. [74], hence overcoming the trouble with a long random needle-like forest. The interdigitated MSC was built basing on MOF-CVD films (up to 6 μm). The films were fabricated generally in three

Fig. 8.7 SEM images of typical of $Cu_{0.56}Co_{2.44}O_4$@MnO_2 core–shell nanoflowers [68].

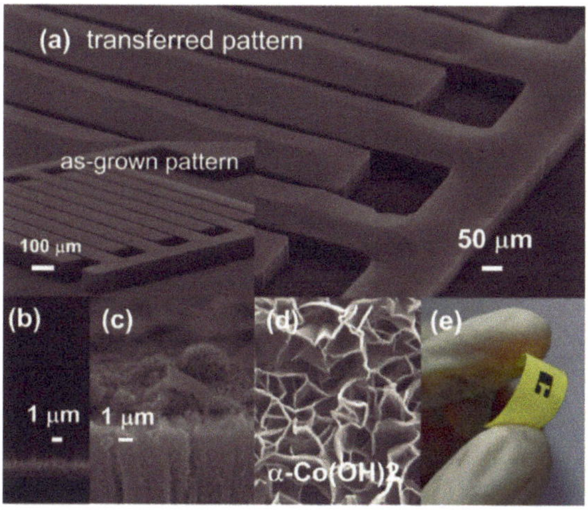

Fig. 8.8 (A–D) SEM Images of the interdigital electrodes and microstructures of as-grown CNT electrodes and (E) MSC on a polymer substrate. *(Reprinted with permission from C.-H. Chen, D.-S. Tsai, W.-H. Chung, K.-Y. Lee, Y.-M. Chen, Y-S. Huang, Electrochemical capacitors of miniature size with patterned carbon nanotube sand cobalt hydroxide, J. Power Sources 205 (2012) 510–515).*

steps: (1) electrodeposition of ZnO films on the metal current collector, (2) conversion of the ZnO into MOF, and (3) carbonization of the MOF to get highly porous carbon electrodes at 800 °C under a flowing environment of N_2 gas (Fig. 8.9).

8.3.3 3D scaffold

Because maintaining a reduced footprint area (<1 cm^2) with the expansion of the electrode thickness is complicated with micrometer separations, especially in terms of avoiding short circuit paths, the 3D scaffold—an array of high-aspect-ratio and dense microstructures (walls, trenches, pillars, tubes) is an alternative. The array consists of supporting nonconducting walls enclosing cells. Into the empty cells of the array, the active material is loaded—by the conformal deposition on the sites of the array walls. Usually, the deposited layer is 10–100 times thicker as opposed to that of nanostructured arrays. This topology offers robustness hence, it is possible to avoid, e.g., a short circuit, collapsing, or mechanical damage, which would take place for free-standing structures of the same height. On the other hand, the array should be carefully designed to avoid overloading of the active material, as well as ample free access of the ions from the electrolyte to the electrode surface. To maximize the electrode surface area of the loaded material, the

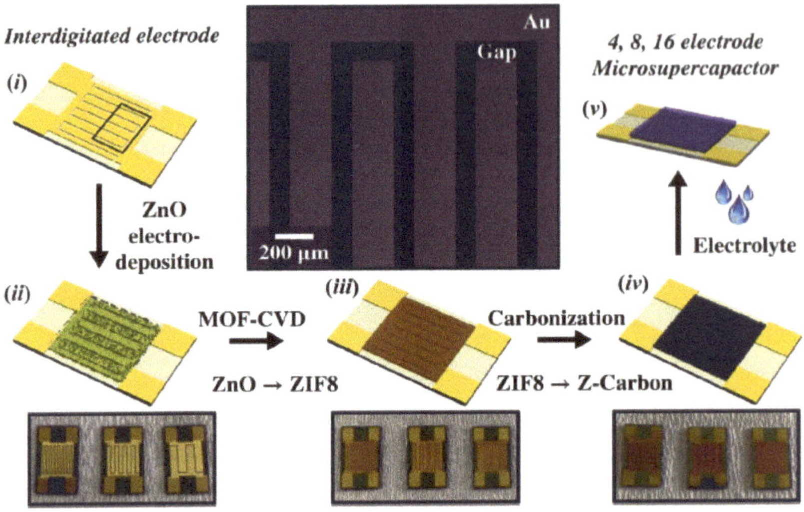

Fig. 8.9 MOF-CVD-based microsupercapacitor—a scheme illustrating the fabrication method. *(Reprinted with permission from Y. Li, H. Xie, J. Li, Y. Bando, Y. Yamauchi, J. Henzie, High performance nanoporous carbon microsupercapacitors generated by a solvent-free MOF-CVD method, Carbon 152 (2019) 688–696).*

strategy is to maximize the total number of empty cells by minimizing its cross-sectional dimension and the width of the wall separating cells. Lethien et al. [75] point the way to get the optimized design of the scaffold by the parameter named an area enhancement factor (AEF) as a function of the geometrical parameters of the 3D scaffold. They often recalled the accessed geometrical surface in case of the relevant two-dimension and three-dimension structure of the scaffold. The authors showed that for the array composed of nanopillars of 100 nm in diameter, the deeper the template, the higher the AEF, while reducing the spacing between two nanopillars improves the density of the template, i.e., the number of pillars per square centimeter. Eustache et al. [76] using a DRIE Bosch process made the scaffold in silicon forming tubal microstructures. The tubes were covered by conformal layers of Al_2O_3 and Pt by the atomic layer deposition, which acts as the insulator and current collector, respectively. Finally, a film of the active material (MnO_2) was electrodeposited. Films of different thicknesses could be made according to the number of deposition pulses, e.g., to obtain the thickness of ~300 nm needed about 3000 electrodeposition pulses. The measured capacitance was 670 mF cm^{-2} with a lifetime up to 15,000 cycles. Lei et al. [77] designed a honeycomb monolith with ultrathin and stiff alumina nano scaffold. The fabrication included nanoindentation, anodization, and a two-step etching process. The cell density was estimated to be 4.65×10^9 cells per square inch. The cell depth was about 25 μm, while cell-wall-thickness was about 40 nm including the current collector made of the deposited SnO_2 film. In the end, the active material (MnO_2) was electrodeposited revealing the capacitance of 137 mF cm^{-2}, in 1 M Na_2SO_4. Wang et al. [78] made the scaffold from Si by etching through a silicon wafer (400 μm thick) using the DRIE technique, and deposited carbon and hydrothermally grown MnO_2 film (15 nm) forming a Si/C/MnO_2 hybrid structure. It showed the capacitance of ~29 mF cm^{-2} in LiCl/PVA gel electrolyte.

The scaffolds without the dead volume being supporting walls were presented as well. Beidaghi et al. [79] form 140 μm thick 39 pillars (Fig. 8.10) having 50 μm diameter each. The pillars were patterned from photosensitive resin (SU-8), and next pyrolyzed to carbonize them, i.e., make them conductive. Finally, the residual material was removed by short exposure of the oxygen plasma, and the carbonized pillars (the current collector) were coated in the electrochemical process by PPy (the MSC electrodes) having up to 5 μm thickness. On the one finger (7 mm long) of the interdigitated structure (2.5 cm^2) as many as 39 pillars were placed. The resulting capacitance was 190 mF cm^{-2} in 0.1 M KCl. The scaffolds without supporting walls allow making higher (thicker) structures with better access of the ions but at the cost of lowering the mass loading per footprint area. In other words, there is a need to keep the balance between the design parameters

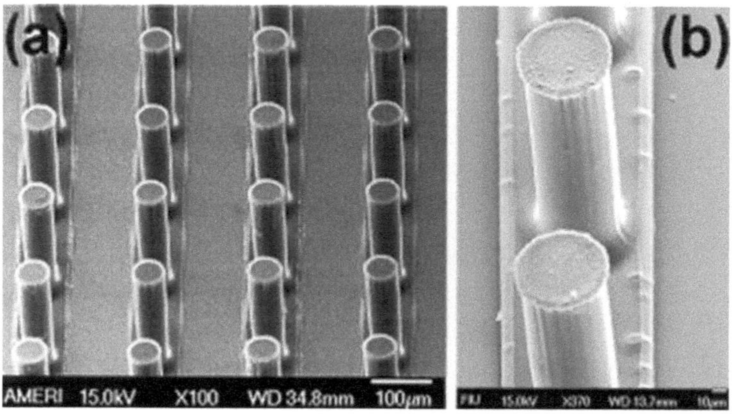

Fig. 8.10 SEM images of the pyrolyzed SU-8 pillars. *(Reprinted with permission from M. Beidaghi, C. Wang, Micro-supercapacitors based on three dimensional interdigital polypyrrole/C-MEMS electrodes, Electrochim. Acta 56 (2011) 9508–9514).*

of the pillars and the deposition technique of the active layer to avoid the short circuit problem. A response to this issue is a micrometer-thick porous carbon electrode having a fully connected uniform pore network (Fig. 8.11) presented by Kim et al. [80]. The authors also used the thick SU-8 resist, which was patterned by the 3D lithography process to form the uniform pore network, and next was carbonized by the pyrolysis. But, unlike the previous work, the obtained carbon electrodes were doped with boron, while the sputtered Au played the role of the current collector. The capacitance of the 3D porous carbon was ~7 mF cm^{-2} and the electrodes showed highly stable performance maintaining 98% of the initial capacitance up to 30,000 cycles in PVA- H_3PO_4 polymer gel electrolyte. The drawback of the

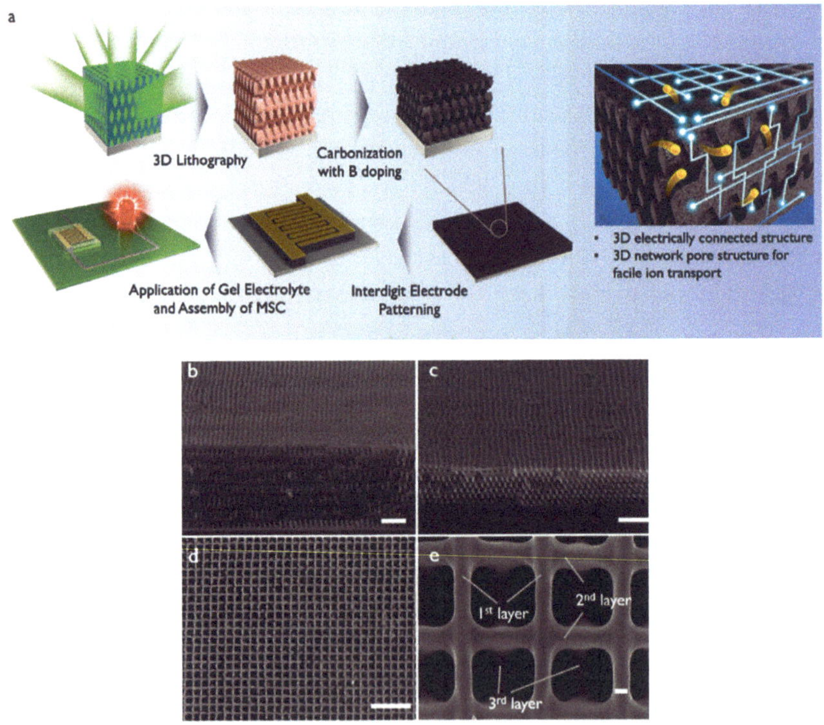

Fig. 8.11 Schematic fabrication of the MSCs composed of porous carbon electrodes (A). (B–E) Cross-sectional SEM images of the 3D electrodes (scale bar = 5 μm for (D) and 200 nm for (E)). *(Reprinted with permission from C. Kim, D.-Y. Kang, J.H. Moon, Full lithographic fabrication of boron-doped 3D porous carbon patterns for high volumetric energy density microsupercapacitors, Nano Energy 53 (2018) 182–188).*

method is the fact that the structure shrink was reported up to 80% due to high temperature (1000 °C) implemented during the pyrolysis. There is little knowledge about the effect itself and the trend governing the change of the designed dimensions of the network with the nanosized pores. Beside SU-8, the other photopolymers can be pyrolyzed to obtain the porous carbon electrodes, however, the electrode thickness is less than 1 μm [81].

Wang et al. [82] form the tubular pillars (80 μm high) composed of ruthenium oxide and carbon nanotubes (Fig. 8.12). A typical electrode is $2 \times 2\ mm^2$, containing 832 pillars of 50 μm in diameter. The distance between the centers of the adjacent pillars is 75 μm. In order to fabricate the pillars, the substrate was etched with inductively coupled plasma (ICP). Next, the surface of the Si substrate was coated with the insulation layer (thermally grown SiO_2), then the metal layer for the current collector (sputtered Au). In the next step, etching by a focused ion beam was performed to make a deep electrical isolation space between the opposed electrodes in the Au layer. Finally, the CNTs and RuO_2 were electrochemically co-deposited on pre-etched 3D microarrays of Si. The made pillars electrodes exhibited capacitance of $\sim$37 mF cm^{-2} in Na_2SO_4 solution. It is shown that the undesired phenomena so-called "grassy prominence" micro-structures—typically observed during ICP processes—enhanced the growth of the ripple-like CNT composite film. As the result, this ensured a good contact of the active material to the Au layer, hence lowering the internal resistance and contributing to the relatively high capacitance and stable cycling (6000 cycles). Gnerlich et al. [83] combined photolithographically patterned polyimide micropillars (5 μm tall) covered by Tobacco mosaic virus and selective deposition of ruthenium oxide (Fig. 8.13). The virus is self-assembled on the pillar surface and its role is to function as a template for the selective RuO_2 coating. Meanwhile, Ni film and partial Au film play the role of a sacrificial metal during the reduction of RuO_4 to RuO_2. Titanium nitride is chemically stable and adheres well to RuO_2. The plating reaction stops when the sacrificial nickel is exhausted. Thanks to this, the active material was deposited by simply submerging the substrate with the micropillars and virus template in the suspension of RuO_4 solution. Hence, the substrate does not require a seed layer or any external electrical current source during the deposition process as typical with electroplating. The coated film is specific to the micro-electrode areas. The capacitance of such an interdigitated electrode has been measured as high as 203 mF cm^{-2} with solid Nafion electrolyte. However, it should be noticed that RuO_4 solution is volatile and a strong oxidizer produces a hazardous vapor at room temperature.

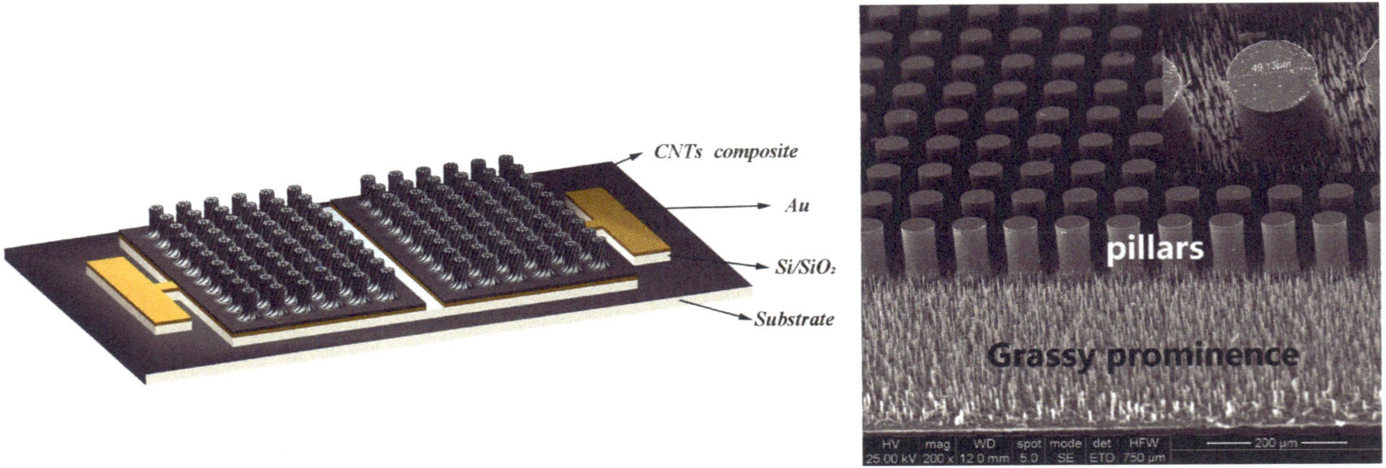

Fig. 8.12 Schematic drawing of the MSC with the tubular pillar of CNT and RuO_2 (left) and the SEM image of as-made electrodes with visible so-called "grassy prominence" (right). *(Reprinted with permission from X. Wang, Y. Yin, C. HaO, Z. You, A high-performance three-dimensional micro supercapacitor based on ripple-like ruthenium oxide-carbon nanotube composite films. Carbon 82 (2015) 436–445).*

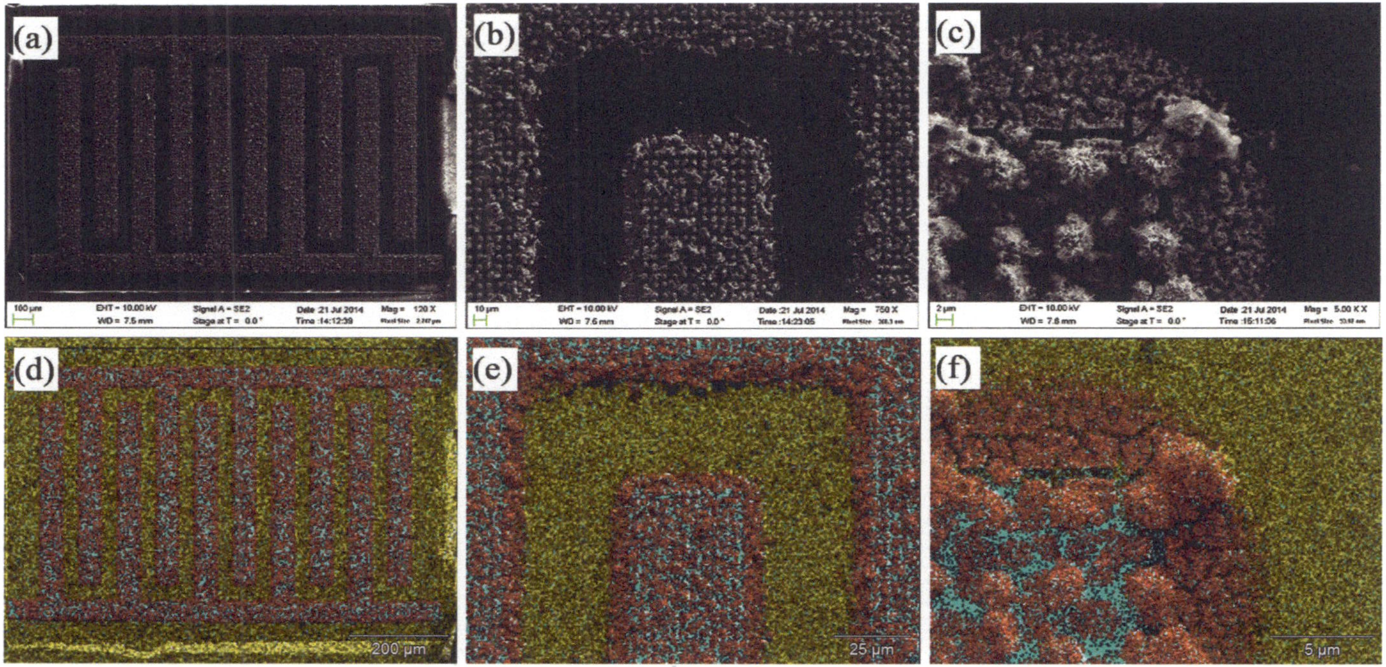

Fig. 8.13 Microsupercapacitor with micropillars covered by the Tobacco virus and RuO_2: SEM and EDX spectral map—red for ruthenium, yellow for carbon, and cyan for titanium. *(With the permission from M. Gnerlich, H. Ben-Yoav, J.N. Culver, D.R. Ketchum, R. Ghodssi, Selective deposition of nanostructured ruthenium oxide using Tobacco mosaic virus for micro-supercapacitors in solid Nafion electrolyte, J. Power Sources 293 (2015) 649–656).*

8.3.4 Microcavities

Maeng et al. [84] propose the reverse technique of mass loading through microcavities etched in the polymer (Parylene-C) substrate via plasma-assisted reactive ion etching. In contrast to the scaffolds being rigid structures, the choice of a polymer as a substrate enables this MSC for emerging flexible applications such as wearable and implantable wireless sensors. The microcavities have a depth and a diameter of 15 μm and form an array of 5616 cavities per single electrode finger ($2 \times 2\ mm^2$) in the interdigitated structure. Subsequently, first metal current collector (Ti/Au), next the active material (PANI fibers) were deposited on extra sidewall surface areas on which the active materials are grown in the form of nanofibers. The fabricated microsupercapacitors exhibited a maximum areal capacitance of $\sim$65 mF cm^{-2} by PVA/H_2SO_4. A similar solution revealing free-standing electrodes were chosen by Shen et al. [85] with a capacitance of $\sim$90 mF cm^{-2} in $NaNO_3$ electrolyte. In the substrate (Si) high aspect-ratio interdigital channels were etched leaving Si walls between fingers. Next, it was necessary to deposit an insulation layer to prevent leakage current through the substrate and a metal layer as a current collector. Then the active material (AC) loading into the channels between the Si walls was followed ensuring perfect separation of two adjacent electrodes. Finally, the Si walls were etched, freeing the self-supported electrode material and the air space was filled with electrolyte. The technology flow is presented briefly in Fig. 8.14. Although the deeper cavity is favorable for achieving a higher capacitance gain, it results in an increased substrate thickness, a prolonged etching time, and an increased manufacturing cost [84].

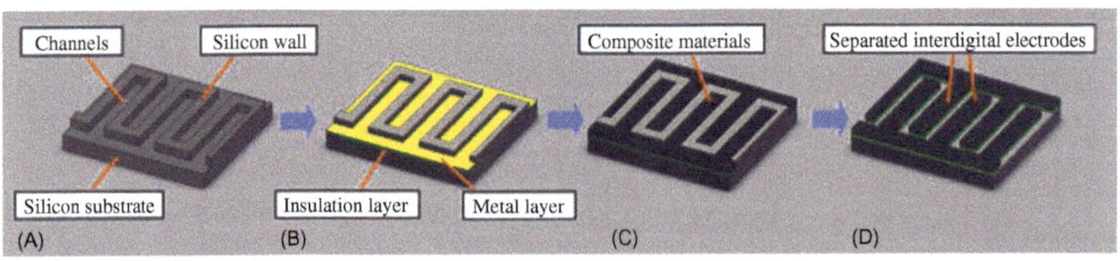

Fig. 8.14 The technology flow for the MSC with free-standing electrodes made of activated carbon (AC). *(Reprinted with permission of C. Shen, X. Wang, W. Zhang, F. Kang, A high-performance three-dimensional micro supercapacitor based on self-supporting composite materials, J. Power Sources 196 (2011) 10465–10471).*

8.3.5 Sponges

Among the 3D concepts have emerged for supercapacitor electrodes are foams, sponges, and aerogels. The foams and aerogels are also classified as building materials for the sponges [86]. The sponges are a kind of scaffolds, but instead of an array, composed of cells enclosed by high aspect ratio walls, and the base is built of the ordered or irregular network. The latest is often observed as a kind of scaffold with polyhedron empty spaces (pores) possessing similar dimensions. The sponge scaffold is basing on conducting or non-conducting materials. The conductive base materials are metals, e.g., commercially available foam sheets (Fig. 8.15), which is de-alloyed as nano-porous metal film [87]. The sponge for supercapacitors is synthesized in the CVD synthesis [88], hydrothermal reaction [89, 90], or obtained by combining the hydrogels with nanocarbons, such as rGO or CNT, which transform the hybrid into the porous free-standing monolith through the drying process [91]. The supercapacitor sponge is acquired also using a die of polyurethane, cellulose, PEDOT (poly (3,4-ethylenedioxythiophene), and others, which sucks the active material into the gaps through the subsequent immersion in a suspension and the process of drying [92–96]. The latest opens the route to compressible supercapacitors (CSCs, Fig. 8.16) with unique properties like the ability to work in different degrees of compression so that it can be easily embedded into smaller space than itself, together with

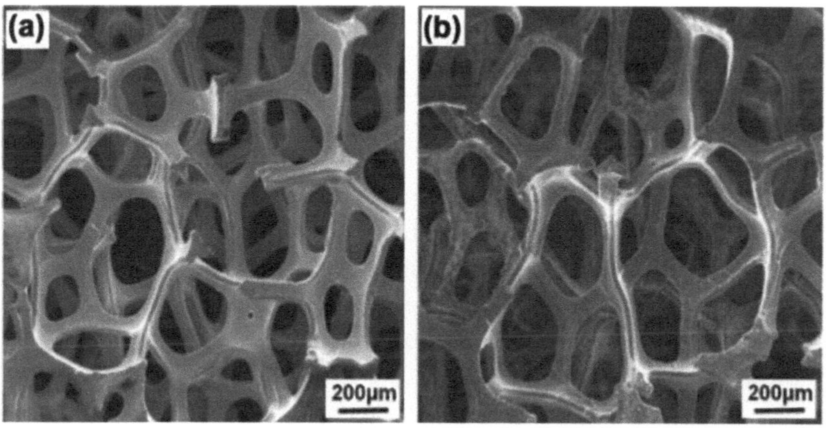

Fig. 8.15 (A) FESEM image of 3D nickel foam substrate. (B) FESEM image of 3D graphene foam without nickel substrate. *(Reprinted with permission from X. Dong, X. Wang, J. Wang, H. Song, X. Li, L. Wang, M.B. Chan-Park, C.M. Li, P. Chen, Synthesis of a MnO2–graphene foam hybrid with controlled MnO2 particle shape and its use as a supercapacitor electrode, Carbon 50 (2012) 4865–4870).*

Fig. 8.16 An example of compressible supercapacitor. *(Reprinted with permission from S.-W. Zhang, B.-S. Yin, C. Liu, Z.-B. Wang, D.-M. Gu, A lightweight, compressible and portable sponge-based supercapacitor for future power supply, Chem. Eng. J. 349 (2018) 509–521).*

good bending performance and extremely low mass [97]. To enhance the performance, besides the basing material and method used to make the sponge for supercapacitor electrodes, the active materials (EDL-like or pseudocapacitive) are electrodeposited [94, 98], and CVD or hydrothermally grown [99, 100] on the basing material. It is important then to optimize the coating technique—undercoating degrades the conductivity of the sponge electrode while overcoating closes the pores of the sponge and prevents the movement of the electrolyte ions. The sponge is barely investigated for micropatterning. Some attempts have been undertaken but not yet well developed, e.g., using slurry printing of an aerogel [101], single-mode fiber lasers [102], or laser cutting of the metal foam [103] or self-healable aerogel composite [104]. The latest was fabricated as a self-healable MSC consisting of MXene-rGO composite aerogel, delivering a large area-specific capacitance of 34.6 mF cm^{-2} at a scan rate of 1 mV s^{-1} and an outstanding cycling performance with capacitance retention up to 91% over 15,000 cycles. Shi et al. [105] proposed another method—to directly print the electrode material (commercial CNT ink) onto the ion-gel film (Fig. 8.17). The film does not have pores and acts as a substrate and gel-electrolyte simultaneously. The depth of the printed ink defined at once the thickness of the electrodes (50 μm). Though the electrodes are also limited in downscaling (mm in range), they offer additional self-healable features and stretchability with the easy method to release the electrodes in the drying process.

8.3.6 Additive manufacturing

The above-described 2D or 3D MSCs required mostly the multiple processes with the specialized machines and manpower with the knowledge to design process flow and to serve the equipment. Although it seems there

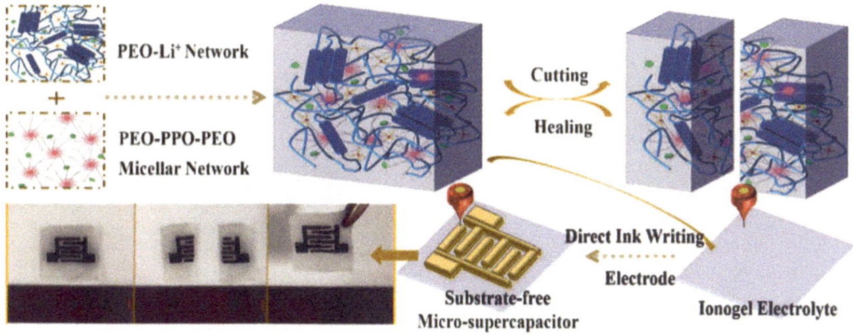

Fig. 8.17 A schematic picture of the fabrication process of the directly printed MSC in the ionogel. *(Reprinted with permission from Y. Shi, Y. Wang, Y. Gu, L. Zheng, S. Ma, X. Xu, Self-healable and stretchable ionogels serve as electrolytes and substrates for integrated all-in-one micro-supercapacitors, Chem. Eng. J. 392 (2020) 123645).*

is one standard route to make an interdigitated MSC, i.e., the subsequent deposition and patterning of the metal current collector, active material, followed by electrolyte filling, the know-how and tricks are hidden behind the special treatment condition of the selected materials. Therefore, additive manufacturing seems to be the simplest and most attainable technology which might develop novel three-dimensional designs of the electrodes. The additive manufacturing as a 3D printing technique also offers a low-cost fabrication protocol for future MSCs and a broad range of materials to be selected, including polymers, metal, and ceramics [59]. It evolved from direct ink writing (DIW) based on extrusion technique and offers far greater possibilities in assembling of spatial structures as developed for microbatteries (Fig. 8.18) [106]. This technique is highly sensitive to the viscoelastic properties of the ink, and has to satisfy several key design requirements. First, the ink should rapidly solidify once extruded from the nozzle. Second, the extruded ink must have good adhesion to the substrate. Then the solidified ink must be mechanically robust enough to maintain its shape under the load of the completed structure and the anticipated external stresses [107]. Besides, the ink for extrusion should possess a high viscosity to avoid slipping off the steep edges of curved surfaces. The micronozzle must be precisely controlled with its displacement to reproduce the design, yet without large agglomerates to achieve adequate resolution [108]. As the micronozzle moves to print the design, the ejected inks travel along the direction of the micronozzle movement, and spread then solidifies forming a line of continuous droplets unlike the discrete droplets of inkjet printing (Fig. 8.19) [109]. The printed structure is expanded in the third direction in a layer-

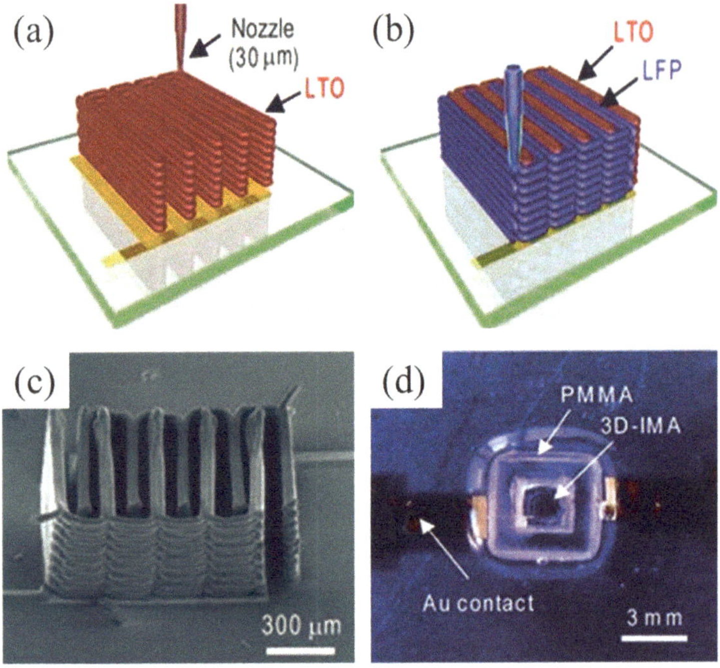

Fig. 8.18 The printed microbattery by extrusion of the inks. *(Reprinted with permission from C. Zhua, T. Liu, F. Qian, W. Chen, S. Chandrasekaran, B. Yao,Y. Song, E. B. Duoss, J. D. Kuntz, C. M. Spadaccini, M. A. Worsley, Y. Li, 3D printed functional nanomaterials for electrochemical energy storage, Nano Today 15 (2017) 107–120).*

by-layer manner by controlling the motion in three-axis directions of the 3D printer. The reproducibility of the design structures depends on the substrate temperature, the extrusion speed, the working distance, and the moving speed of the printing nozzle enabling to achieve 100-μm thick electrodes. The heating of the substrate also advances adhesion between printed layers and to improve the structural integrity without structure delamination or distortion [110]. Additionally, the dimensions of printed electrodes vary as a function of the micronozzle-basal distance or driving gas pressure. For example, as the applied gas pressure changes from 10 to 30 KPa, the height rises from 6.10 to 9.97 μm, and the width grows from 162.88 to 304.78 μm [111]. Meanwhile, Liu et al. [112] presented that the areal capacitance also depends on the number of extruded layers, e.g., the areal capacitance increased from 17.5 to 153.6 mF cm^{-2} with increasing the number of printed layers from 2 to 20. The porous hierarchical structure of the electrode materials was investigated with the highly conductive and

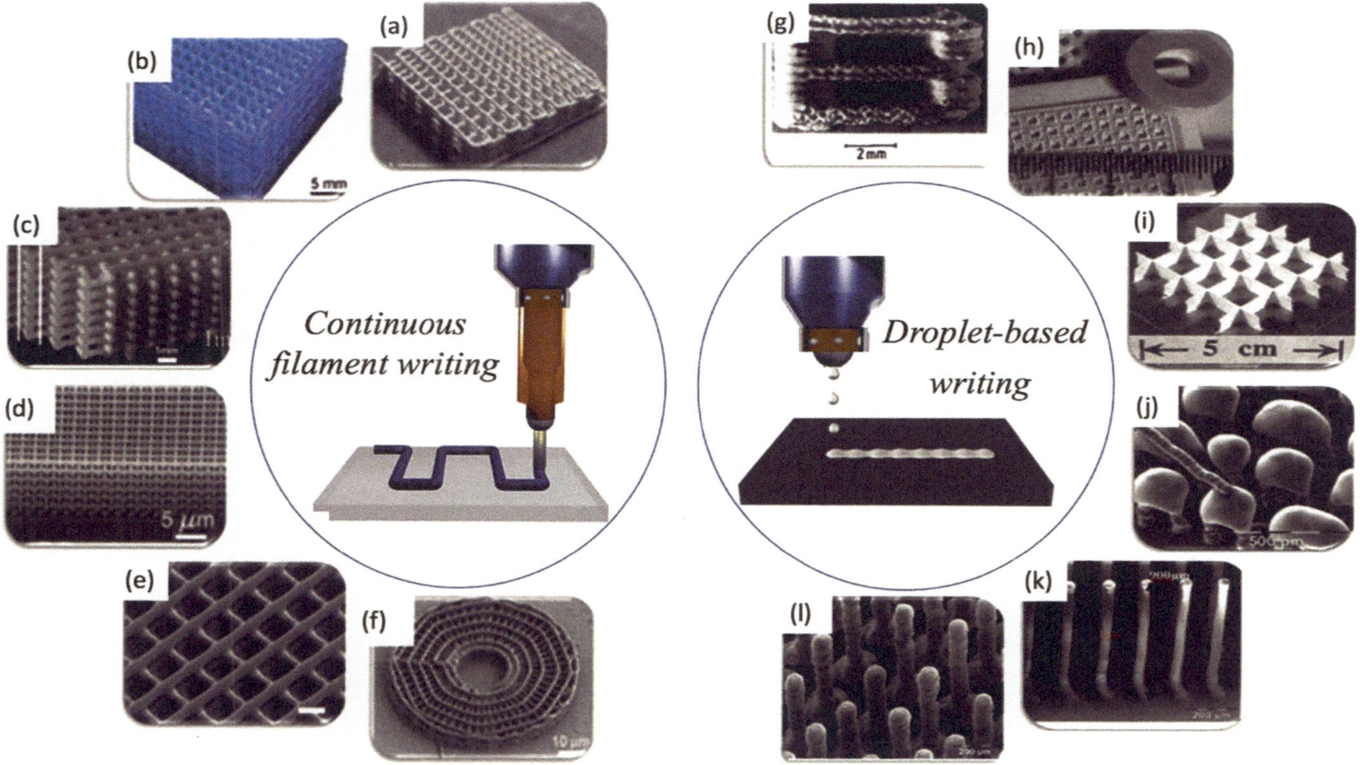

Fig. 8.19 Continuous writing vs droplet-based writing with the application of various inks. *(Reprinted with permission from M. Wei, F. Zhang, W. Wang, P. Alexandridis, C. Zhou, G. Wu, 3D direct writing fabrication of electrodes for electrochemical storage devices, J. Power Sources 354 (2017) 134–147).*

ion-permeable PEDOT: PSS, which facilitates the penetration and transfer of electrolyte ions in between the inner surface of electrode materials. Similar results were obtained by Sun et al. [113] who confirmed the increase of capacitance with the number of extruded layers (Fig. 8.20). Moreover, they show that the conductivity of rGO electrodes increases from ~58 S cm^{-1} for a single layer up to ~114 S cm^{-1} for the film consisting of four layers, where the single rGO layer is ~1 μm thick. It should be underlined here that the stacked printed films gave a different tendency from the previously reported results, where the decrease of the capacitance with electrode thickness was caused by the increased electrode resistance and the reduced electrolyte penetration and ion transfer for thick and dense electrode [114].

Yu et al. [111] form free-standing interdigitated MSC extruding 8 wt% CNTs slurry through a micronozzle controlled by an air-powered fluid dispenser. The electrodes were fabricated at a rate of 6 mm s^{-1}. Though the heated-up plate promotes removing the solvent, it does not leave the pure active material. Therefore the MSC electrodes need to be annealed at 120 °C for a night using a vacuum oven. Then MSC is filled with gel electrolyte (PVA-H_3PO_4) and detached from the glass base by the treatment of the glass with C_4F_8 plasma which makes its surface hydrophobic. The metal current collector was attached with an adhesive copper tape assisted by silver paint and reinforced with the Kapton tapes to ensure good electrical contact. The measured capacitance of the MSC is about 4 mF cm^{-2}, which is

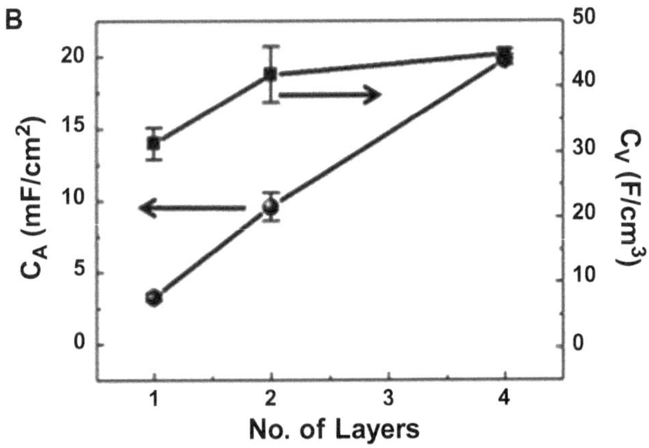

Fig. 8.20 Specific areal (CA) and volumetric (CV) capacitance vs the number of printed layers. *(Reprinted with permission from G. Sun, J. An, C.K. Chua, H. Pang, J. Zhang, P. Chen, Layer-by-layer printing of laminated graphene-based interdigitated microelectrodes for flexible planar micro-supercapacitors, Electrochem. Commun. 51 (2015) 33–36).*

remarkable taking into account the thickness of the electrodes (27 μm). Wang et al. [115] made a framework consisting of graphene and metal oxide. Although both materials work for electrochemical performance, the graphene support also plays the role of a current collector. First using GO-based ink, the graphene framework was printed by layer-by-layer extruding method on the already patterned silver paths, and frozen in liquid nitrogen, subsequently freeze-dried for 12 h. The freezing prevents the structure design from deformation caused by free drying in the air, and enables to make 825 μm thick electrodes (Fig. 8.21). The printed GO was reduced at 700 °C under Ar/H_2 gas flow for 6 h. Next, the pseudocapacitive nanosheets of metal oxide (NiO nanorods and MnO_2 nanosheets) were electrochemically and selectively deposited on the graphene by soaking the structure in the solution containing pseudocapacitive additives.

Another technique, Selective Laser Melting (SLM) was chosen by Zhao et al. [116] to obtain the interdigitated MSC electrodes, but different from the standard comb-type architecture coming from planar MSC (Fig. 8.22). By the SLM in the metal alloy, the vertical posts were fabricated at the top and the bottom electrode with the adjacent alignment. Each post has a 500 μm diameter and an 8 mm height. Next, the posts were coated with a polypyrrole (PPy) thin-film via electro-polymerization, and filled with

Fig. 8.21 The selected frame of Suppl. Movie presenting printing process of GO-ink on the Ag path. *(Reprinted with permission from T. Wang, L. Li, X. Tian, H. Jin, K. Tang, S. Hou, H. Zhou, X. Yu, 3D-printed interdigitated graphene framework as superior support of metal oxide nanostructures for remarkable micro-pseudocapacitors, Electrochim. Acta 319, (2019) 245–252).*

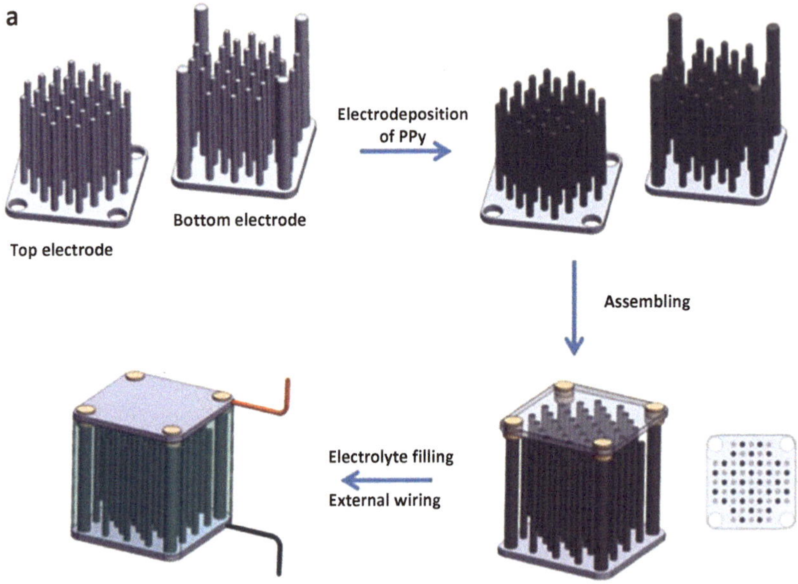

Fig. 8.22 The SLM-based MSC—schematic of the fabrication and electrodes configuration. *(Reprinted with permission from C. Zhao, C. Wang, R. Gorkin, S. Beirne, K. Shu, G.G. Wallace, Three dimensional (3D) printed electrodes for interdigitated supercapacitors, Electrochem. Commun. 41 (2014) 20–23).*

PVA–H_3PO_4 electrolyte showing the volumetric capacitance (2.4 F cm^{-3}) comparable to the 3D electrodes patterned by the lithography.

Seol et al. [117] did an end-to-end printing for two types of microsupercapacitors based on the EDL (activated carbon) and pseudocapacitance (graphene-Mn_3O_4). They printed the substrate, current collector, active layers, and a gel polymer electrolyte. The substrate was printed from the mixture of acrylonitrile butadiene styrene (ABS) and PVA by a fused deposition modeling (FDM). The substrate consists of the 3D-printed double layer of ABS and PVA. Subsequently, the metal current collector was printed by the Ag ink to build the interdigitated planar patterns, followed by the printing of the active material for the electrodes, and electrolyte (0.5 M $LiClO_4$ in PVA) on the top. The line width was 1 mm, while the space between the electrodes was 1.5 mm. The entire manufacturing was performed in a sequential single-instrument process having the potential for the integration of such MSC with other functional systems. For successful printing, the right recipe for the printable inks is crucial, including the proper viscosity, particle size, shear strength, and software settings, which are used to build the virtual model of the printed structure. If the viscosity is too low, the pattern rigidness becomes poor, which causes electrical shorts between the electrodes. If the viscosity is too high, clogging occurs resulting in poor printing stability. On the other hand, the high viscosity is beneficial to maintain the printed structure, which is required for printing multilayer 3D architectures, aside from the elastic modulus and stress yield due to the high concentration of the ink [118].

8.3.7 Stamping

The stamping offers rapid and scalable production of flexible MSCs with the potential of low-cost. This strategy affords 3D-printed stamps (Fig. 8.23) from rigid or polymer materials [119] with arbitrary shapes, i.e., pad or rolling stamps. The active material slurries are transformed into free-standing electrodes. The stamps can be designed to get various scaled up patterns allowing for multiplication of MSCs in seconds. Moreover, the bank of MSC cells using stamping was presented as well, wherein nine MSCs were in series connected with the electrode thickness maintained to be 20 μm [120].

Zhang et al. [121] demonstrated the facile method to stamp MSCs of arbitrary shapes exhibiting high areal capacitance of 61 mF cm^{-2}. For the process, they printed different stamp shapes and coated the stamp base with

c)

Fig. 8.23 The stamp templates used to make printed stamps of MSCs using so-called letterpress stamp method. *(Reprinted with permission from J. Coelho, M.P. Kremer, S. Pinilla, V. Nicolosi, An outlook on printed microsupercapacitors: technology status, remaining challenges, and opportunities, Curr. Opin. Electrochem. 21 (2020) 69–75, ISSN 2451-9103, https://doi.org/10.1016/j.coelec.2019.12.004).*

the MXene ink, followed by the firm pressing onto paper substrates and cold rolling process. Due to the analogy to the traditional paper press printing, the method is called the letterpress stamp method. After attaching Ag wires and casting the gel electrolyte, the devices were dried naturally at room temperature and become ready for testing. However, the uniformity of the finger electrodes needs to be further improved, by lowering the lateral dimension of the electrodes as well as inter-space to enhance the areal capacitance and rate performance. While stamps with a smaller finger gap can be designed, the as-stamped MSCs could be short-circuited if the gap is too small. To avoid this issue, Esfahani et al. [17] used another mold-like approach, in which they combined the stamp technique with blade coating. The authors stamped the patterns of the electrodes and a current collector in the top layer (parafilm) of the substrate, forming the channels which the depth was related to the top layer thickness. Hence, the space between the interdigitated electrodes was maintained by the walls of the parafilm. Into the imprinted channels, subsequently, a fine silver powder for the current collector and active material paste (graphene-based) were transferred (Fig. 8.24). The prepared MSCs display a large capacitance of ~14 mF cm^{-2} and capacitance retention of 85% after 25,000 cycles. The similar mold-like strategy was showed by Chen et al. [122]. On the conductive glass substrate, the imprinted channels reflecting the electrode design were stamped using an insulating paste to enable the selective Ni electroplating and electrodeposition of the active material (MnO_2). The resulting MSC has a capacitance of ~4 mF cm^{-2} (Fig. 8.25). To improve the integrity and resolution of transferring electrodes of MSCs, Wang et al. [123] combined the stamping with the gravure printing. First, the ink is applied to the base of the stamp followed by the gravure printing on the substrate leaving the designed patterns with the active material. Here, the ink for gravure printing is essential, which is strongly correlated with the adhesion between the ink and the substrate as well the ink and the stamp mold. Meanwhile, the gravure printing works as the inverse of the filling process, which includes emptying a filled cell by displacing liquid with air [124].

8.4 Encapsulation and packaging

The majority of the published works on microsupercapacitors are nonencapsulated and tested in the open air or a controlled glove-box atmosphere. Only a few projects address this issue. Meanwhile, the packaging of the MSC including its encapsulation becomes more complex when liquid electrolytes

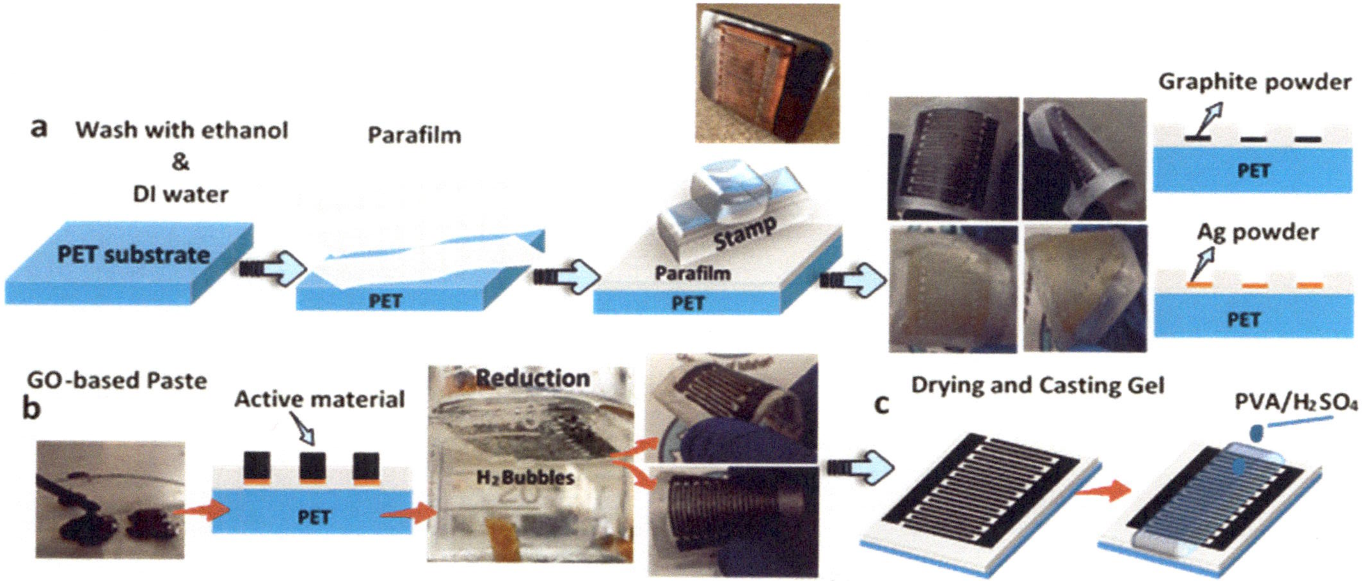

Fig. 8.24 MSC printed in a top layer of the substrate. *(Reprinted with permission from M.Z. Esfahani, M. Khosravi, Stamp-assisted flexible graphene-based micro-supercapacitors, J. Power Sources 462 (2020) 228166).*

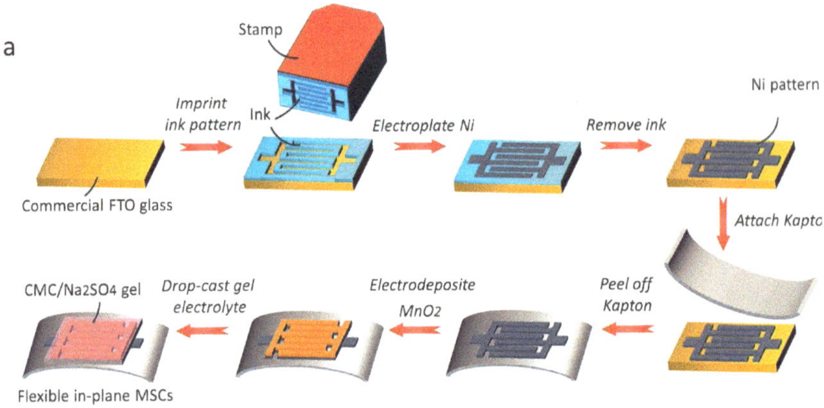

Fig. 8.25 The stamp and stamped insulating layer with the channels for the selective electroplating of Ni. *(Reprinted with permission from Y. Chen, X. Li, Z. Bi, G. Li, X. He, X. Gao, Stamp-assisted printing of nanotextured electrodes for high-performance flexible planar micro-supercapacitors, Chem. Eng. J. 353 (2018) 499–506).*

are used. Hence, one of the solutions is to build the MSC based on solid or gel electrolyte, integrating on the PCB [125, 126] at the cost of lowering the power performance. The whole electronics industry relies on the semiconducting methods to package electronics and mounting them on PCBs. Therefore, the earliest strategy was presented by Durou et al. [127] who propose hermetic sealing using the bonding process. Although it is a reliable technique, the use can be properly chosen except for textile electronics and all types of materials requiring high temperature or high voltage for the application. Toan et al. [73] demonstrated that this method is well adapted to the electrodes based on the native material of the substrate such as Si forest. The sophisticated method by micromachining was presented by In et al. [128], applying an origami structure. In this method, the electrodes and a metal current collector are patterned by the conventional photolithography. The etching process fabricated one of the halves of MSCs electrodes to be suspended, while another one was placed on the rigid substrate. The suspended part is made to move mechanically to bond opposed electrodes together (Fig. 8.26). The other idea is to make parylene film sealing [12, 84] or encapsulation in PDMS [129] and protective coating (e.g., Ecoflex) [130]. Additionally, there is another challenge: wireless packaging of the MSC [131]. It is especially important in clinical applications, where wires are first of all discomfort to the patient, and constitute the primary source of failure and infection. It is one of the key factors to successfully deploy

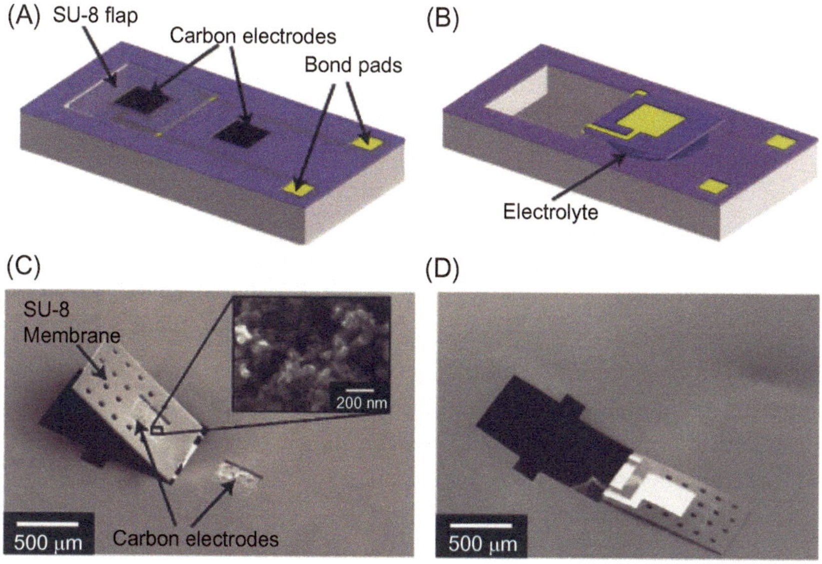

Fig. 8.26 Origami-like MSC. *(Reprinted with permission from H.J. In, S. Kumar, Y. Shao-Horn, G. Barbastathis, Origami fabrication of nanostructured three-dimensional devices: electrochemical capacitors with carbon electrodes, Appl. Phys. Lett. 88 (2006) 083104).*

the MSC into the health-related market, and some attempts have been already presented utilizing the wireless powering transfer [131] or the MSC integration with triboelectric nanogenerators [132].

8.5 Summary

The intensified research in materials for energy storage goes in parallel with revealing a wide spectrum of techniques to combine these materials for the microsupercapacitors. The current trend is to preserve the superb conductivity and morphology of the raw materials at once with relative ease and low cost of processing to respond to the severe demand for industrial level manufacture. The techniques still need to secure high scalability and throughput, without which the commercial boom of the microsupercapacitors will not happen. There are numerous issues to be solved. One of them is to improve the resolution in printing to seriously compromise device uniformity, yield, and miniaturization. Second, significant efforts must be placed toward the development of suitable suspensions to prevent the clog printing nozzles and the required performance of the as-made devices in accord with the nonuniform morphology (aggregates present inside electrodes). Third, the combination of the novel materials such as MXenes with additive manufacturing might incredibly boost the performance of the microsupercapacitors. However, the technique is still limited by a lower through output in terms of the volume and area than conventional technologies. Fourth, the sponge materials might address the mentioned above issue, but there is no adequate way to make them patternable in a convenient easy way preserving the fine resolution. For the capacitive performance enhancement, the scaling down is especially important as demonstrated for both electric-double-layer and pseudocapacitive mechanisms of the energy storage. Finally, full integration of the microsupercapacitors in circuitry, and daily life applications are still a challenging process to avoid the depletion of natural resources and make the technology flow environment friendly. Despite all these shortcomings, significant breakthroughs were achieved making the MSCs promising energy storage devices.

References

[1] Y.Q. Jiang, Q. Zhou, L. Lin, Planar MEMS supercapacitor using carbon nanotube forests, in: 2009 IEEE 22nd International Conference on Micro Electro Mechanical Systems, Sorrento, 2009, pp. 587–590.
[2] N. Liu, Y. Gao, Recent progress in micro-supercapacitors with in-plane interdigital electrode architecture, Small 13 (2017) 1701989.
[3] N.A. Kyeremateng, T. Brousse, D. Pech, Microsupercapacitors as miniaturized energy storage components for on-chip electronics, Nat. Nanotechnol. 12 (2017) 7–15.
[4] F. Bu, W. Zhou, Y. Xu, Y. Du, C. Guan, W. Huang, Recent developments of advanced micro-supercapacitors: design, fabrication and applications, NPJ Flex. Electr. 31 (2020) 1–16.
[5] Z. Liu, H.I. Wang, A. Narita, Q. Chen, Z. Mics, D. Turchinovich, M. Kläui, M. Bonn, K. Müllen, Photoswitchable micro-supercapacitor based on a diarylethene-graphene composite film, Am. Chem. Soc. 139 (2017) 9443–9446.
[6] J. Li, A. Levitt, N. Kurra, K. Juan, N. Noriega, X. Xiao, X. Wang, H. Wang, H.N. Alshareef, Y. Gogotsi, MXene-conducting polymer electrochromic microsupercapacitors, Energy Storage Mater. 20 (2019) 455–461.
[7] D. Pech, M. Brunet, H. Durou, P. Huang, V. Mochalin, Y. Gogotsi, P.-L. Taberna, P. Simon, Ultrahigh-power micrometre-sized supercapacitors based on onion-like carbon, Nat. Nanotechnol. 5 (2010) 651–654.
[8] J.-P. Randin, E. Yeager, Differential capacitance study on the edge orientation of pyrolytic graphite and glassy carbon electrodes, J. Electroanal. Chem. Interfacial Electrochem. 58 (1975) 313–322.
[9] C.-H. Kim, S.-I. Pyun, J.-H. Kim, An investigation of the capacitance dispersion on the fractal carbon electrode with edge and basal orientations, Electrochim. Acta 48 (2003) 3455–3463.
[10] W. Yuan, Y. Zhou, Y. Li, C. Li, H. Peng, J. Zhang, Z. Liu, L. Dai, G. Shi, The edge- and basal-plane-specific electrochemistry of a single-layer graphene sheet, Sci. Rep. 3 (2013) 2248.

[11] J. Yoo, S. Byun, C.-W. Lee, C.-Y. Yoo, J. Yu, Precisely geometry controlled micro-supercapacitors for ultrahigh areal capacitance, volumetric capacitance, and energy density, Chem. Mater. 30 (2018) 3979–3990.

[12] K.U. Laszczyk, D.N. Futaba, K. Kobashi, K. Hata, T. Yamada, A. Sekiguchi, The limitation of electrode shape on the operational speed of a carbon nanotube based micro-supercapacitor, Sustain. Energy Fuels 1 (2017) 1282–1286.

[13] W. Liu, C. Lu, X. Wang, R.Y. Tay, B.K. Tay, High-performance microsupercapacitors based on two-dimensional graphene/manganese dioxide/silver nanowire ternary hybrid film, ACS Nano 9 (2) (2015) 1528–1542.

[14] Z.-S. Wu, K. Parvez, X. Feng, K. Müllen, Photolithographic fabrication of high-performance photolithographic fabrication of high-performance all-solid-state graphene-based planar microsupercapacitors with different interdigital fingers, J. Mater. Chem. A 2 (2014) 8288–8293.

[15] J. Zhu, B. Huang, C. Zhao, H. Xu, S. Wang, Y. Chen, L. Xie, L. Chen, Benzoic acid-assisted substrate-free synthesis of ultrathin nanosheets assembled two-dimensional porous Co3O4 thin sheets with 3D hierarchical micro-/nano-structures and enhanced performance as battery-type materials for supercapacitors, Electrochim. Acta 313 (2019) 194–204.

[16] M.K. Hota, Q. Jiang, Y. Mashraei, K.N. Salama, H.N. Alshareef, Fractal electrochemical microsupercapacitors, Adv. Electron. Mater. 1700185 (2017).

[17] M.Z. Esfahani, M. Khosravi, Stamp-assisted flexible graphene-based micro-supercapacitors, J. Power Sources 462 (2020) 228166.

[18] S. Kumar, S. Telpande, V. Manikandan, P. Kumar, A. Misra, Novel electrode geometry for high performance CF/Fe2O3 based planar solid state micro-electrochemical capacitors, Nanoscale 12 (2020) 19438.

[19] X. Tian, M. Shi, X. Xu, M. Yan, L. Xu, A. Minhas-Khan, C. Han, L. He, L. Mai, Arbitrary shape engineerable spiral micropseudocapacitors with ultrahigh energy and power densities, Adv. Mater. 27 (2015) 7476–7482.

[20] P. Zhuang, Y. Sun, L. Li, M. Oliver, L. Chee, P. Dong, L. Pei, H. Chu, Z. Sun, J. Shen, M. Ye, P.M. Ajayan, FIB-patterned nano-supercapacitors: minimized size with ultrahigh performances, Adv. Mater. 32 (2020) 1908072.

[21] C.-C. Liu, D.-S. Tsai, W.-H. Chung, K.-W. Li, K.-Y. Lee, Y.-S. Huang, Electrochemical micro-capacitors of patterned electrodes loaded with manganese oxide and carbon nanotubes, J. Power Sources 196 (13) (2011) 5761–5768.

[22] Y. Gogotsi, A. Nikitin, H. Ye, W. Zhou, J.E. Fischer, B. Yi, H.C. Foley, M.W. Barsoum, Nanoporous carbide-derived carbon with tunable pore size, Nat. Mater. 2 (2003) 591–594.

[23] S. Makino, Y. Yamauchi, W. Sugimoto, Synthesis of electro-deposited ordered mesoporous RuOx using lyotropic liquid crystal and application toward micro-supercapacitors, J. Power Sources 227 (2013) 153–160.

[24] P. Hang, D. Pech, R. Lin, J.K. McDonough, M. Brunet, P.-L. Taberna, Y. Gogotsi, P. Simon, On-chip micro-supercapacitors for operation in a wide temperature range, Electrochem. Commun. 36 (2013) 53–56.

[25] C. Shen, X. Wang, W. Zhang, F. Kang, Direct prototyping of patterned nanoporous carbon: a route from materials to on-chip devices, Sci. Rep. 3 (2013) 2294.

[26] M. Beidaghi, C. Wang, Micro-supercapacitors based on interdigital electrodes of reduced graphene oxide and carbon nanotube composites with ultrahigh power handling performance, Adv. Funct. Mater. 22 (2012) 4501–4510.

[27] K.U. Laszczyk, K. Kobashi, S. Sakurai, A. Sekiguchi, D.N. Futaba, T. Yamada, K. Hata, Lithographically integrated microsupercapacitors for compact, high performance, and designable energy circuits, Adv. Energy Mater. 1500741 (2015).

[28] N. Wang, J. Liu, Y. Zhao, M. Hu, G. Shan, High-performance asymmetric microsupercapacitors based on electrodeposited MnO2 and N-doped graphene, Nanotechnology 30 (2019) 235403.

[29] S.-K. Kim, H.-J. Koo, A. Lee, P.V. Braun, Selective wetting-induced micro-electrode patterning for flexible micro-supercapacitors, Adv. Mater. 26 (2014) 5108–5112.

[30] J.-H. Sung, S.-J. Kim, K.-H. Lee, Fabrication of all-solid-state electrochemical microcapacitors, J. Power Sources 133 (2) (2004) 312–319.

[31] L. Wei, N. Nitta, G. Yushin, Lithographically patterned thin activated carbon films as a new technology platform for on-chip devices, ACS Nano 7 (2013) 6498–6506.

[32] X. Wang, B.D. Myers, J. Yan, G. Shekhawat, V. Dravid, P.S. Lee, Manganese oxide micro-supercapacitors with ultra-high areal capacitance, Nanoscale 5 (2013) 4119.

[33] Z.-S. Wu, Y.-Z. Tan, S. Zheng, S. Wang, K. Parvez, J. Qin, X. Shi, C. Sun, X. Bao, X. Feng, K. Müllen, Bottom-up fabrication of sulfur-doped graphene films derived from sulfur-annulated nanographene for ultrahigh volumetric capacitance micro-supercapacitors, J. Am. Chem. Soc. 139 (2017) 4506–4512.

[34] M.S. Kim, B. Hsia, C. Carraro, R. Maboudian, Flexible micro-supercapacitors from photoresist-derived carbon electrodes on flexible substrates, in: 2014 IEEE 27th International Conference on Micro Electro Mechanical Systems (MEMS), San Francisco, CA, 2014, pp. 389–392.

[35] J.B. In, B. Hsia, J.-H. Yoo, S. Hyun, C. Carraro, R. Maboudian, C.P. Grigoropoulos, Facile fabrication of flexible all solid-state micro-supercapacitor by direct laser writing of porous carbon in polyimide, Carbon 83 (2015) 144–151.

[36] H.-C. Huang, C.-J. Chung, C.-T. Hsieh, P.-L. Kuo, H. Teng, Laser fabrication of all-solid-state microsupercapacitors with ultrahigh energy and power based on hierarchical pore carbon, Nano Energy 21 (2016) 90–105.

[37] W. Gao, N. Singh, L. Song, Z. Liu, A.L.M. Reddy, L. Ci, R. Vajtai, Q. Zhang, B. Wei, P.M. Ajayan, Direct laser writing of micro-supercapacitors on hydrated graphite oxide films, Nat. Nanotechnol. 6 (2011) 496–500.

[38] M. El-Kady, R. Kaner, Scalable fabrication of high-power graphene micro-supercapacitors for flexible and on-chip energy storage, Nat. Commun. 4 (2013) 1475.

[39] T.M. Dinh, K. Armstrong, D. Guay, D. Pech, High-resolution on-chip supercapacitors with ultra-high scan rate ability, J. Mater. Chem. A 2 (2014) 7170.

[40] E. Kim, B.-J. Lee, K. Maleski, Y. Chae, Y. Lee, Y. Gogotsi, C.W. Ahn, Microsupercapacitor with a 500 nm gap between MXene/CNT electrodes, Nano Energy 81 (2021) 105616.

[41] N. Kamboj, T. Purkait, M. Das, S. Sarkar, K.S. Hazra, R.S. Dey, Ultralong cycle life and outstanding capacitive performance of a 10.8 V metal free micro-supercapacitor with highly conducting and robust laser-irradiated graphene for an integrated storage device, Energ. Environ. Sci. 12 (2019) 2507.

[42] S. Park, H. Lee, Y.-J. Kim, P.S. Lee, Fully laser-patterned stretchable microsupercapacitors integrated with soft electronic circuit components, NPG Asia Mater. 10 (2018) 959–969.

[43] G.D. Martin, S.D. Hoath, I.M. Hutchings, Inkjet printing—the physics of manipulating liquid jets and drops, J. Phys. Conf. Ser. 105 (2008), 012001.

[44] H. Li, J. Liang, Recent development of printed micro-supercapacitors: printable materials, printing technologies, and perspectives, Adv. Mater. 32 (2020) 1805864.

[45] S.S. Delekta, A.D. Smith, J. Li, M. Östling, Inkjet printed highly transparent and flexible graphene micro-supercapacitors, Nanoscale 9 (2017) 6998.

[46] D. Soltman, V. Subramanian, Inkjet-printed line morphologies and temperature control of the coffee ring effect, Langmuir 24 (2008) 2224–2231.

[47] J.A. Lim, W.H. Lee, H.S. Lee, J.H. Lee, Y.D. Park, K. Cho, Self-organization of ink-jet-printed triisopropylsilylethynyl pentacene via evaporation-induced flows in a drying droplet, Adv. Funct. Mater. 18 (2008) 229–234.

[48] C.-W. Wu, B. Unnikrishnan, I.-W.P. Chen, S.G. Harroun, H.-T. Chang, C.-C. Huang, Excellent oxidation resistive MXene aqueous ink for micro-supercapacitor application micro-supercapacitor application, Energy Storage Mater. 25 (2020) 563–571.

[49] D. Pech, M. Brunet, P. Taberna, P. Simon, N. Fabre, F. Mesnilgrente, V. Conedéra, H. Durou, Elaboration of a microstructured inkjet-printed carbon electrochemical capacitor, J. Power Sources 195 (4) (2010) 1266–1269.

[50] S.S. Delekta, M. Östling, J. Li, Wet transfer of inkjet printed graphene for microsuper-capacitors on arbitrary substrates, ACS Appl. Energy Mater. 2 (1) (2019) 158–163.

[51] P. Giannakou, M.G. Masteghin, R.C.T. Slade, S.J. Hinder, M. Shkunov, Energy storage on demand: ultra-high-rate and high-energy-density inkjet-printed NiO micro-supercapacitors, J. Mater. Chem. A 7 (2019) 21496–21506.

[52] Y. Wang, Y.-Z. Zhang, D. Dubbink, J.E. ten Elshof, Inkjet printing of δ-MnO2 nanosheets for flexible solid-state micro-supercapacitor, Nano Energy 49 (2018) 481–488.

[53] H. Huang, X. Chu, H. Su, H. Zhang, Y. Xie, W. Deng, N. Chen, F. Liu, H. Zhang, B. Gu, W. Deng, W. Yang, Massively manufactured paper-based all-solid-state flexible micro-supercapacitors with sprayable MXene conductive inks, J. Power Sources 415 (2019) 1–7.

[54] H. Huang, J. He, Z. Wang, H. Zhang, L. Jin, N. Chen, Y. Xie, X. Chu, B. Gu, W. Deng, W. Yang, Scalable, and low-cost treating-cutting-coating manufacture platform for MXene-based on-chip micro-supercapacitors, Nano Energy 69 (2020) 104431.

[55] P. Sundriyal, S. Bhattacharya, Scalable micro-fabrication of flexible, solid-state, inexpensive, and high-performance planar micro-supercapacitors through inkjet printing, ACS Appl. Energy Mater. 2 (3) (2019) 1876–1890.

[56] L. Qin, Q. Tao, X. Liu, M. Fahlman, J. Halim, P.O.Å. Persson, J. Rosen, F. Zhang, Polymer-MXene composite films formed by MXene-facilitated electrochemical polymerization for flexible solid-state microsupercapacitors, Nano Energy 60 (2019) 734–742.

[57] Q. Jiang, N. Kurra, H.N. Alshareef, Marker pen lithography for flexible and curvilinear on-chip energy storage, Adv. Funct. Mater. 25 (2015) 4976–4984.

[58] Z. Liu, Z.-S. Wu, S. Yang, R. Dong, X. Feng, K. Müllen, Ultraflexible in-plane micro-supercapacitors by direct printing of solution-processable electrochemically exfoliated graphene, Adv. Mater. 28 (2016) 2217–2222.

[59] K. Fu, Y. Yao, J. Dai, L. Hu, Progress in 3D printing of carbon materials for energy related applications, Adv. Mater. 29 (2017) 1603486.

[60] M.F. El-Kady, M. Ihnsa, M. Li, J.Y. Hwang, M.F. Mousavi, L. Chaney, A.T. Lech, R.B. Kaner, Engineering three-dimensional hybrid supercapacitors and microsupercapacitors for high-performance integrated energy storage, PNAS 112 (4) (2015) 4233–4238.

[61] L. Xia, L. Yu, D. Hua, G.Z. Chen, Electrolytes for electrochemical energy storage, Mater. Chem. Front. 1 (2017) 584.

[62] Y. Shao, M.F. El-Kady, J. Sun, Y. Li, Q. Zhang, M. Zhu, H. Wang, B. Dunn, R.B. Kaner, Design and mechanisms of asymmetric supercapacitors, Chem. Rev. 118 (2018) 9233–9280.

[63] X.H. Lu, T. Zhai, X.H. Zhang, Y.Q. Shen, L.Y. Yuan, B. Hu, L. Gong, J. Chen, Y.H. Gao, J. Zhou, Y.X. Tong, Z.L. Wang, WO3-x@au@MnO2 core-shell nanowires on carbon fabric for high-performance flexible supercapacitors, Adv. Mater. 24 (2012) 938.

[64] Liu, et al., Hierarchical three-dimensional ZnCo2O4 nanowire arrays/carbon cloth anodes for a novel class of high-performance flexible lithium-ion batteries, Nano Lett. 12 (6) (2012) 3005–3011.

[65] C. Zhou, Y. Zhang, Y. Li, J. Liu, Construction of high-capacitance 3D CoO@polypyrrole nanowire array electrode for aqueous asymmetric supercapacitor, Nano Lett. 13 (5) (2013) 2078–2085.

[66] X. Ma, S. Feng, L. He, M. Yan, X. Tian, Y. Li, C. Tang, X. Hong, L. Mai, Rapid, all dry microfabrication of three-dimensional Co3O4/Pt nanonetworks for high-performance microsupercapacitors, Nanoscale 9 (2017) 11765.
[67] H. Hu, Z. Pei, H. Fan, C. Ye, 3D interdigital au/MnO 2/au stacked hybrid electrodes for on-chip microsupercapacitors, Small 12 (22) (2016) 3059–3069.
[68] F. Li, M. Huang, J. Wang, J. Qu, Y. Li, L. Liu, V.K. Bandari, Y. Hong, B. Sun, M. Zhu, F. Zhu, Y.X. Zhang, O.G. Schmidt, On-chip 3D interdigital micro-supercapacitors with ultrahigh areal energy density, Energy Storage Mater. 27 (2020) 17–24.
[69] F. Thissandier, P. Gentile, N. Pauc, E. Hadji, A. Le Comte, O. Crosnier, G. Bidan, S. Sadki, T. Brousse, Highly N-doped silicon nanowires as a possible alternative to carbon for on-chip electrochemical capacitors, Electrochemistry 81 (10) (2013) 777–782.
[70] D. Aradilla, D. Gaboriau, G. Bidan, P. Gentile, M. Boniface, D. Dubal, P. Gómez-Romero, J. Wimberg, T.J.S. Schubert, S. Sadki, An innovative 3-D nanoforest heterostructure made of polypyrrole coated silicon nanotrees for new high performance hybrid micro-supercapacitors, J. Mater. Chem. A 3 (2015) 13978–13985.
[71] C.-H. Chen, D.-S. Tsai, W.-H. Chung, K.-Y. Lee, Y.-M. Chen, Y.-S. Huang, Electrochemical capacitors of miniature size with patterned carbon nanotube sand cobalt hydroxide, J. Power Sources 205 (2012) 510–515.
[72] C.C. Liu, D.S. Tsai, D. Susanti, W.C. Yeh, Y.S. Huang, F.J. Liu, Planar ultracapacitors of miniature interdigital electrode loaded with hydrous RuO2 and RuO2 nanorods, Electrochim. Acta 55 (20) (2010) 5768–5774.
[73] N.V. Toan, T.T.K. Tuoi, J. Li, N. Inomata, T. Ono, Liquid and solid states on-chip micro-supercapacitors using silicon nanowire-graphene nanowall-pani electrode based on microfabrication technology, Mater. Res. Bull. 131 (2020) 1109772.
[74] Y. Li, H. Xie, J. Li, Y. Bando, Y. Yamauchi, J. Henzie, High performance nanoporous carbon microsupercapacitors generated by a solvent-free MOF-CVD method, Carbon 152 (2019) 688–696.
[75] C. Lethien, J. Le Bideau, T. Brousse, Challenges and prospects of 3D micro-supercapacitors for powering the internet of things, Energ. Environ. Sci. 12 (2019) 96.
[76] E. Eustache, C. Douard, R. Retoux, C. Lethien, T. Brousse, MnO2 thin films on 3D scaffold: microsupercapacitor electrodes competing with "bulk" carbon electrodes, Adv. Energy Mater. (2015), 1500680.
[77] Z. Lei, L. Liu, H. Zhao, F. Liang, S. Chang, L. Li, Y. Zhang, Z. Lin, J. Kröger, Y. Lei, Nanoelectrode design from microminiaturized honeycomb monolith with ultrathin and stiff nanoscaffold for high-energy microsupercapacitors, Nat. Commun. 11 (2020) 299.
[78] Y. Wang, L. Sun, D. Xiao, H. Du, Z. Yang, X. Wang, L. Tu, C. Zhao, F. Hu, B. Lu, Silicon-based 3D all-solid-state micro-supercapacitor with superior performance, ACS Appl. Mater. Interfaces 12 (2020) 43864–43875.
[79] M. Beidaghi, C. Wang, Micro-supercapacitors based on three dimensional interdigital polypyrrole/C-MEMS electrodes, Electrochim. Acta 56 (2011) 9508–9514.
[80] C. Kim, D.-Y. Kang, J.H. Moon, Full lithographic fabrication of boron-doped 3D porous carbon patterns for high volumetric energy density microsupercapacitors, Nano Energy 53 (2018) 182–188.
[81] M.S. Kim, B. Hsia, C. Carraro, R. Maboudian, Flexible micro-supercapacitors with high energy density from simple transfer of photoresist-derived porous carbon electrodes, Carbon 74 (2014) 163–169.
[82] X. Wang, Y. Yin, C. HaO, Z. You, A high-performance three-dimensional micro supercapacitor based on ripple-like ruthenium oxide-carbon nanotube composite films, Carbon 82 (2015) 436–445.

[83] M. Gnerlich, H. Ben-Yoav, J.N. Culver, D.R. Ketchum, R. Ghodssi, Selective deposition of nanostructured ruthenium oxide using Tobacco mosaic virus for micro-supercapacitors in solid Nafion electrolyte, J. Power Sources 293 (2015) 649–656.

[84] J. Maeng, Y.-J. Kim, C. Meng, P.P. Irazoqui, Three-dimensional microcavity array electrodes for high capacitance all-solid-state flexible microsupercapacitors, ACS Appl. Mater. Interfaces 8 (2016) 13458–13465.

[85] C. Shen, X. Wang, W. Zhang, F. Kang, A high-performance three-dimensional micro supercapacitor based on self-supporting composite materials, J. Power Sources 196 (2011) 10465–10471.

[86] V. Chabot, D. Higgins, A. Yu, X. Xiao, Z. Chen, J. Zhang, A review of graphene and graphene oxide sponge: material synthesis and applications to energy and the environment, Energ. Environ. Sci. 7 (2014) 1564.

[87] X. Lang, A. Hirata, T. Fujita, M. Chen, Nanoporous metal/oxide hybrid electrodes for electrochemical supercapacitors, Nat. Nanotechnol. 6 (2011) 232–236.

[88] A.J. Cortés-López, E. Muñoz-Sandoval, F. López-Urías, Efficient carbon nanotube sponges production boosted by acetone in CVD-synthesis, Carbon 135 (2018) 145–156.

[89] J. Ji, L.L. Zhang, H. Ji, Y. Li, X. Zhao, X. Bai, X. Fan, F. Zhang, R.S. Ruoff, Nanoporous Ni(OH)2 thin film on 3D ultrathin-graphite foam for asymmetric supercapacitor, ACS Nano 7 (2013) 6237–6243.

[90] L. Yu, G. Zhang, C. Yuan, X.W. Lo, Hierarchical NiCo2O4@MnO2 core–shell heterostructured nanowire arrays on Ni foam as high-performance supercapacitor electrodes, Chem. Commun. 49 (2013) 137–139.

[91] Y. Li, Z. Kang, X. Yan, S. Cao, M. Li, Y. Guo, Y. Huan, X. Wen, Y. Zhang, A three-dimensional reticulate CNT-aerogel for a high mechanical flexibility fiber supercapacitor, Nanoscale 10 (2018) 9360–9368.

[92] J. Xu, Z. Tan, W. Zeng, G. Chen, S. Wu, Y. Zhao, K. Ni, Z. Tao, M. Ikram, H. Ji, Y. Zhu, A hierarchical carbon derived from sponge-templated activation of graphene oxide for high-performance supercapacitor electrodes, Adv. Mater. 28 (2016) 5222–5228.

[93] J. Ge, H.-B. Yao, W. Hu, X.-F. Yu, Y.-X. Yan, L.-B. Mao, H.-H. Li, S.-S. Li, S.-H. Yu, Facile dip coating processed graphene/MnO2 nanostructured sponges as high performance supercapacitor electrodes, Nano Energy 2 (2013) 505–513.

[94] W. Chen, R.B. Rakhi, L. Hu, X. Xie, Y. Cui, H.N. Alshareef, High-performance nanostructured supercapacitors on a sponge, Nano Lett. 11 (2011) 5165–5172.

[95] X. He, W. Yang, X. Mao, L. Xu, Y. Zhou, Y. Chen, Y. Zhao, Y. Yang, J. Xu, All-solid state symmetric supercapacitors based on compressible and flexible free-standing 3D carbon nanotubes (CNTs)/poly(3,4-ethylenedioxythiophene) (PEDOT) sponge electrodes, J. Power Sources 376 (2018) 138–146.

[96] I.K. Moon, S. Yoon, J. Oh, 3D highly conductive silver nanowire@PEDOT:PSS composite sponges for flexible conductors and their all-solid-state supercapacitor applications, Adv. Func. Interf. 4 (2017) 1700860.

[97] S.-W. Zhang, B.-S. Yin, C. Liu, Z.-B. Wang, D.-M. Gu, A lightweight, compressible and portable sponge-based supercapacitor for future power supply, Chem. Eng. J. 349 (2018) 509–521.

[98] G.-W. Yang, C.-L. Xu, H.-L. Li, Electrodeposited nickel hydroxide on nickel foam with ultrahigh capacitance, Chem. Commun. 48 (2008) 6537–6539.

[99] X. Dong, X. Wang, J. Wang, H. Song, X. Li, L. Wang, M.B. Chan-Park, C.M. Li, P. Chen, Synthesis of a MnO2–graphene foam hybrid with controlled MnO2 particle shape and its use as a supercapacitor electrode, Carbon 50 (2012) 4865–4870.

[100] I.K. Moon, S. Yoon, J. Oh, Three-dimensional hierarchically mesoporous ZnCo2O4 nanowires grown on graphene/sponge foam for high-performance, flexible, all-solid-state supercapacitors, Chem. A Eur. J. 23 (2017) 597.
[101] L. Li, F. Lu, C. Wang, F. Zhang, W. Liang, S. Kuga, Z. Dong, Y. Zhao, Y. Huang, M. Wu, Flexible double-cross-linked cellulose-based hydrogel and aerogel membrane for supercapacitor separator, J. Mater. Chem. A 6 (2018) 24468–24478.
[102] R. Baumann, P. Herwig, A. Wetzig, E. Beyer, Investigations of corrosion resistance of laser separated open cell metal, Adv. Eng. Mater. 19 (10) (2017) 1700107.
[103] M. Wei, X. Wu, Y. Yao, S. Yu, R. Sun, C.P. Wong, Toward high micro-supercapacitive performance by constructing graphene-supported NiMoS4 hybrid materials on 3D current collectors, ACS Sustain. Chem. Eng. 7 (2019) 19779–19786.
[104] Y. Yue, N. Liu, Y. Ma, S. Wang, W. Liu, C. Luo, H. Zhang, F. Cheng, J. Rao, X. Hu, J. Su, Y. Gao, Highly self-healable 3D microsupercapacitor with MXene–graphene composite aerogel, ACS Nano 12 (2018) 4224–4232.
[105] Y. Shi, Y. Wang, Y. Gu, L. Zheng, S. Ma, X. Xu, Self-healable and stretchable ionogels serve as electrolytes and substrates for integrated all-in-one micro-supercapacitors, Chem. Eng. J. 392 (2020) 123645.
[106] C. Zhua, T. Liu, F. Qian, W. Chen, S. Chandrasekaran, B. Yao, Y. Song, E.B. Duoss, J.D. Kuntz, C.M. Spadaccini, M.A. Worsley, Y. Li, 3D printed functional nanomaterials for electrochemical energy storage, Nano Today 15 (2017) 107–120.
[107] P.-.Y. Chen, M. Liu, Z. Wang, R.H. Hurt, I.Y. Wong, From flatland to spaceland: higher dimensional patterning with two-dimensional materials, Adv. Mater. 29 (2017) 1605096.
[108] X. Tang, H. Zhou, Z. Cai, D. Cheng, P. He, P. Xie, D. Zhang, T. Fan, Generalized 3D printing of graphene-based mixed-dimensional hybrid aerogels, ACS Nano 12 (2018) 3502–3511.
[109] M. Wei, F. Zhang, W. Wang, P. Alexandridis, C. Zhou, G. Wu, 3D direct writing fabrication of electrodes for electrochemical storage devices, J. Power Sources 354 (2017) 134–147.
[110] W. Yu, B.Q. Li, S. Ding, H. Liu, 3D printing of interdigitated electrode for all-solid-state microsupercapacitors, J. Micromech. Microeng. 28 (2018) 105014.
[111] W. Yu, H. Zhou, B.Q. Li, S. Ding, 3D printing of carbon nanotubes-based micro-supercapacitors, ACS Appl. Mater. Interfaces 9 (2017) 4597–4604.
[112] Y. Liu, B. Zhang, Q. Xu, Y. Hou, S. Seyedin, S. Qin, G.G. Wallace, S. Beirne, J.M. Razal, J. Chen, Development of graphene oxide/polyaniline inks for high performance flexible microsupercapacitors via extrusion printing, Adv. Funct. Mater. 28 (2018) 1706592.
[113] G. Sun, J. An, C.K. Chua, H. Pang, J. Zhang, P. Chen, Layer-by-layer printing of laminated graphene-based interdigitated microelectrodes for flexible planar micro-supercapacitors, Electrochem. Commun. 51 (2015) 33–36.
[114] Z.-S. Wu, Z. Liu, K. Parvez, X. Feng, K. Müllen, Ultrathin printable graphene supercapacitors with AC line-filtering performance, Adv. Mater. 27 (2015) 3669–3675.
[115] T. Wang, L. Li, X. Tian, H. Jin, K. Tang, S. Hou, H. Zhou, X. Yu, 3D-printed interdigitated graphene framework as superior support of metal oxide nanostructures for remarkable micro-pseudocapacitors, Electrochim. Acta 319 (2019) 245–252.
[116] C. Zhao, C. Wang, R. Gorkin, S. Beirne, K. Shu, G.G. Wallace, Three dimensional (3D) printed electrodes for interdigitated supercapacitors, Electrochem. Commun. 41 (2014) 20–23.
[117] M.-L. Seol, I. Nam, E.L. Ribeiro, B. Segel, D. Lee, T. Palma, H. Wu, D. Mukherjee, B. Khomam, C. Hill, J.W. Han, M. Meyyappan, All-printed in-plane supercapacitors by sequential additive manufacturing process, ACS Appl. Energy Mater. 3 (2020) 4965–4973.

[118] X. Yun, B. Lu, Z. Xiong, B. Jia, B. Tang, H. Mao, T. Zhang, X. Wang, Direct 3D printing of a graphene oxide hydrogel for fabrication of a high areal specific capacitance microsupercapacitor, RSC Adv. 9 (2019) 29384–29395.
[119] L. Jia, L. Zheng, W. Wang, Y. Shi, Q. Zhang, X. Xu, Controllable design of coaxial MnO2/polyaniline for asymmetric supercapacitors and stamping flexible micro-device, Mater. Lett. 252 (2019) 80–83.
[120] F. Li, J. Qu, Y. Li, J. Wang, M. Zhu, L. Liu, J. Ge, S. Duan, T. Li, V.K. Bandari, M. Huang, F. Zhu, O.G. Schmidt, Stamping fabrication of flexible planar micro-supercapacitors using porous graphene inks, Adv. Sci. 7 (2020) 2001561.
[121] C. Zhang, M.P. Kremer, A. Seral-Ascaso, S.-H. Park, N. McEvoy, B. Anasori, Y. Gogotsi, V. Nicolosi, Stamping of flexible, coplanar micro-supercapacitors using MXene inks, Adv. Funct. Mater. 28 (2018) 1705506.
[122] Y. Chen, X. Li, Z. Bi, G. Li, X. He, X. Gao, Stamp-assisted printing of nanotextured electrodes for high-performance flexible planar micro-supercapacitors, Chem. Eng. J. 353 (2018) 499–506.
[123] H. Wang, Y. Song, L. Miao, J. Wan, X. Chen, X. Cheng, H. Guo, H. Zhang, Stamp-assisted gravure printing of micro-supercapacitors with general flexible substrates, in: 2019 IEEE 32nd International Conference on Micro Electro Mechanical Systems (MEMS), Seoul, Korea (South), 2019, pp. 950–953.
[124] J. Cen, R. Kitsomboonloha, V. Subramanian, Cell filling in gravure printing for printed electronics, Langmuir 30 (45) (2014) 13716–13726.
[125] K.-H. Lee, S.-S. Lee, D.B. Ahn, J. Lee, D. Byun, S.-Y. Lee, Ultrahigh areal number density solid-state on-chip microsupercapacitors via electrohydrodynamic jet printing, Sci. Adv. 6 (2020), eaaz1692.
[126] B. Asbani, B. Bounor, K. Robert, C. Douard, L. Athouël, C. Lethien, J. Le Bideau, T. Brousse, Reflow soldering-resistant solid-state 3D micro-supercapacitors based on ionogel electrolyte for powering the internet of things, J. Electrochem. Soc. 167 (2020) 100551.
[127] H. Durou, D. Pech, D. Colin, P. Simon, P.-L. Taberna, M. Brunet, Wafer-level fabrication process for fully encapsulated micro-supercapacitors with high specific energy, Microsyst. Technol. 18 (2012) 467–473.
[128] H.J. In, S. Kumar, Y. Shao-Horn, G. Barbastathis, Origami fabrication of nanostructured three-dimensional devices: electrochemical capacitors with carbon electrodes, Appl. Phys. Lett. 88 (2006), 083104.
[129] J. Pu, X. Wang, R. Xu, S. Xu, K. Komvopoulos, Highly flexible, foldable, and rollable microsupercapacitors on an ultrathin polyimide substrate with high power density, Microsyst. Nanoeng. 4 (2018) 1–11.
[130] H. Kim, J. Yoon, G. Lee, S.-H. Paik, G. Choi, D. Kim, B.-M. Kim, G. Zi, J.S. Ha, Encapsulated, high-performance, stretchable array of stacked planar micro-supercapacitors as waterproof wearable energy storage devices, ACS Appl. Mater. Interfaces 8 (2016) 16016–16025.
[131] J. Park, D.B. Ahn, J. Kim, E. Cha, B.-S. Bae, S.-Y. Lee, J.-U. Park, Printing of wirelessly rechargeable solid-state supercapacitors for soft, smart contact lenses with continuous operations, Sci. Adv. 5 (2019), eaay0764.
[132] Q. Jianga, C. Wub, Z. Wang, A.C. Wang, J.-H. Hec, Z.L. Wang, H.N. Alshareefa, MXene electrochemical microsupercapacitor integrated with triboelectric nanogenerator as a wearable self-charging power unit, Nano Energy 45 (2018) 266–272.

CHAPTER 9

Fiber-shaped micro-supercapacitors

Shengli Zhai[a,b], Zhi Li[b], and Yuan Chen[a]
[a]School of Chemical and Biomolecular Engineering, The University of Sydney, Sydney, NSW, Australia
[b]Department of Chemical and Materials Engineering, University of Alberta, Edmonton, AB, Canada

9.1 History of fiber-shaped supercapacitors

Supercapacitors (SCs) are electrochemical energy storage devices that bridge the gap between electrolytic capacitors and rechargeable batteries. They store energy through electrochemical double-layer capacitance by reversible adsorption/desorption of electrolyte ions at the surface or inside the pores of electrode and pseudocapacitance through surface faradic redox reactions or ion intercalation at electrode surfaces [1, 2]. Compared to rechargeable batteries, they have the advantage of fast charge/discharge rate, high power density, long service life, high stability and safety, environmental friendliness, wide operating temperature window, and low cost [1, 2]. Recently, their energy density is gradually catching up that of rechargeable batteries when intercalative and extrinsic pseudocapacitive materials are used [3–5]. Therefore, SCs are regarded as promising energy solutions for emerging wearable electronic devices. Among various types of SCs, fiber-shaped SCs (FSCs) with elongated (including yarn/wire/cable-like) structures have attracted significant interests because of their unique characteristics. First, their high mechanical flexibility enables structural integrity after long-term and repetitive deformations; second, they can be easily scaled-up by self-integration to meet a wide range of power requirements; third, they can be packed into small space with diverse shapes which ensures great design versatility; and forth, they can be integrated with other one-dimensional (1D) devices in multifunctional systems [5, 6].

In 2003, Baughman et al. demonstrated the first FSC by twisting two conductive carbon nanotube (CNT) composite fiber electrodes together and using poly(vinyl alcohol) (PVA)/H_3PO_4 as both the gel electrolyte and separator [7]. The FSC was further woven into an existing textile, demonstrating its application potential in powering wearable electronics. However, the electrochemical energy storage performance of the FSC is far from

Microsupercapacitors
https://doi.org/10.1016/B978-0-08-102888-9.00009-4

satisfactory due to the limited surface area of CNT composite fiber electrodes. Therefore, the FSC did not initially attract significant attention for a few years. However, since 2011, the electrochemical energy storage performance of FSCs has increased significantly, and FSCs are attracting more and more interest. For example, in 2011, Wang et al. deposited a pseudocapacitive material, MnO_2, on fiber substrates with pre-grown ZnO_2 nanowires, yielding a high-performance FSC [8]. Wang et al. also demonstrated the concept of self-powered 1D devices by integrating energy generators (i.e., a nanogenerator and solar cell) with an energy storage component (i.e., an FSC) on one fiber [9]. It should be noted that the concept of self-powering has already been used in consumer electronics. For example, Citizen Eco-Drive watches have a self-powering unit (a solar cell connected with a battery) that can last for over 10 years. In 2012, multiple FSCs were self-integrated into an energy storage unit by connecting them in series and parallel to increase their working voltage window and current, respectively. The energy storage unit can fulfill diverse power and energy requirements [10]. In summary, these pioneering studies during 2011–12 demonstrated three main approaches, which can be used to enhance the energy storage performance of FSCs: adding pseudocapacitive materials, integrating with energy generators, and self-integration.

In 2013, the appearance of microscale FSCs [11, 12] and the introduction of graphene-based fibers [13, 14] greatly enriched the family of FSCs and the research on FSCs has ushered in a spurt of development since then. From 2014, multifunctional FSCs have been demonstrated with expanded application possibilities. For example, FSCs with an electrochromic function can indicate their charge states in real-time [15]. Other features, such as self-healability [16] sensitivity to environmental parameters [17, 18] and shape-recoverability [19], have also been demonstrated. In 2015, long FSCs were woven into electrochemical energy storage textiles via commercial textile machines for the first time, opening the possibility of creating energy storage textiles [20]. Further, the discovery of new electrode materials makes it possible to improve the energy density of FSCs. For example, MXenes with both high electrical conductivity (e.g., up to 10,000 S cm^{-1}) and specific capacitance (up to 1500 F cm^{-3}) [21], have been incorporated into various carbon-based fibers to demonstrate FSCs with ultrahigh energy densities [22–25]. In 2019, our group combined a high energy capacity battery electrode (zinc deposited on carbon fibers) with and a high-power capacitive electrode (graphene fibers) to show a fiber-shaped hybrid energy storage device, achieving both high density and power density

[26]. Fiber-shaped hybrid energy storage devices with some advantages of both batteries and supercapacitors are becoming a promising direction in the development of FSCs (Fig. 9.1).

9.2 Fiber-shaped micro-supercapacitors (FMSCs)

FSCs typically have a diameter ranging from tens to hundreds of micrometers and a length from millimeters to meters. According to their size, FSCs can be roughly categorized into two sub-branches: fiber-shaped micro-SCs (denoted as FMSCs, typically less than 10 mm in length) which are used to power microelectronics (Fig. 9.2A) and long FSCs that can be knitted/woven into textiles for wearable energy storage textiles (Fig. 9.2B). While in this chapter, we only focus on FMSCs, while a summary of long FSCs can be found in several recently published review articles and books [6, 27–30].

9.2.1 Device configurations of FMSCs

FMSCs consist of a positive electrode, a negative electrode, a separator, and an electrolyte confined into a 1D structure. As shown in Fig. 9.3, FMSCs may have three different configurations, including (1) the parallel type by placing two fiber electrodes closely in parallel, (2) the two-ply type by twisting two fiber electrodes together, and (3) the coaxial type by wrapping a shell electrode on the outer surface of a core fiber electrode. Regardless of their configurations, two electrodes in one FMSC should be separated either by a separator or a gel/solid electrolyte to prevent short-circuiting.

The parallel type FMSCs are easy to be fabricated and scaled up into integrated energy storage units. Fiber electrodes can be integrated on planar substrates by connecting them in series and/or parallel to achieve different power or energy requirements. The two-ply FMSCs are also easy to be fabricated; however, they require fiber electrodes to have higher mechanical characteristics. The coaxial type FMSCs can enable efficient contacts between two electrodes and potentially provide better structural stability. However, their fabrication process is more complicated. Overall, the device configurations may vary depending on the structures and properties of fiber electrodes, and assembly methods.

9.2.2 Design considerations for FMSCs

Electrode materials and their microstructures: First, electrode materials with high intrinsic energy storage capacity are desirable. It is beneficial to push the

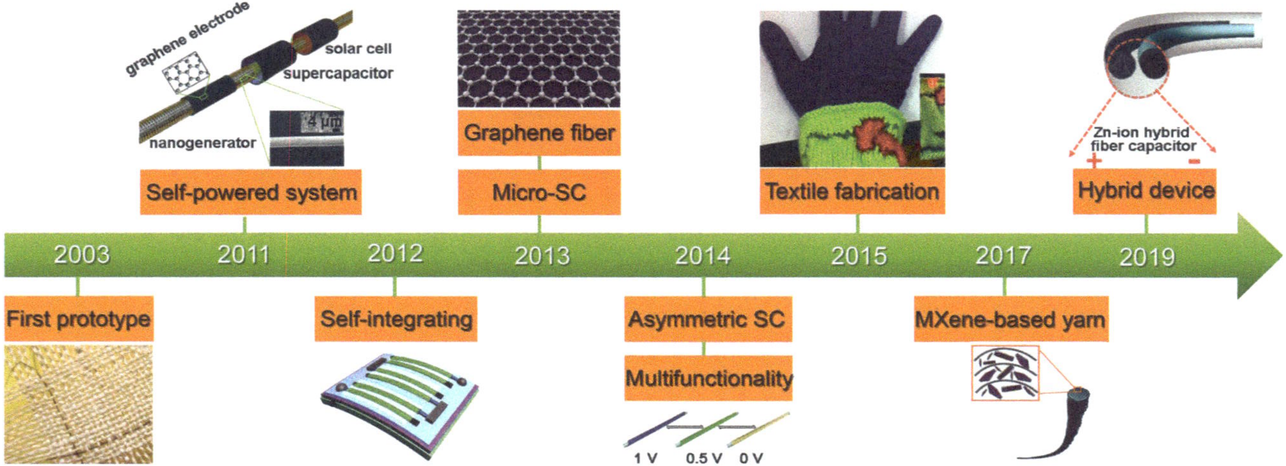

Fig. 9.1 A brief chronicle of the development of FSCs. *(Image for "First prototype": Reproduced with permission from A.B. Dalton, S. Collins, E. Munoz, J.M. Razal, V.H. Ebron, J.P. Ferraris, J.N. Coleman, B.G. Kim, R.H. Baughman, Nature 423 (2003) 703.. Copyright, 2003, Springer Nature. Image for "Self-powered system": Reproduced with permission J. Bae, Y.J. Park, M. Lee, S.N. Cha, Y.J. Choi, C.S. Lee, J.M. Kim, Z.L. Wang, Adv. Mater. 23 (2011) 3446–3449. Copyright, 2011, Wiley-VCH. Image for "Self-integrating": Reproduced with permission from X. Xiao, T. Li, P. Yang, Y. Gao, H. Jin, W. Ni, W. Zhan, X. Zhang, Y. Cao, J. Zhong, ACS Nano 6 (2012) 9200–9206. Copyright 2012, American Chemical Society (ACS). Image for "Multifunctionality": Reproduced with permission from X. Chen, H. Lin, J. Deng, Y. Zhang, X. Sun, P. Chen, X. Fang, Z. Zhang, G. Guan, H. Peng, Adv. Mater. 26 (2014) 8126–8132. Copyright, 2014, Wiley-VCH. Image for "Textile fabrication": Reproduced with permission Y. Huang, H. Hu, Y. Huang, M. Zhu, W. Meng, C. Liu, Z. Pei, C. Hao, Z. Wang, C. Zhi, ACS Nano 9 (2015) 4766–4775. Copyright, 2015, ACS. Image for "MXene-based yarn": Reproduced with permission from Z. Wang, S. Qin, S. Seyedin, J. Zhang, J. Wang, A. Levitt, N. Li, C. Haines, R. Ovalle-Robles, W. Lei, Small, 14 (2018) 1802225. Copyright, 2017, Wiley-VCH. Image for "hybrid device": Reproduced with permission from X. Zhang, Z. Pei, C. Wang, Z. Yuan, L. Wei, Y. Pan, A. Mahmood, Q. Shao, Y. Chen, Small 15 (2019) 1903817. Copyright, 2019, Wiley-VCH.)*

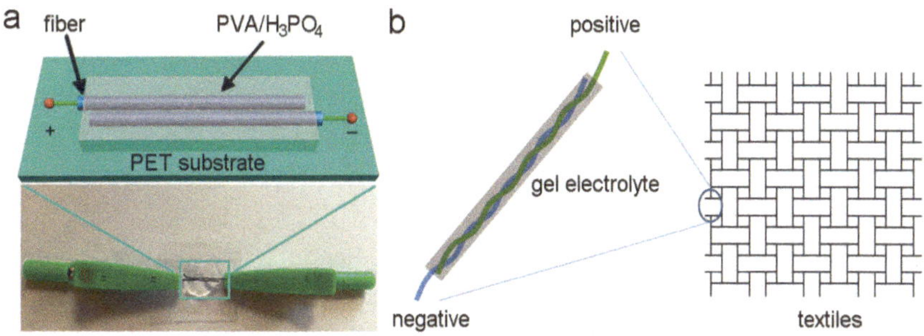

Fig. 9.2 Schematic illustration of (A) FMSCs and (B) long FSCs for wearable energy storage. *(Image for "FMSCs": Reproduced with permission from S. Zhai, H.E. Karahan, L. Wei, X. Chen, Z. Zhou, X. Wang, Y. Chen, Energy Storage Mater. 9 (2017) 221–228. Copyright, 2017, Elsevier. Image for "Long FSCs": Reproduced with permission from S. Zhai, L. Wei, H.E. Karahan, X. Chen, C. Wang, X. Zhang, J. Chen, X. Wang, Y. Chen, Energy Storage Mater. 19 (2019) 102–123. Copyright, 2016, Elsevier.)*

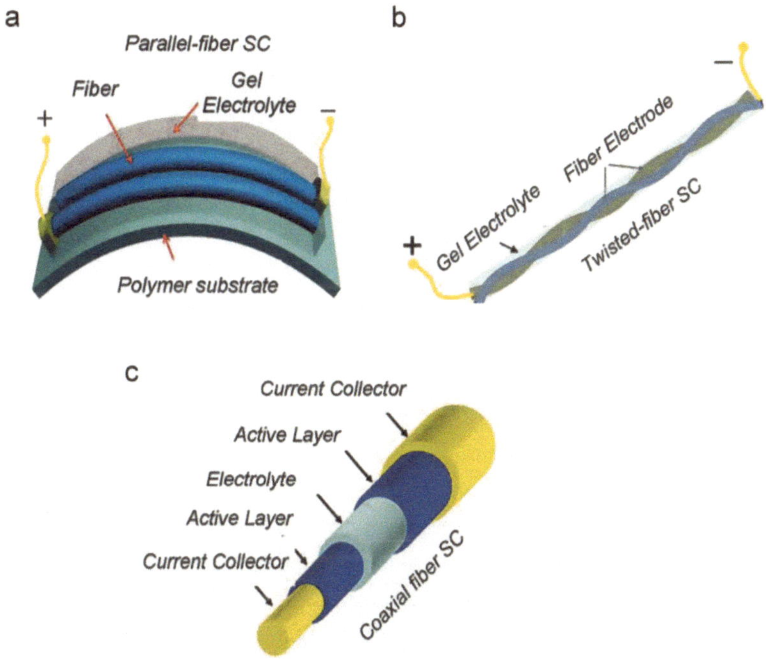

Fig. 9.3 Schematic illustration of three major device configurations of FMSCs: (A) parallel type, (B) two-ply type, and (C) coaxial type. *(Reproduced with permission from D. Yu, Q. Qian, L. Wei, W. Jiang, K. Goh, J. Wei, J. Zhang, Y. Chen, Chem. Soc. Rev. 44 (2015) 647–662. Copyright 2014, Royal Society of Chemistry (RSC).)*

performances of electrode materials to their theoretical limits so that FMSCs can have a high energy storage capacity in their limited volume. Second, the microstructures of electrodes should also be carefully tailored. High intrinsic energy storage capacity does not always guarantee excellent electrochemical performance. Creating 3D porous structures may allow efficient access of electrolyte ions.

Electrical conductivity: For FMSCs, electrons need to travel a long distance along the axial direction to reach external circuits. In contrast, for conventional planar devices, electrons only transfer across the thickness of electrode materials, ranging from several nanometers to hundreds of micrometers, which is several orders smaller than that of FMSCs. Therefore, both electrode materials and current collectors should have high electrical conductivity in FMSCs. Further, the contact resistance between electrode materials and current collectors should also be minimized.

Macrostructures of fiber electrodes: As microscale power sources, the limited volume is one key constraint for FMSCs. Electrode materials may be directly deposited on conductive fibers to form a core-sheath structure, which often increases the power density of FMSCs. Alternatively, electrode materials and other materials may form freestanding fiber electrodes via bottom-up synthesis methods. In core-sheath fiber electrodes, the cores typically work as current collectors without contributing to the total capacitance. In comparison, freestanding fibers may concurrently serve as current collectors and electrode materials. Therefore, freestanding fibers often deliver higher energy storage performance in FMSCs.

Electrolytes: Most of the recently reported FMSCs use polymeric gel electrolytes other than liquid electrolytes to avoid possible leakage of liquids. Gel electrolytes offer several advantages, such as serving as separators to simplify device fabrication, offering mechanical flexibility, better safety with non-flammability and low toxicity, and wide working temperatures [31, 32].

Mechanical characteristics: Long FSCs require excellent mechanical properties to withstand mechanical deformations taking place during the textile fabrication. In comparison, the requirements for the mechanical properties of FMSCs are less stringent. For powering flexible electronics, FMSCs still need to possess enough mechanical strength to maintain their structural and performative stability under various mechanical deformation conditions. This can normally be achieved by either using mechanically robust fiber electrode scaffolds or attaching fiber electrodes on flexible substrates, such as plastic sheets (Fig. 9.4).

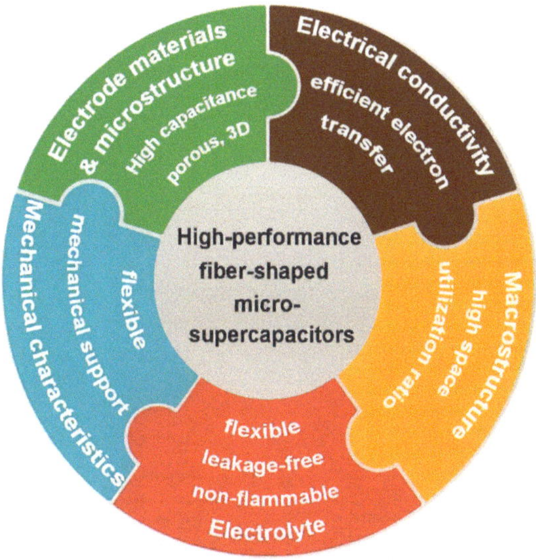

Fig. 9.4 Design considerations to create high-performance 1D FMSCs.

9.3 FMSCs with different fiber scaffolds

The general requirements of fiber scaffolds for FMSCs include high electrical conductivity to enable efficient electron transfer and mechanical strength and flexibility to withstand various deformations. Metal wires and different carbon-based fibers are the most often used scaffolds to fabricate FMSCs (see Fig. 9.5 and Table 9.1).

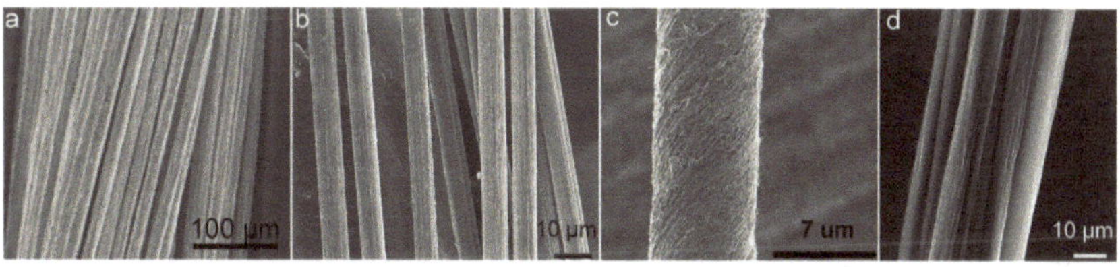

Fig. 9.5 SEM images of the surface of (A) stainless steel (SS) fibers, (B) carbon fibers, (C) multi-walled CNT fibers, and (D) graphene fibers. *(Panel A: Reproduced with permission from J. Sun, Y. Huang, C. Fu, Y. Huang, M. Zhu, X. Tao, C. Zhi, H. Hu, J. Mater. Chem. A 4 (2016) 14877–14883. Copyright, 2016, RSC; Panel B: reproduced with permission from S. Zhai, W. Jiang, L. Wei, H.E. Karahan, Y. Yuan, A.K. Ng, Y. Chen, Mater. Horiz. 2 (2015) 598–605. Copyright, 2015, RSC; panel C: reproduced with permission from M. Miao, Carbon 49 (2011) 3755–3761. Copyright, 2011, Elsevier.)*

Table 9.1 Physical properties and costs of various metal wires and carbon-based fibers [5].

Type	Specific surface area ($m^2\ g^{-1}$)	Density ($g\ cm^{-3}$)	Electrical conductivity ($10^3\ S\ m^{-1}$)	Tensile strength (GPa)	Failure strain (%)	Cost (US$/kg)
SS fibers	Low	7.90	1450	2.75	2.5	3–30
Ni wire	Low	8.90	14,300	NA	NA	20–50
Cu wire	Low	8.96	59,600	NA	NA	8–30
CNT fiber	100–210	0.54–1.30	30–100	0.10–3.00	2.5–13.0	NA
Graphene-based fiber	24–2210	0.29–1.75	0.78–111	0.10–1.08	0.5–10.1	NA
Carbon fiber	1–10	1.75–2.00	58–140	2.70–7.00	0.7–2.4	10–60

9.3.1 FMSCs based on metal wires

Due to the correlation between bending strain and thickness, many materials would become flexible when they are thin [33]. Therefore, metal wires (e.g., stainless steel (SS), nickel (Ni), copper (Cu), and titanium (Ti)) with diameters down to several micrometers are flexible, and they have been used to fabricate FMSCs. Metal wires have the advantages of high electrical conductivity, which helps to achieve excellent rate capability for FMSCs. They are mechanically robust and can withstand harsh deformation conditions. Their low-cost and commercial availability can enable large-scale device production. In 2016, Duan et al. prepared a nanoarray electrode consisting of CuO nanowires grown on copper wires, which currently provided high electron transfer pathways and a large surface area to anchor active electrode materials [34]. FeCo-layered double hydroxide nanoplatelets were further deposited on CuO nanowires to introduce pseudocapacitance. A solid-state, asymmetric FMSC was fabricated by pairing this electrode with another electrode made of a Cu wire coated with activated carbon in PVA/KOH gel electrolyte (see Fig. 9.6A). The FMSC delivered a high energy density of 1.857 mWh cm^{-3}, high rate capability, and high cycling stability. In 2016, Dai et al. demonstrated a solid-state FMSC using a Ti wire sheathed with radially aligned TiO_2 nanotubes as a positive electrode and a CNT fiber yarn as a negative electrode [35]. The CNT fiber yarn was wrapped around the outer surface of the positive electrode, forming a coaxial-helix device structure. The FMSC showed a high capacitance of 1.84 mF cm^{-2} and excellent performance stability under various bending states.

However, metal wires also have several critical drawbacks. They are usually heavy, and some of them can be easily corroded in acidic or alkaline electrolytes, which restrict their lifetime. Therefore, many efforts have also been made to find alternatives to metal wires to fabricate FMSCs.

9.3.2 FMSCs based on carbon-based fibers

Carbon-based fibers usually have high electrical conductivity, excellent mechanical characteristics, low weight, chemical stability, and environmental friendliness. Besides, some carbon-based fibers can also contribute to the energy storage by either electrical double-layer capacitance and/or pseudocapacitance. Therefore, carbon-based fibers are regarded as promising candidates to fabricate FMSCs. To date, carbon-based fibers used to build FMSCs are mainly carbon fibers, CNT fibers, and graphene fibers.

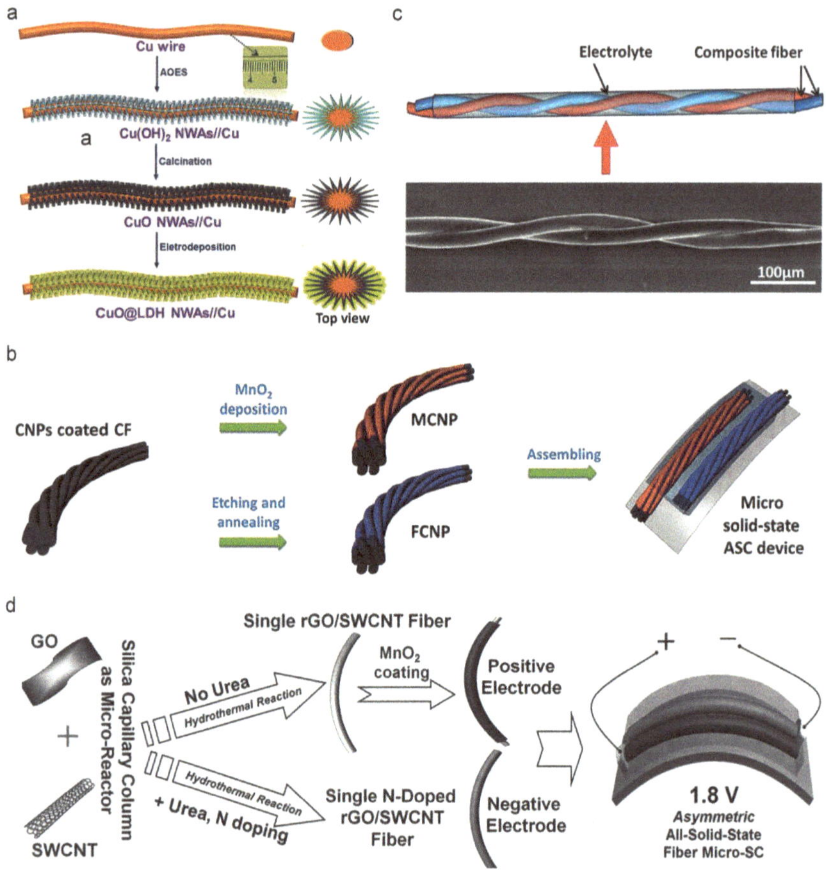

Fig. 9.6 Various FMSCs based on different fiber scaffolds. Schematic illustration for the fabrication of (A) hierarchical CuO@CoFe layer double hydroxide nanowire arrays supported on a Cu wire, which was used to assemble an asymmetric FMSC with an activated carbon-coated Cu wire electrode. (B) An asymmetric FMSC using functionalized carbon nanoparticles coated carbon fibers as a negative electrode, MnO_2 and carbon nanoparticle coated carbon fiber as the positive electrode. (C) A symmetric FMSC assembled by twisting two aligned polyaniline-deposited CNT fiber electrodes. (D) An asymmetric FMSC using MnO_2 deposited graphene/CNT fiber electrode and nitrogen-doped graphene/CNT fiber electrode. *(Panel A: Reproduced with permission from Z. Li, M. Shao, L. Zhou, R. Zhang, C. Zhang, J. Han, M. Wei, D.G. Evans, X. Duan, Nano Energy 20 (2016) 294–304. Copyright, 2016, Elsevier; panel B: reproduced with permission from H. Jin, L. Zhou, C.L. Mak, H. Huang, W.M. Tang, H.L.W. Chan, J. Mater. Chem. A 3 (2015) 15633–15641. Copyright, 2015, RSC; panel C: reproduced with permission from Z. Cai, L. Li, J. Ren, L. Qiu, H. Lin, H. Peng, J. Mater. Chem. A 1 (2013) 258–261. Copyright, 2013, RSC; panel D: Reproduced with permission from D. Yu, K. Goh, Q. Zhang, L. Wei, H. Wang, W. Jiang, Y. Chen, Adv. Mater. 26 (2014) 6790–6797. Copyright, 2014, RSC.)*

9.3.2.1 FMSC based on carbon fibers

Carbon fibers, spun from pitch or polyacrylonitrile, are commercially available in large quantities [36]. As tabulated in Table 9.1, carbon fibers have both high electrical conductivity and excellent mechanical characteristics (e.g., tensile strength up to 7 GPa). In 2014, Zhou et al. reported an all-solid-state FMSC, which was fabricated by two placing MnO_2 deposited carbon fiber electrodes in parallel (see Fig. 9.6B) [10]. The two electrodes were separated by PVA/H_3PO_4 gel electrolyte. The assembled device yielded a high capacitance of 2.5 F cm^{-3} and an energy density of 2.2×10^{-4} Wh cm^{-3}. Assembling asymmetric FMSCs to expand the working voltage window is one of the most popular approaches to increase the energy density of FMSCs. In 2015, Chan et al. demonstrated an asymmetric FMSC using MnO_2 deposited carbon fiber as a positive electrode and carbon nanoparticle coated carbon fiber as a negative electrode [37]. The assembled FMSC operated at an expanded voltage of 1.8 V (typically 1 V for symmetric SCs in aqueous electrolytes) and exhibited a high volumetric capacitance of 4.6 F cm^{-3} and an energy density of 2.1 mWh cm^{-3}.

9.3.2.2 FMSCs based on CNT fibers

Individual CNT exhibits remarkable electrical conductivity and tensile strength. Various methods have been demonstrated to assemble individual CNT into CNT fibers, such as dry-spinning from CNT forests [38] or CNT aerogels [39], wet-pinning from liquid CNT dispersions [40] and twisting CNT thin sheets [41]. Among them, the dry-spinning method is the most commonly used one for FMSC fabrication. As shown in Table 9.1, CNT fibers show a high electrical conductivity of up to 100 S cm^{-1} and excellent tensile strength in a range of 0.1–3.0 GPa. Besides, CNT fibers also have a much higher specific area in comparison to carbon fibers (Table 9.1). Therefore, CNT fibers can concurrently serve as current collectors and directly contribute to energy storage via double-layer capacitance. A higher surface area is also favorable for enabling a high mass loading of active electrode materials. In 2013, Peng et al. reported a high-performance FMSC, which was fabricated by twisting two identical, polyaniline-deposited CNT fibers together (Fig. 9.6C) [42]. The FMSC showed a specific length capacitance of 263 F cm^{-1}. Besides conductive polymers, other pseudocapacitive materials (e.g., MnO_2) have also been deposited on CNT fibers to fabricate FMSCs [11].

9.3.2.3 FMSCs based on graphene-based fibers

Graphene has excellent mechanical, electrical, and electrochemical properties. Graphene-based fibers can be synthesized via various bottom-up approaches, including wet-spinning [43] or hydrothermal synthesis [44] methods from graphene oxide (GO) dispersions. With the incorporation of a spacer and/or conductivity enhancer, graphene-based fibers can concurrently possess large surface area (up to 2200 $m^2\ g^{-1}$), high electrical conductivity (up to $\sim$100 k $S\ m^{-1}$) and tensile strength (up to 1 GPa), as tabulated in Table 9.1. As mentioned in Section 9.2.2, the small volume is a key constrain of FMSCs. Electrode materials can be directly mixed with GO to form graphene-based hybrid fibers via bottom-up synthesis approaches. Therefore, graphene-based fibers can not only serve as current collectors but also directly contribute to energy storage. In 2014, we demonstrated an FMSC with excellent capacitive performance using hydrothermally synthesized GO/CNT hybrid fibers [44]. With the further incorporation of pseudocapacitive RuO_2, the FMSC achieved one of the highest volumetric capacitances of 199 F cm^{-3} and an ultrahigh energy density of 27.3 mWh cm^{-3} [45]. Other pseudocapacitive materials (e.g., MnO_2, MoS_2) and/or heteroatoms (e.g., N, S) have also been used to enhance the capacitive performance of graphene-based hybrid fibers and created high-performance FMSCs (see Fig. 9.6D) [37, 46, 47].

9.4 Conclusion and outlook

An overview of recent progress related to FMSCs was presented. We first provided a brief chronicle of the development of FSCs. Then, the specific design considerations of FMSCs were highlighted. Afterward, representative FMSCs based on different fiber scaffolds used are described. Despite the fast progress in the last few years, the development of FMSCs still faces several key challenges.

Improving energy storage performance: It is desired to improve the energy storage performance of FMSCs further to realize practical applications. Two types of electrode materials are attractive to increase energy storage capacity: intercalative pseudocapacitive materials and battery materials in nanoscale. Intercalative pseudocapacitive materials can provide higher energy storage capacity via rapid ion intercalation/deintercalation (e.g., MXenes with H^+ ion intercalation). Battery materials in nanoscale can deliver large extrinsic pseudocapacitance because they can fully expose their reaction sites with short ion diffusion distances. From the device configuration aspect, the combination of one battery type electrode with high energy storage capacity and one capacitive electrode with high power density can enable FMSCs with higher energy and power densities.

Managing trade-offs between energy storage and multifunctionality: In the last few years, various additional functions have been integrated into FMSCs, such as electrochromic capability, self-healability, and shape memorability. Multifunctional FMSCs bring added limitations on material selections and device configurations. More complex electrodes/devices also require additional components and fabrication steps, which often have negative impacts on the energy storage performance of FMSCs. New research is needed to achieve an optimum balance between energy storage and other functions in FMSCs.

Standardizing electrochemical performance testing metrics: The electrochemical performance parameters of FMSCs have been reported based on their weight, length, area, or volume. In consideration of their small volume, footprint, and limited masses of materials used, volume or area-based parameters would be more meaningful to evaluate their performance. FMSCs are capacitive devices. Therefore, they are expected to deliver energy quickly. It is meaningless to assess their performance at very low scan rates or current densities. Considering that the performances of commercial Li-ion batteries are often evaluated under the charging/discharging rate of 1C, FMSCs should be assessed at a much higher rate. For example, FMSCs should be fully charged in 6 min, equivalent to the charging/discharging rate of 10 °C. Besides, the details of testing conditions and geometric parameters are often missing in many current studies. They should be provided to allow a fair comparison among different studies. It is essential to have standardized testing metrics to move the research field forward.

Exploring packaging materials: In most of the current studies, FMSCs fabricated using gel electrolytes (mostly PVA-based gels) have been applied directly to power electronics without packaging materials. However, PVA-based gels are not stable in hot or humid environments. Thus, suitable packaging materials are required to enable FMSCs to work stably under different environmental conditions. Flexible packing materials which are compatible with FMSCs have seldom been reported.

Achieving scalable device manufacture: To date, most FMSCs are manually assembled in research labs, and the scalability of their fabrication methods is rarely studied. Scalable production of delicately designed FMSCs in a cost-effective manner is complicated, which typically involves the coating of active electrode materials on 1D fiber scaffolds, the pairing of 1D fiber electrodes in different configurations, and encapsulation of completed 1D devices. It is essential to explore novel scalable fabrication methods for FMSCs to achieve commercialization.

References

[1] P. Simon, Y. Gogotsi, Nat. Mater. 7 (2008) 845.
[2] Y. Wang, Y. Song, Y. Xia, Chem. Soc. Rev. 45 (2016) 5925–5950.
[3] J. Liu, J. Wang, C. Xu, H. Jiang, C. Li, L. Zhang, J. Lin, Z.X. Shen, Adv. Sci. 5 (2018) 1700322.
[4] V. Augustyn, P. Simon, B. Dunn, Energy Environ. Sci. 7 (2014) 1597–1614.
[5] S. Zhai, H.E. Karahan, C. Wang, Z. Pei, L. Wei, Y. Chen, Adv. Mater. 32 (2020) 1902387.
[6] D. Yu, Q. Qian, L. Wei, W. Jiang, K. Goh, J. Wei, J. Zhang, Y. Chen, Chem. Soc. Rev. 44 (2015) 647–662.
[7] A.B. Dalton, S. Collins, E. Munoz, J.M. Razal, V.H. Ebron, J.P. Ferraris, J.N. Coleman, B.G. Kim, R.H. Baughman, Nature 423 (2003) 703.
[8] J. Bae, M.K. Song, Y.J. Park, J.M. Kim, M. Liu, Z.L. Wang, Angew. Chem. Int. Ed. 50 (2011) 1683–1687.
[9] J. Bae, Y.J. Park, M. Lee, S.N. Cha, Y.J. Choi, C.S. Lee, J.M. Kim, Z.L. Wang, Adv. Mater. 23 (2011) 3446–3449.
[10] X. Xiao, T. Li, P. Yang, Y. Gao, H. Jin, W. Ni, W. Zhan, X. Zhang, Y. Cao, J. Zhong, ACS Nano 6 (2012) 9200–9206.
[11] J. Ren, L. Li, C. Chen, X. Chen, Z. Cai, L. Qiu, Y. Wang, X. Zhu, H. Peng, Adv. Mater. 25 (2013) 1155–1159.
[12] L. Li, C. Chen, J. Xie, Z. Shao, F. Yang, J. Nanomater. 2013 (2013).
[13] T. Huang, B. Zheng, L. Kou, K. Gopalsamy, Z. Xu, C. Gao, Y. Meng, Z. Wei, RSC Adv. 3 (2013) 23957–23962.
[14] X. Li, T. Zhao, Q. Chen, P. Li, K. Wang, M. Zhong, J. Wei, D. Wu, B. Wei, H. Zhu, Phys. Chem. Chem. Phys. 15 (2013) 17752–17757.
[15] X. Chen, H. Lin, J. Deng, Y. Zhang, X. Sun, P. Chen, X. Fang, Z. Zhang, G. Guan, H. Peng, Adv. Mater. 26 (2014) 8126–8132.
[16] Y. Huang, Y. Huang, M. Zhu, W. Meng, Z. Pei, C. Liu, H. Hu, C. Zhi, ACS Nano 9 (2015) 6242–6251.
[17] C. Choi, J.M. Lee, S.H. Kim, S.J. Kim, J. Di, R.H. Baughman, Nano Lett. 16 (2016) 7677–7684.
[18] H. Wang, Z. Liu, J. Ding, X. Lepró, S. Fang, N. Jiang, N. Yuan, R. Wang, Q. Yin, W. Lv, Adv. Mater. 28 (2016) 4998–5007.
[19] J. Deng, Y. Zhang, Y. Zhao, P. Chen, X. Cheng, H. Peng, Angew. Chem. Int. Ed. 54 (2015) 15419–15423.
[20] Y. Huang, H. Hu, Y. Huang, M. Zhu, W. Meng, C. Liu, Z. Pei, C. Hao, Z. Wang, C. Zhi, ACS Nano 9 (2015) 4766–4775.
[21] B. Anasori, M.R. Lukatskaya, Y. Gogotsi, Nat. Rev. Mater. 2 (2017) 1–17.
[22] Z. Wang, S. Qin, S. Seyedin, J. Zhang, J. Wang, A. Levitt, N. Li, C. Haines, R. Ovalle-Robles, W. Lei, Small 14 (2018) 1802225.
[23] S. Seyedin, E.R.S. Yanza, J.M. Razal, J. Mater. Chem. A 5 (2017) 24076–24082.
[24] Q. Yang, Z. Xu, B. Fang, T. Huang, S. Cai, H. Chen, Y. Liu, K. Gopalsamy, W. Gao, C. Gao, J. Mater. Chem. A 5 (2017) 22113–22119.
[25] K. Krishnamoorthy, P. Pazhamalai, S. Sahoo, S.-J. Kim, J. Mater. Chem. A 5 (2017) 5726–5736.

[26] X. Zhang, Z. Pei, C. Wang, Z. Yuan, L. Wei, Y. Pan, A. Mahmood, Q. Shao, Y. Chen, Small 15 (2019) 1903817.
[27] S. Zhai, H.E. Karahan, L. Wei, Q. Qian, A.T. Harris, A.I. Minett, S. Ramakrishna, A.-K. Ng, Y. Chen, Energy Storage Mater. 3 (2016) 123–139.
[28] S. Zhai, L. Wei, H.E. Karahan, X. Chen, C. Wang, X. Zhang, J. Chen, X. Wang, Y. Chen, Energy Storage Mater. 19 (2019) 102–123.
[29] H. Peng, Fiber-Shaped Energy Harvesting and Storage Devices, Springer, 2015.
[30] Y. Zhou, C.H. Wang, W. Lu, L. Dai, Adv. Mater. 32 (2020) 1902779.
[31] X. Lu, M. Yu, G. Wang, Y. Tong, Y. Li, Energy Environ. Sci. 7 (2014) 2160–2181.
[32] S. Senthilkumar, Y. Wang, H. Huang, J. Mater. Chem. A 3 (2015) 20863–20879.
[33] H. Lee, M. Kim, I. Kim, H. Lee, Adv. Mater. 28 (2016) 4541–4548.
[34] Z. Li, M. Shao, L. Zhou, R. Zhang, C. Zhang, J. Han, M. Wei, D.G. Evans, X. Duan, Nano Energy 20 (2016) 294–304.
[35] T. Chen, L. Dai, Energy Storage Mater. 2 (2016) 21–26.
[36] S. Zhai, W. Jiang, L. Wei, H.E. Karahan, Y. Yuan, A.K. Ng, Y. Chen, Mater. Horiz. 2 (2015) 598–605.
[37] H. Jin, L. Zhou, C.L. Mak, H. Huang, W.M. Tang, H.L.W. Chan, J. Mater. Chem. A 3 (2015) 15633–15641.
[38] X. Zhang, Q. Li, Y. Tu, Y. Li, J.Y. Coulter, L. Zheng, Y. Zhao, Q. Jia, D.E. Peterson, Y. Zhu, Small 3 (2007) 244–248.
[39] Y.-L. Li, I.A. Kinloch, A.H. Windle, Science 304 (2004) 276–278.
[40] L.M. Ericson, H. Fan, H. Peng, V.A. Davis, W. Zhou, J. Sulpizio, Y. Wang, R. Booker, J. Vavro, C. Guthy, Science 305 (2004) 1447–1450.
[41] W. Ma, L. Song, R. Yang, T. Zhang, Y. Zhao, L. Sun, Y. Ren, D. Liu, L. Liu, J. Shen, Nano Lett. 7 (2007) 2307–2311.
[42] Z. Cai, L. Li, J. Ren, L. Qiu, H. Lin, H. Peng, J. Mater. Chem. A 1 (2013) 258–261.
[43] L. Kou, T. Huang, B. Zheng, Y. Han, X. Zhao, K. Gopalsamy, H. Sun, C. Gao, Nat. Commun. 5 (2014) 1–10.
[44] D. Yu, K. Goh, H. Wang, L. Wei, W. Jiang, Q. Zhang, L. Dai, Y. Chen, Nat. Nanotechnol. 9 (2014) 555.
[45] S. Zhai, C. Wang, H.E. Karahan, Y. Wang, X. Chen, X. Sui, Q. Huang, X. Liao, X. Wang, Y. Chen, Small 14 (2018) 1800582.
[46] C. Wang, S. Zhai, Z. Yuan, J. Chen, X. Zhang, Q. Huang, Y. Wang, X. Liao, L. Wei, Y. Chen, Electrochim. Acta 305 (2019) 493–501.
[47] D. Yu, K. Goh, Q. Zhang, L. Wei, H. Wang, W. Jiang, Y. Chen, Adv. Mater. 26 (2014) 6790–6797.

CHAPTER 10

Recent research and issues

Karolina Urszula Laszczyk
Faculty of Microsystem Electronics and Photonics, Wroclaw University of Science and Technology, Wrocław, Poland

10.1 Introduction

There are some issues and challenges related to the development of biomedical MSCs. First of all, the performance enhancement requires understanding diverse scientific parameters such as the mechanism of diffusion, electron migration, chemical reactions, heat transfer, and ion penetration into the pore networks, which were found to vary upon the operational voltage [1], ions [2, 3], electrolytes [4–6], the active material building the electrodes, and the way the material is synthesized [7–9], finally leading to the strict electrode-electrolyte relationship to consider when assembling the MSCs and SCs in general [10–12]. This all could be applied with limited and specialized in situ techniques. The scanning tunneling microscope (STM) and atomic force microscope (AFM) offer to map the electric double layer (EDL) structure locally with atomic-level resolution [13, 14]. Meanwhile, the neutron scattering [15], electrochemical quartz crystal microbalance [16], X-ray reflectivity [17], X-ray photoelectron spectroscopy [18], elucidating the structure and composition of the electric double layer. The nuclear magnetic resonance (NMR) spectroscopy has been used to study the local environments of anions in supercapacitor electrodes and to quantify changes in the populations of adsorbed species during charging [19]. The surface plasmon resonance (SPR) enables monitoring of the electrochemical reactions of small molecules at electrode surfaces and the visualization of the spatial distribution of surface charge and local electric field [20]. Due to the wide span of materials to be used and their combinations, all of them have not been verified experimentally. Meanwhile, simulation models play a significant role in drawing new perspectives and findings that might lead to the solution [21, 22] but they do not represent the simplified environment. For example, modeling of the charged molecules requires density functional theory (DFT), which does not describe anions adequately. On the other hand, molecular dynamics (MD) simulations do not readily deal with the

Microsupercapacitors
https://doi.org/10.1016/B978-0-08-102888-9.00005-7

partial charge transfer between anions and cations of an ionic liquid (IL) [23]. Therefore combining the in situ characterizations and theoretical simulations with electrochemical characterizations allows to find a more complementary view on the charge storage. It would help in the chemical construction of the electrodes for the particular electrolyte, e.g., by regulating functional groups to balance the active site, wettability, and electrical conductivity [24, 25].

On the other hand, the performance aspects such as a wide potential window (up to 3.7–4.1 V [26]), hence a high energy density, were achieved by the use of asymmetric electrodes of the MSCs. Then, the cathode and anode are composed of various materials, e.g., EDL-based and pseudocapacitive [27, 28], or battery-type and capacitive materials simultaneously [29, 30]. There are also applicable abundant elements such as Li, Na, Mg, or K, which are inserted/extracted into/from the electrode, or metal nitrides offering high electronic conductivities [31].

Another important challenge is the multifunctionality that gives the perspective of the off-grid network for the Internet of Things (IoT) through a self-charging function of the MSCs. It offers new functionality without the necessity to form an additional wire and component connection within the electronic circuit, which might be of high interest for biomedical and implantable devices. Nonetheless, almost all of these issues demand novel materials and the compatible technology as well as application to the existing models. For example, new perspectives open metal-organic frameworks (MOFs) also refer to the coordination polymers, porous ionic liquids, or hybrid electrolytes. The first material can be applied by making use of controllable morphology, high specific surface area, copious pores, and multifunctionalities [32], while the second might lead to the synthesis of other porous liquids, of which guest-uptake properties [33] may result in novel functions in the field of energy storage. The third example, i.e., the hybrid electrolytes, relies on mixing the different types, such as aqueous and organic, aqueous, and gel polymers [34].

Although there are many interesting approaches to develop MSCs, in this chapter the multifunctionality and biomedical applications are of special consideration. The multifunctionality facilitates developing autonomous electronic devices and off-grid energy storage systems based on available and renewable energies. Biomedical MSCs offer long cyclability and operational stability. Yet both features can be combined into one device opening the route to develop novel continuous power for advanced sustainable electronics and medical examination systems for sensing and actuating.

As mentioned above, one key disadvantage of existing fabrication processes for μSCs is their poor compatibility with major components of wearable devices, such as textile, paper, and plastics.

10.2 Multifunctional microsupercapacitors

"Multifunctional" means the material or device having an additional function, other than assigned due to its application. In the case of the energy storage electrochemical cell, it can be the recovery to its shape after deformation, change in microenvironment polarity, cleavage or rebonding of covalent bonds, changing color and luminescence, and self-healing behavior. This additional functionality indicates the intrinsic property of the materials used for electrodes, electrolytes, or additives. The majority of these materials are so-called stimuli-responsive polymers [35]. The existence, form, and mechanism of responsive polymer materials can be embodied in molecular chains, colloidal particles, self-assembled ordered aggregates, (hydro)gels, thin films, liquid crystal (LC) elastomers, and bulk materials.

Among the stimuli-responsive functions applied for the SCs or MSCs, there are (1) changing the color as the result of electrochemical reduction/oxidation processes, i.e., electrochromism; (2) self-charging, by piezoelectric or triboelectric effect; (3) harvesting electrical energy by thermal-to-electrical energy conversion techniques, such as the Seebeck effect, thermally driven ion diffusion, and temperature-dependent electrochemical redox potential; (4) photo self-charging, i.e., photovoltaic electricity generation; (5) returning to the original shape and removing all plastic deformations after being bent, fold, etc., i.e., shape-memory effect, as well as (6) restoring the configuration integrity and electrical properties after mechanical damage, i.e., self-healability, (7) protection from the temperature increase, by use of thermally responsive polymers.

Various transition metal oxides (WO_3, $W_{18}O_{49}$, NiO) and conducting polymers (PANI) were used as colayered films of the electrochromic electrodes for SCs [36–38]. Electrochromic SCs change their colors while they store energy; hence to make it visible, the SCs require transparent and conductive current collectors. Besides Ti foil [39], crystal materials, such as indium tin oxide (ITO)-coated glass or fluorine-doped tin oxide (FTO)-coated glass, are used as the current collector [36]. But they are brittle, which prevents the development of flexible electrochromic electrodes. Nonetheless, flexible silver grid/polyethylene terephthalate (PET) was presented [40] as well as poly(3,4-ethylenedioxythiophene)–polystyrenesulfonate

(PEDOT:PSS) [38] or in general CNTs, nanowires grids, and organic fluids were proposed [41]. The electrochromic electrodes have multipurpose applications: chemical and biosensing, gas sensing, mechanical sensing, smart windows for light transmittance and thermal regulation, entertainment, and technology resembling the microsupercapacitors. However, there is an issue to use them for energy storage, related to the most important operating parameters of an electrochromic device, i.e., color efficiency and the switching time. The first is irreversibly proportional to charge density (the electronic charge inserted into or extracted from the electrochromic material per unit area). The second is defined as the time required for a system to reach 90% of its full optical modulation and also the less the charge is acquired, the faster the response becomes [42]. Therefore a high coloration efficiency and high speed of the switching processes require a low charge density that stays opposite to the energy storage, which needs high charge density. There are few electrochromic MSCs presented using the V_2O_5 nanoribbons and exfoliated graphene [43], Mxene-PEDOT heterostructures [44], or a solid-state hybrid organic-inorganic PEDOT:PSS–MoO_3 [45] offering a color-switching time of few seconds. The examples of the basic architecture of electrochromic SCs are presented in Fig. 10.1.

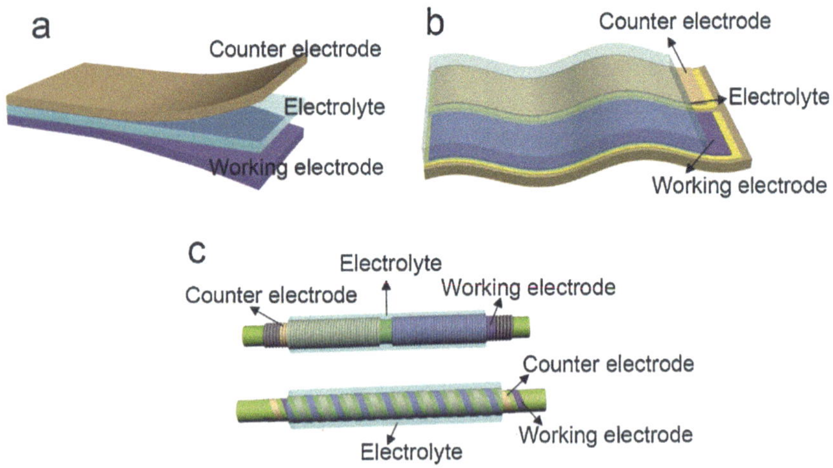

Fig. 10.1 Schematic designs of flexible wearable electrochromic devices with (A) layered design, (B) lateral design, and (C) fiber design. *(Reprinted with permission from Z. Wang, X. Wang, S. Cong, F. Geng, Z. Zhao, Fusing electrochromic technology with other advanced technologies: a new roadmap for future development, Mater. Sci. Eng. R 140 (2020) 100524).*

Self-charging SCs, also found as nanogenerator-based energy storage devices relying on the piezoelectric and triboelectric nanogenerators, solve issues of continuous energy supply of the microelectronic devices [46]. One of the first proposals was applied for Li-ion batteries and SCs [47–49]. In the case of the piezoelectric SCs, the role of the multifunctionality is played by a polymer or piezo-electrolyte separator. The mechanical strain along the polarized film causes anion and cation migration toward the interface of the positive and negative electrodes. As a consequence, the electronic double layer or pseudocapacitance is formed, and the energy can be stored electrochemically. For example, instead of insulating cellulose as the separator, a gel electrolyte enriched with piezoelectric compound ($BaTiO_3$ powder) [50] or a well-polarized film is applied based on poly(vinylidene fluoride) (PVDF) [49, 51–53]. The triboelectric-based charging goes by an ambient mechanical motion. Due to a pulse signal generated and a low current output, the triboelectric effect is not suitable to drive microelectronic device directly, hence it is integrated with the MSCs (Fig. 10.2) [54–57].

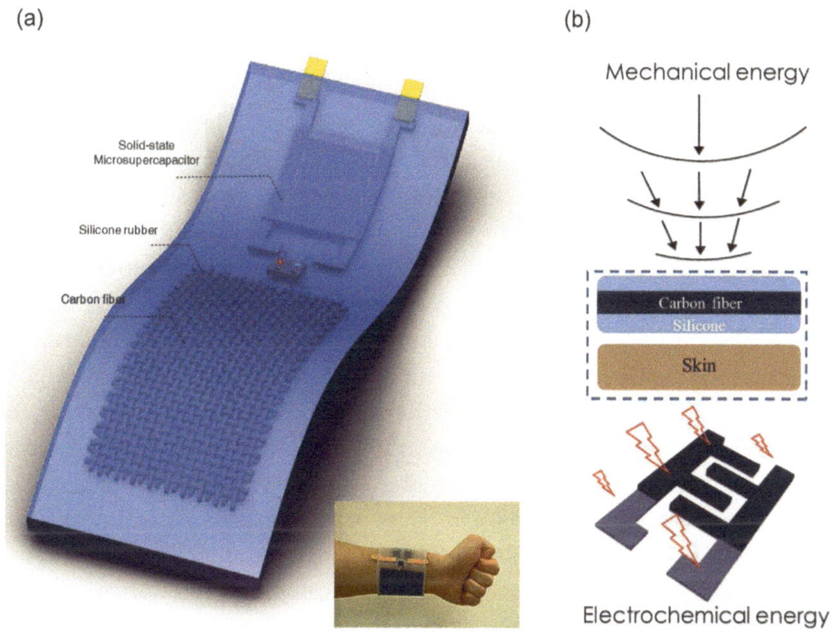

Fig. 10.2 Wearable self-charging system based on the integration of a triboelectric harvester and a microsupercapacitor. *(Reprinted with permission from Q. Jiang, C. Wu, Z. Wang, A.C. Wang, H. He Jr, Z.L. Wang, H.N. Alshareef, MXene electrochemical microsupercapacitor integrated with triboelectric nanogenerator as a wearable self-charging power unit, Nano Energy 45 (2018) 266–272).*

Particularly, the charging takes place when two materials are brought into contact through rubbing, pressing, or friction. There are a series of material compositions, e.g., cotton and glass, silk and paper, and many others, also reported for the pair of the same materials. Moreover, it is influenced by environmental parameters (e.g., humidity, load stress, etc.) as well as material surface properties (e.g., microstructure patterns, surface roughness, etc.) [58]. However, it remains a huge challenge to achieve self-charging SC with versatile capabilities in recharging with a high specific capacity, and yet transferring it on microscale, especially in the case of these SCs. The mechanical (compressional) force is the key to make them serve as self-charging devices [59]. In the case of using the triboelectric effect, the formation of a water film at the electrode interfaces may deteriorate charge transfer or induce charge dissipation. If not well encapsulated, the contamination of surfaces by dust and particulates also deteriorates its performance and durability [60].

The thermal and photonic self-charging MSCs could become useful in environments with large temperature differences ("low-grade" heat, i.e., less than 100–200 °C [61]) and sunlight exposure. It is significant in the context of off-grid energy storage systems due to ubiquitous unwanted energy losses. Thermally self-charging MSCs can absorb heat when the temperature rises and releases electricity during the discharging process. They use the conversion techniques including the electron diffusion by the temperature gradient (the Seebeck effect) [61], temperature-dependent electrochemical redox potential [62], and thermally driven ion diffusion caused by the Soret effect [63, 64]. In the case of the Seebeck effect, the architecture of electrodes must contain hundreds of p- and n-type thermoelectric legs, yet connected electrically in series and thermally in parallel increasing the complexity of device fabrication as well as its electrical resistance [65]. In general, to develop high-performance thermoelectric devices, the requirements are not only adequate thermoelectric material, but also contacts, thermal interfaces, mechanical properties, interconnects, and packaging technique [65]. The Soret effect of a redox-free electrolyte, i.e., from ionic charge carriers, is similarly to the electronic Seebeck one. Due to thermodiffusion of ions in the electrolyte, a gradient of an ionic concentration is created, where a corresponding thermo-voltage is governed by the temperature difference across the material and the ionic Seebeck coefficient [64]. Temperature-dependent electrochemical redox potential (also known as thermogalvanic effect) is found in electrolytes with redox couples, such as ferricyanide/ferrocyanide [62]. Although this approach is promising, its disadvantages are the low voltage,

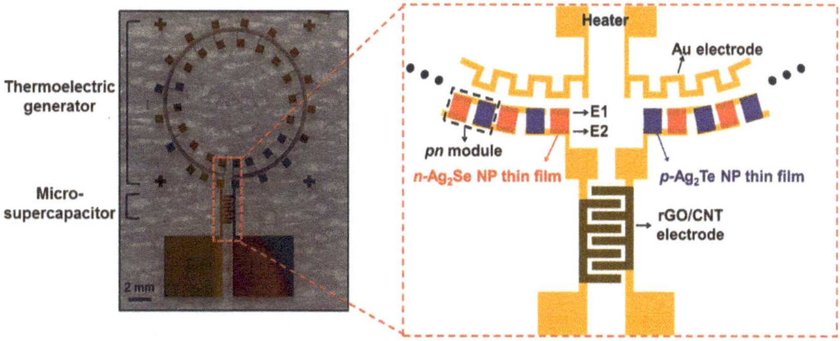

Fig. 10.3 Optical image (*left*) and a schematic illustration (*right*) of the integrated MSC and thermoelectric generator. *(Reprinted with permission from S. Yang, K. Cho, S. Kim, Energy devices generating and storing electricity from finger and solar thermal energy, Nano Energy 69 (2020) 104458).*

less than 0.5 V [66, 67] and difficulty to fabricate planar portable and flexible electronic devices, Because vertically stacked electrodes architecture with a thick electrolyte layer (~few millimeter) are preferable. Thus integration of MSC and thermoelectric generator is one of the solutions (Fig. 10.3) [68, 69].

Photonic self-charging is implemented by either the integration of the solar cells and supercapacitors or using the supercapacitor itself as a photodetector. In the integration strategy, various photoactive materials were demonstrated, such as perovskites [70, 71], silicon [72], CNTs [73], and many others. Meanwhile, for the heteroelectrodes, the preferred materials are photoresponsive and electrocapacitive, e.g., based on WO_3 [74], ZnO [75], graphene/Fe_2O_3 [76], and conducting polymers by the adequate molecule design to improve energy conversion efficiency and sensitivity to light [77, 78]. Moreover, the other pseudocapacitive materials, e.g., V_2O_5 and Ni(OH), have also been extensively studied as active materials in photoelectrochemical systems [74]. However, photo-charging performances are still limited by the voltage response (<0.5 V) and its improvement must be addressed to the photovoltaic material. On the other hand, the photodetecting SC with long cyclability is not an easy task, since most of the efficient photovoltaic materials do not possess such features directing to electrode separation and electrolyte leakage under bending. To solve this problem, the researchers proposed a wrapping fiber structure [79, 80].

The shape-memory effect enables a piece of material to recover to its initial shape or size after deformations when subjected to various stimuli such as temperature, light, magnetic fields, and electric currents, and many others. Such materials are usually shape memory alloys (SMAs) or polymers (SMPs) [81]. However, the shape-memory effect has been demonstrated for the composites loaded with nanoparticles [82], pseudocapacitive flakes [83], or nanowires [84, 85]. The recent examples are dominated by fiber-shaped microscale SCs [86, 87], which could be adapted to transferring onto various types of substrates thanks to the freestanding property of the developed polypyrolle (PPy)-based electrode. It should be noted here, in some cases, high-aspect-ratio nanoparticles, yet characterized by outstanding mechanical and electrical properties, can achieve a similar goal. Cao et al. [88] demonstrated a sticky-note SC with easy and repeated adhesive performance. The sticky-note SC could be attached to various substrates including cloths, glasses, papers, plastics, and metals. Up to 200 attaching/removing cycles on different substrates, the capacitance of the supercapacitor note can be well maintained at above 99%. To achieve such an advance at once with flexibility and well performance, the authors employed a novel kind of sticky aligned CNT array electrode, which does not possess a shape-memory feature. Instead, the vertically aligned CNT structure with ordered nanoscale CNT junctions formed in the top region enabled conductivity, high porosity, and more importantly superior repeated adhesive properties with the well-retained CNT junction structure. Moreover, due to the high electrical conductivity, the sticky-note supercapacitors can be effectively connected in series by directly stacking them to increase the output voltage without the use of extraconducting wires. Most of the examples for multifunctional energy storage devices are macroscale supercapacitors possessing fibrous or sandwiched-shape formed by electrodeposition and manual assembly. Nevertheless, the poor compatibility between the active materials and processing technologies remains a great challenge to integrate various kinds of stimuli into modern functional devices without sacrificing the device performance. Especially, this task requires all the constituent parts including electrodes, electrolyte, and separator to withstand the deformability without degradation of the electrical and chemical properties. Meanwhile, the existing fabrication processes for MSCs offer poor compatibility with major components of wearable devices such as textiles, papers, or plastics. Emerging technologies like 3D printing [89], direct ink writing (DIW) [90], and stamping [91–93] have the potential to make functional inks directly printed onto soft materials as the substrates (see a chapter dedicated to MSC technology). These technologies can find an application for self-healable MSCs, which open the perspective to enhance the MSC structural safety and extend lifetime. The latest is mostly affected by the gradual degradation of the integrity between the MSC components. Achieving the materials capable of self-healing and restoration of electrical conductivity after the damage was a milestone for flexible and stretchable electronics, and nowadays we can observe the transfer of self-healable materials for SCs and MSCs. One of the first examples was the SC that applies mechanical and electrical self-healing through spreading the active material films onto self-healing substrates composed of a supramolecular network and hierarchical flower-like TiO_2 nanostructures. After the damage took place, lateral movement of the self-healing composite layer brings the separated areas of the single-walled CNT layer into contact, hence enabling the restoration of the device's configuration and conductivity [94]. Other examples use carboxylated polyurethane (PU) shell [95] or the gel-like materials [96] as the constructing base for the whole cell, at once playing a role of the electrolyte, separator, and electrodes incorporating the active material for the charge storage. By analogy to aqueous and ion gel electrolytes, the reported were gel-like materials' hydrogels with the high ionic conductivity but the limited voltage range to 1 V. The material comprised a lot of water and a small amount of polymer [97, 98] and ion gels, which immobilize ionic liquids with polymers [99].

The thermal runaway poses a significant safety issue. It can happen when the SC or MSC works at high environmental temperatures, and they also can generate a lot of heat themselves. If the heat cannot be dissipated efficiently, the lifespan is shortened and the current leakage or even the fire might occur. Therefore a thermo-reversible self-protective or thermo-switchable SCs and MSCs were proposed. The active role was played by thermosensitive or thermally responsive polymers, which are loaded on either the electrode surface or electrolytes. Those materials block the ion transport at elevated temperature, meanwhile enable to restore completely the ion accessibility to the electrodes at room temperature, as well as excellent reversibility upon heating–cooling. For example, upon changing the temperature to 40 °C, the polymer collapses and inhibits ion transport, resulting in the weakened power delivery. As the temperature decreases to 20 °C the polymer dissolves and releases the accessible ion channels for the normal charge-discharge process [100]. Another example is to utilize the thermodynamic behavior of a smart electrolyte that can reversibly transform between a liquid state and a hydrogel state upon heating up to 60 °C and cooling cycles [101, 102].

10.3 Microsupercapacitors for the medical purpose

Thanks to the small size, long cyclability, flexible and stretchable electrodes, the MSCs provide an opportunity for supplementary application in medical examinations and powering the medical implants. However, to be used for medical purposes, the MSCs need first of all biocompatible materials with a low safety risk in the diagnosis process. Moreover, if the biodegradable electronics is applied, monitoring in terms of real-time in vivo detection becomes easy and the repeated surgery can be avoided [103]. Although the research on the MSCs with biocompatible materials for electrodes, electrolytes, and also separators [104, 105] has been expanding for the last decade, there are still few examples of the medical MSCs reported in the scientific literature. For example, biosupercapacitor (bSC) proposed by Mosa et al. [106] consists of biophilized reduced graphene oxide electrodes in a calf serum as an electrolyte. The bSC achieved an energy density up to 1.8 mW h cm^{-3}, which is 3–11 times higher than commercially available thin-film electrochemical capacitors. The electrode material showed no signs of cell toxicity at concentrations 160 times higher than those used in the demonstrated bSC. Moreover, the encapsulated bSC showed no adverse effects on cells after 4 days of continuous charge/discharge cycles.

Another important role of the MSC in the wearable electronic devices is to present the circuits to monitor personal health status in real-time. For example, the MSC was used to power continuous monitoring performance and sensitivities for ions and glucose in sweat (Fig. 10.4), or incorporated into the soft lens without obstructing vision. In both examples, the MSCs were combined with wireless information transfer technology to achieve real-time monitoring; yet in the case of the soft lens, the MSC could be charged thanks to the inductive wireless system [107, 108].

In addition to the implantation of permanent and biodegradable devices, the digestive system may be an alternative to modulate cellular and organ function without the need for implantation. Wang et al. [109] demonstrated a fully edible supercapacitor made from food-grade materials that supports killing bacteria (through the electric current) and powering medical devices (an endoscope). The edible MSC possesses electrodes consisting of activated charcoal from dietary supplements bound with egg whites, an edible gold leaf (edible bakery artisan decoration) as the current collector (Fig. 10.5). Both components demonstrate high electrical conductivity and chemical stability, and to prevent the short circuit, a sheet of seaweed is used as a separator. As an electrolyte, a sports drink has high concentrations of sodium,

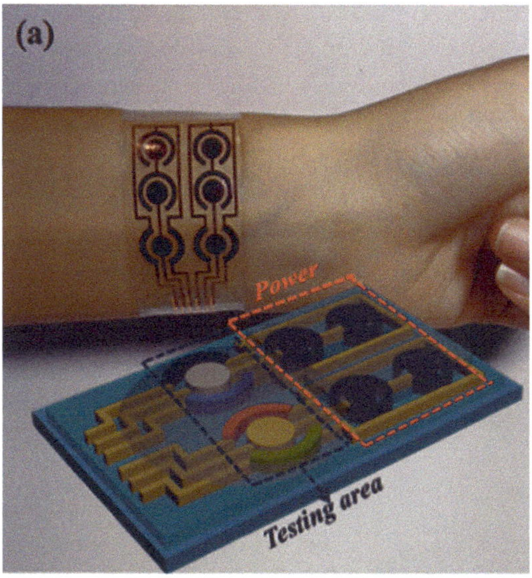

Fig. 10.4 Schematic illustration of the glucose sensor with the integrated MSCs. *(Reprinted with permission from Y. Lu, K. Jiang, D. Chen, G. Shen, Wearable sweat monitoring system with integrated micro-supercapacitors, Nano Energy 58 (2019) 624–632).*

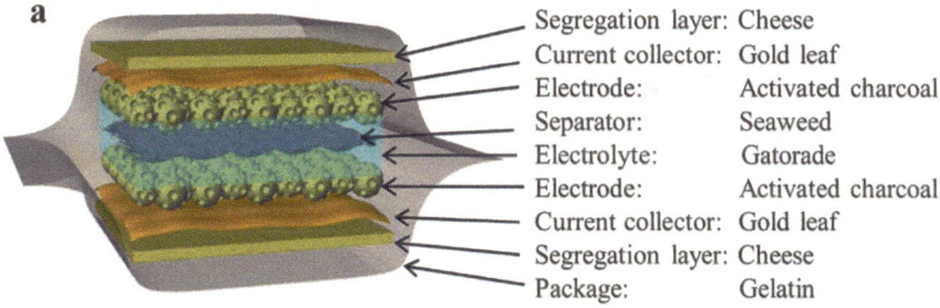

Fig. 10.5 Schematic illustration presenting an edible MSC. *(Reprinted with permission from X. Wang, W. Xu, P. Chatterjee, C. Lv, J. Popovich, Z. Song, L. Dai, M.Y.S. Kalani, S.E. Haydel, H. Jiang, Food-materials-based edible supercapacitors, Adv. Mater. Technol. 1 (2016) 1600059).*

potassium, and citrate, and yet high ionic conductivity. The complete capacitor was sealed inside gelatin sheets, with cheese slices used to segregate the electrolyte from the gelatin. The edible supercapacitor retains about 92% of capacitance after 1000 cycles with a low leakage current and the degradation mainly results from the electrolyte absorption by the gelatin sheet. In order to improve the longevity of gelatin in the digestive fluids, the authors proposed the surface modification and crosslinking of gelatin. The device was also demonstrated to be charged wirelessly. Another edible MSC for the endoscopy, made from food constituents by template-imprinting strategy, was demonstrated by Gao et al. [110]. The MSC's electrode and electrolyte can be ingestible, swallowable, and absorbed or degraded by the human body. The authors used gelatin to enhance the integrity and robustness of the jelly as the substrate, together with gold leaf as the current collector, activated charcoal (AC) as the electrodes, and an agar aqueous solution as the electrolyte, respectively. The two MSCs connected in series could light a red LED. Meanwhile, the MSCs in the capsule can be charged even in the simulated stomach gastric juice for 28 min.

References

[1] S. Kondrat, C.R. Perez, V. Presser, Y. Gogotsi, A.A. Kornyshev, Effect of pore size and its dispersity on the energy storage in nanoporous supercapacitors, Energy Environ. Sci. 5 (2012) 6474–6479.
[2] P. Ratajczak, M.E. Suss, F. Kaasik, F. Béguin, Carbon electrodes for capacitive technologies, Energy Storage Mater. 16 (2019) 126–145.
[3] P. Fayon, J.M.H. Thomas, A. Trewin, Structure and properties of a nanoporous supercapacitor, J. Phys. Chem. C 120 (2016) 25880–25891.
[4] D. Jiang, J. Wu, Unusual effects of solvent polarity on capacitance for organic electrolytes in a nanoporous electrode, Nanoscale 6 (2014) 5545–5550.
[5] A. Karatrantos, S. Khan, T. Ohba, Q. Cai, The effect of different organic solvents on sodium ion storage in carbon nanopores, Phys. Chem. Chem. Phys. 20 (2018) 6307–6315.
[6] C. Prehal, C. Koczwara, H. Amenitsch, V. Presser, O. Paris, Salt concentration and charging velocity determine ion charge storage mechanism in nanoporous supercapacitors, Nat. Commun. 9 (2018) 4145.
[7] J. Yang, H. Su, C. Lian, Y. Shang, H. Liu, J. Wu, Understanding surface charge regulation in silica nanopores, Phys. Chem. Chem. Phys. 22 (2020) 15373–15380.
[8] C. Merlet, C. Pean, B. Rotenberg, P.A. Madden, B. Daffos, P.L. Taberna, P. Simon, M. Salanne, Highly confined ions store charge more efficiently in supercapacitors, Nat. Commun. 4 (2013) 2701.
[9] D. Wang, Z. Lu, L. Xu, Microstructure design of porous nanocarbons for ultrahigh energy and power density supercapacitors in ionic liquid electrolyte, J. Mater. Sci. 55 (2020) 7477–7491.
[10] A.C. Forse, C. Merlet, J.M. Griffin, C.P. Grey, New perspectives on the charging mechanisms of supercapacitors, J. Am. Chem. Soc. 138 (2016) 5731–5744.
[11] X. Wang, M. Salari, D. Jiang, J.C. Varela, B. Anasori, D.J. Wesolowski, S. Dai, M.W. Grinstaff, Y. Gogotsi, Electrode material–ionic liquid coupling for electrochemical energy storage, Nat. Rev. Mater. 5 (2020) 787–808.
[12] A.A. Lee, D. Vella, A. Goriely, S. Kondrat, Capacitance-power-hysteresis trilemma in nanoporous supercapacitors, Phys. Rev. X 6 (2016), 021034.
[13] I. Siretanu, D. Ebeling, M.P. Andersson, S.L. Stipp, A. Philipse, M.C. Stuart, D.D.E. Van, F. Mugele, Direct observation of ionic structure at solid-liquid interfaces: a deep look into the stern layer, Sci. Rep. 4 (2014) 4956.

[14] M. Gnahm, C. Berger, M. Arkhipova, H. Kunkel, T. Pajkossy, G. Maas, D.M. Kolb, The interfaces of Au(111) and Au(100) in a hexaalkyl-substituted guanidinium ionic liquid: an electrochemical and in situ STM study, Phys. Chem. Chem. Phys. 14 (2012) 10647–10652.
[15] Y. Lauw, M.D. Horne, T. Rodopoulos, V. Lockett, B. Akgun, W.A. Hamilton, A.R.J. Nelson, Structure of [C4mpyr][NTf2] room-temperature ionic liquid at charged gold interfaces, Langmuir 28 (2012) 7374–7381.
[16] J.M. Griffin, A.C. Forse, W.Y. Tsai, P.L. Taberna, P. Simon, C.P. Grey, In situ NMR and electrochemical quartz crystal microbalance techniques reveal the structure of the electrical double layer in supercapacitors, Nat. Mater. 14 (2015) 812–819.
[17] E. Sloutskin, M.O. Benjamin, L. Tamam, I. Kuzmenko, A.T. Gog, M. Deutsch, Surface layering in ionic liquids: an X-ray reflectivity study, J. Am. Chem. Soc. 127 (2005) 7796–7804.
[18] G. Olivieri, A. Goel, M.A. Brown, Molecular level insight on the adsorption of carboxylic acids to oxide nanoparticles in aqueous solution by X-ray photoelectron spectroscopy, Chem. Commun. 52 (2016) 9040–9043.
[19] J.M. Griffin, A.C. Forse, H. Wang, N.M. Trease, P.L. Taberna, P. Simon, C.P. Grey, Ion counting in supercapacitor electrodes using NMR spectroscopy, Faraday Discuss. 176 (2014) 49–68.
[20] X. Luo, S. Denga, P. Wang, Temporal–spatial-resolved mapping of the electrical double layer changes by surface plasmon resonance imaging, RSC Adv. 8 (2018) 28266–28274.
[21] H. Tao, C. Lian, H. Liu, Multiscale modeling of electrolytes in porous electrode: from equilibrium structure to non-equilibrium transport, Green Energy Environ. 5 (2020) 303–321.
[22] B.A. Ali, N.K. Allam, A first-principles roadmap and limits to design efficient supercapacitor electrode materials, Phys. Chem. Chem. Phys. 21 (2019) 17494.
[23] E. Jonsson, Ionic liquids as electrolytes for energy storage applications—a modelling perspective, Energy Storage Mater. 25 (2020) 827–835.
[24] X. Su, J. Ye, Y. Zhu, Advances in in-situ characterizations of electrode materials for better supercapacitors, J. Energy Chem. 54 (2021) 242–253.
[25] Y. Wu, Y. Zhang, Y. Liu, P. Cui, S. Chen, Z. Zhang, J. Fu, E. Xie, Boosting the electrochemical performance of graphene-based on-chip micro-supercapacitors by regulating the functional groups, ACS Appl. Mater. Interfaces 12 (2020) 42933–42941.
[26] C.F. Liu, Y.C. Liu, T.Y. Yi, C.C. Hu, Carbon materials for high-voltage supercapacitors, Carbon 145 (2019) 529–548.
[27] W. Liu, M. Zhang, M. Li, B. Li, W. Zhang, G. Li, M. Xiao, J. Zhu, A. Yu, Z. Chen, Advanced electrode materials comprising of structure-engineered quantum dots for high-performance asymmetric micro-supercapacitors, Adv. Energy Mater. 10 (2020) 1903724.
[28] C. Couly, M. Alhabeb, K.L. Van Aken, N. Kurra, L. Gomes, A.M. Navarro-Suárez, B. Anasori, H.N. Alshareef, Y. Gogotsi, Asymmetric flexible MXene-reduced graphene oxide microsupercapacitor, Adv. Electron. Mater. 4 (2018) 1700339.
[29] W. Zuo, R. Li, C. Zhou, Y. Li, J. Xia, J. Liu, Battery-supercapacitor hybrid devices: recent progress and future prospects, Adv. Sci. 4 (2017) 1600539.
[30] P. Zhang, Y. Li, G. Wang, F. Wang, S. Yang, F. Zhu, X. Zhuang, O.G. Schmidt, X. Feng, Zn Li, hybrid micro-supercapacitors with ultrahigh areal energy density and long-term durability, Adv. Mater. 31 (2018) 1806005.
[31] F. Wang, X. Wu, X. Yuan, Z. Liu, Y. Zhang, L. Fu, Y. Zhu, Q. Zhou, Y. Wu, W. Huang, Latest advances in supercapacitors: from new electrode materials to novel device designs, Chem. Soc. Rev. 46 (2017) 6816.

[32] T. Mehtab, G. Yasin, M. Arif, M. Shakeel, R.M. Korai, M. Nadeem, N. Muhammad, X. Lu, Metal-organic frameworks for energy storage devices: batteries and supercapacitors, J. Energy Storage 21 (2019) 632–646.
[33] L. Ma, C.J.E. Haynes, A.B. Grommet, A. Walczak, C.C. Parkins, C.M. Doherty, L. Longley, A. Tron, A.R. Stefankiewicz, T.D. Bennett, J.R. Nitschke, Coordination cages as permanently porous ionic liquids, Nat. Chem. 12 (2020) 270–275.
[34] W. Ye, H. Wang, J. Ning, Y. Zhong, Y. Hu, New types of hybrid electrolytes for supercapacitors, J. Energy Chem. 57 (2021) 219–232.
[35] K. Wang, K. Amin, Z. An, Z. Cai, H. Chen, H. Chen, Y. Dong, X. Feng, W. Fu, J. Gu, Y. Han, D. Hu, R. Hu, D. Huang, F. Huang, F. Huang, Y. Huang, J. Jin, X. Jin, Q. Li, T. Li, Z. Li, Z. Li, J. Liu, J. Liu, S. Liu, H. Peng, A. Qin, X. Qing, Y. Shen, J. Shi, X. Sun, B. Tong, B. Wang, H. Wang, L. Wang, S. Wang, Z. Wei, T. Xie, C. Xu, H. Xu, Z.-K. Xu, B. Yang, Y. Yu, X. Zeng, X. Zhan, G. Zhang, J. Zhang, M.Q. Zhang, X.-Z. Zhang, X. Zhang, Y. Zhang, Y. Zhang, C. Zhao, W. Zhao, Y. Zhou, Z. Zhou, J. Zhu, X. Zhu, B.Z. Tang, Advanced functional polymer materials, Mater. Chem. Front. 4 (2020) 1803–1915.
[36] G. Cai, X. Wang, M. Cui, P. Darmawan, J. Wang, A.L.S. Eh, P.S. Lee, Ultra-large optical modulation of electrochromic porous WO3 film and the local monitoring of redox activity, Nano Energy 12 (2015). 258-26.
[37] X. Chen, H. Lin, J. Deng, Y. Zhang, X. Sun, P. Chen, X. Fang, Z. Zhang, G. Guan, H. Peng, Electrochromic fiber-shaped supercapacitors, Adv. Mater. 26 (2014) 8126–8132.
[38] Y. Tian, S. Cong, W. Su, H. Chen, Q. Li, F. Geng, Z. Zhao, Synergy of W18O49 and polyaniline for smart supercapacitor electrode integrated with energy level indicating functionality, Nano Lett. 14 (2014) 2150–2156.
[39] P. Yang, P. Sun, L. Du, Z. Liang, W. Xie, X. Cai, L. Huang, S. Tan, W. Mai, Quantitative analysis of charge storage process of tungsten oxide that combines pseudocapacitive and electrochromic properties, J. Phys. Chem. C 119 (2015) 16483–16489.
[40] G. Cai, P. Darmawan, M. Cui, J. Wang, J. Chen, S. Magdassi, P.S. Lee, Highly stable transparent conductive silver grid/PEDOT:PSS electrodes for integrated bifunctional flexible electrochromic supercapacitors, Adv. Energy Mater. 6 (2016) 1501882.
[41] Z. Wang, X. Wang, S. Cong, F. Geng, Z. Zhao, Fusing electrochromic technology with other advanced technologies: a new roadmap for future development, Mater. Sci. Eng. R 140 (2020) 100524.
[42] P. Yang, P. Sun, W. Mai, Electrochromic energy storage devices, Mater. Today 19 (2016) 394–402.
[43] P. Zhang, F. Zhu, F. Wang, J. Wang, R. Dong, X. Zhuang, O.G. Schmidt, X. Feng, Stimulus-responsive micro-supercapacitors with ultrahigh energy density and reversible electrochromic window, Adv. Mater. 29 (2017) 1604491.
[44] J. Li, A. Levitt, N. Kurra, K. Juan, N. Noriega, X. Xiao, X. Wang, H. Wang, H.N. Alshareef, Y. Gogotsi, MXene-conducting polymer electrochromic microsupercapacitors, Energy Storage Mater. 20 (2019) 455–461.
[45] M. Yoonessi, A. Borenstein, M.F. El-Kady, C.L. Turner, H. Wang, A.Z. Stieg, L. Pilon, Hybrid transparent PEDOT:PSS molybdenum oxide battery-like supercapacitors, ACS Appl. Energy Mater. 2 (7) (2019) 4629–4639.
[46] K. Zhao, Y. Wang, L. Han, Y. Wang, X. Luo, Z. Zhang, Y. Yang, Nanogenerator-based self-charging energy storage devices, Nano-Micro Lett. 11 (19) (2019) 1–19.
[47] X. Xue, S. Wang, W. Guo, Y. Zhang, Z.L. Wang, Hybridizing energy conversion and storage in a mechanical-to-electrochemical process for self-charging power cell, Nano Lett. 12 (2012) 5048–5054.
[48] L.L. Xing, Y.X. Nie, X.Y. Xue, Y. Zhang, PVDF mesoporous nanostructures as the piezo-separator for a self-charging power cell, Nano Energy 10 (2014) 44–52.

[49] A. Ramadoss, B. Saravanakumar, S.W. Lee, Y.S. Kim, S.J. Kim, Z.L. Wang, Piezoelectric-driven self-charging supercapacitor power cell, ACS Nano 9 (2015) 4337–4345.
[50] D. Zhou, F. Wang, X. Zhao, J. Yang, H. Lu, L.-Y. Lin, L.-Z. Fan, Self-chargeable flexible solid-state supercapacitors for wearable electronics, ACS Appl. Mater. Interfaces 12 (2020) 44883–44891.
[51] A.J. Lovinger, Ferroelectric polymers, Science 220 (1983) 1115–1121.
[52] H. He, Y. Fu, T. Zhao, X. Gao, L. Xing, Y. Zhang, X. Xue, All-solid-state flexible self-charging power cell basing on piezo-electrolyte for harvesting/storing body-motion energy and powering wearable electronics, Nano Energy 39 (2017) 590–600.
[53] S. Sahoo, K. Krishnamoorthy, P. Pazhamala, V.K. Mariappan, S. Manoharan, S.-J. Kim, High performance self-charging supercapacitors using a porous PVDF-ionic liquid electrolyte sandwiched between two-dimensional graphene electrodes, J. Mater. Chem. A 7 (2019) 21693–21703.
[54] J. Luo, F.R. Fan, T. Jiang, Z. Wang, W. Tang, C. Zhang, M. Liu, G. Cao, Z.L. Wang, Integration of micro-supercapacitors with triboelectric nanogenerators for a flexible self-charging power unit, Nano Res. 8 (12) (2015) 3934–3943.
[55] K. Dong, Y.-C. Wang, J. Deng, Y. Dai, S.L. Zhang, H. Zou, B. Gu, B. Sun, Z.L. Wang, A highly stretchable and washable all-yarn-based self-charging knitting power textile composed of fiber triboelectric nanogenerators and supercapacitors, ACS Nano 11 (9) (2017) 9490–9499.
[56] X. Feng, K. Wang, S. Li, Triboelectric nanogenerator and integration with electrochemical microsupercapacitor, in: Proceedings—2019 4th International Conference on Control, Robotics and Cybernetics, CRC 2019 September 2019, Tokyo, Japan, Article number 9058810, 2019, pp. 169–173.
[57] Q. Jiang, C. Wu, Z. Wang, A.C. Wang, H. He Jr., Z.L. Wang, H.N. Alshareef, MXene electrochemical microsupercapacitor integrated with triboelectric nanogenerator as a wearable self-charging power unit, Nano Energy 45 (2018) 266–272.
[58] S. Pan, Z. Zhang, Fundamental theories and basic principles of triboelectric effect: a review, Friction 7 (1) (2019) 2–17.
[59] A. Rasheed, W. He, Y. Qian, H. Park, D.J. Kang, Flexible supercapacitor-type rectifier-free self-charging power unit based on a multifunctional polyvinylidene fluoride–ZnO–rGO piezoelectric matrix, ACS Appl. Mater. Interfaces 12 (2020) 20891–20900.
[60] Q. Zhou, K. Lee, K.N. Kim, J.G. Park, J. Pan, J. Bae, J.M. Baik, T. Kim, High humidity- and contamination-resistant triboelectric nanogenerator with superhydrophobic interface, Nano Energy 57 (2019) 903–910.
[61] S. Horike, Q. Wei, K. Kirihara, M. Mukaida, T. Sasaki, Y. Koshiba, T. Fukushima, K. Ishida, Outstanding electrode-dependent seebeck coefficients in ionic hydrogels for thermally chargeable supercapacitor near room temperature, ACS Appl. Mater. Interfaces 12 (2020) 43674–43683.
[62] R. Hu, B.A. Cola, N. Haram, J.N. Barisci, S. Lee, S. Stoughton, G. Wallace, C. Too, M. Thomas, A. Gestos, M.E. dela Cruz, J.P. Ferraris, A.A. Zakhidov, R.H. Baughman, Harvesting waste thermal energy using a carbon-nanotube-based thermo-electrochemical cell, Nano Lett. 10 (2010) 838–846.
[63] H. Wang, D. Zhao, Z.U. Khan, S. Puzinas, M.P. Jonsson, M. Berggren, X. Crispin, Ionic thermoelectric figure of merit for charging of supercapacitors, Adv. Electron. Mater. 3 (2017) 1700013.
[64] D. Zhao, H. Wang, Z.U. Khan, J.C. Chen, R. Gabrielsson, M.P. Jonsson, M. Berggren, X. Crispin, Ionic thermoelectric supercapacitors, Energy Environ. Sci. 9 (2016) 1450–1457.

[65] R. He, G. Schierning, K. Nielsch, Thermoelectric devices: a review of devices, architectures, and contact optimization, Adv. Mater. Technol. 3 (2018) 1700256.
[66] H. Lim, Y. Shi, Y. Qiao, Thermally chargeable supercapacitor based on nickel-coated nanoporous carbon, Int. J. Green Energy 15 (2018) 53–56.
[67] H. Lim, L. Weiyi, X. Chen, Y. Qiao, Effects of ion concentration on thermally-chargeable double-layer supercapacitors, Nanotechnology 24 (2013) 465401.
[68] K. Yang, K. Cho, S. Yang, Y. Park, S. Kim, A laterally designed all-in-one energy device using a thermoelectric generator-coupled micro supercapacitor, Nano Energy 60 (2019) 667–672.
[69] S. Yang, K. Cho, S. Kim, Energy devices generating and storing electricity from finger and solar thermal energy, Nano Energy 69 (2020) 104458.
[70] Z. Liu, Y. Zhong, B. Sun, X. Liu, J. Han, T. Shi, Z. Tang, G. Liao, Novel integration of perovskite solar cell and supercapacitor based on carbon electrode for hybridizing energy conversion and storage, ACS Appl. Mater. Interfaces 9 (2017) 22361.
[71] J. Liang, G. Zhu, C. Wang, P. Zhao, Y. Wang, Y. Hu, L. Ma, Z. Tie, J. Liu, Z. Jin, An all-inorganic perovskite solar capacitor for efficient and stable spontaneous photocharging, Nano Energy 52 (2018) 239.
[72] A.P. Cohn, W.R. Erwin, K. Share, L. Oakes, A.S. Westover, R.E. Carter, R. Bardhan, C.L. Pint, All silicon electrode photocapacitor for integrated energy storage and conversion, Nano Lett. 15 (2015) 2727.
[73] J. Xu, H. Wu, L. Lu, S.F. Leung, D. Chen, X. Chen, Z. Fan, G. Shen, D. Li, Integrated photo-supercapacitor based on bi-polar TiO2 nanotube arrays with selective one-side plasma-assisted hydrogenation, Adv. Funct. Mater. 24 (2014) 1840.
[74] M. Zhu, Y. Huang, Y. Huang, Z. Pei, Q. Xue, H. Li, H. Geng, C. Zhi, Capacitance enhancement in a semiconductor nanostructure-based supercapacitor by solar light and a self-powered supercapacitor–photodetector system, Adv. Funct. Mater. 26 (2016) 4481–4490.
[75] B.D. Boruah, A. Misra, Voltage generation in optically sensitive supercapacitor for enhanced performance, ACS Appl. Energy Mater. 2 (2019) 278–286.
[76] S. Gu, Z. Lou, L. Li, Z. Chen, X. Ma, G. Shen, Fabrication of flexible reduced graphene oxide/Fe2O3 hollow nanospheres based on-chip micro-supercapacitors for integrated photodetecting applications, Nano Res. 9 (2016) 424–434.
[77] Y. Yin, K. Feng, C. Liu, S. Fan, A polymer supercapacitor capable of self-charging under light illumination, J. Phys. Chem. C 119 (2015) 8488–8491.
[78] Z. Zhang, X. Chen, P. Chen, G. Guan, L. Qiu, H. Lin, Z. Yang, W. Bai, Y. Luo, H. Peng, Integrated polymer solar cell and electrochemical supercapacitor in a flexible and stable fiber format, Adv. Mater. 26 (2014) 466–470.
[79] T. Chen, L. Qiu, Z. Yang, Z. Cai, J. Ren, H. Li, H. Lin, X. Sun, H. Peng, An integrated "energy wire" for both photoelectric conversion and energy storage, Angew. Chem. 51 (2012) 11977–11980.
[80] Z. Yang, J. Deng, H. Sun, J. Ren, S. Pan, H. Peng, Self-powered energy fiber: energy conversion in the sheath and storage in the core, Adv. Mater. 26 (2014) 7038–7042.
[81] K. Wang, Y.-G. Jia, C. Zhao, X.X. Zhu, Multiple and two-way reversible shape memory polymers: design strategies and applications, Prog. Mater. Sci. 105 (2019) 100572.
[82] P. Miaudet, A. Derré, M. Maugey, C. Zakri, P.M. Piccione, R. Inoubli, P. Poulin, Shape and temperature memory of nanocomposites with broadened glass transition, Science 318 (2007) 1294–1296.
[83] Y. Huang, M. Zhu, Z. Pei, Q. Xue, Y. Huang, C. Zhi, A shape memory supercapacitor and its application in smart energy storage textiles, J. Mater. Chem. A 4 (2016) 1290–1297.

[84] M. Zhu, Y. Huang, Y. Huang, H. Li, Z. Wang, Z. Pei, Q. Xue, H. Geng, C. Zhi, A highly durable, transferable, and substrate-versatile high-performance all-polymer micro-supercapacitor with plug-and-play function, Adv. Mater. 29 (2017) 1605137.
[85] G. Zhou, L. Xu, G. Hu, L. Mai, Y. Cui, Nanowires for electrochemical energy storage, Chem. Rev. 119 (2019) 11042–11109.
[86] J. Deng, Y. Zhang, Y. Zhao, P. Chen, X. Cheng, H. Peng, A shape-memory supercapacitor fiber, Angew. Chem. 54 (2015) 15419–15423.
[87] J. Zhong, J. Meng, Z. Yang, P. Poulin, N. Koratkar, Shape memory fiber supercapacitors, Nano Energy 17 (2015) 330–338.
[88] J. Cao, Y. Zhao, Y. Xu, Y. Zhang, B. Zhang, H. Peng, Sticky-note supercapacitors, J. Mater. Chem. A 6 (2018) 3355–3360.
[89] K. Fu, Y. Yao, J. Dai, L. Hu, Progress in 3D printing of carbon materials for energy related applications, Adv. Mater. 29 (2017) 1603486.
[90] Z. Liu, Z.-S. Wu, S. Yang, R. Dong, X. Feng, K. Müllen, Ultraflexible in-plane micro-supercapacitors by direct printing of solution-processable electrochemically exfoliated graphene, Adv. Mater. 28 (2016) 2217–2222.
[91] F. Li, J. Qu, Y. Li, J. Wang, M. Zhu, L. Liu, J. Ge, S. Duan, T. Li, V.K. Bandari, M. Huang, F. Zhu, O.G. Schmidt, Stamping fabrication of flexible planar micro-supercapacitors using porous graphene inks, Adv. Sci. 7 (2020) 2001561.
[92] C. Zhang, M.P. Kremer, A. Seral-Ascaso, S.-H. Park, N. McEvoy, B. Anasori, Y. Gogotsi, V. Nicolosi, Stamping of flexible, coplanar micro-supercapacitors using MXene inks, Adv. Funct. Mater. 28 (2018) 1705506.
[93] Y. Chen, X. Li, Z. Bi, G. Li, X. He, X. Gao, Stamp-assisted printing of nanotextured electrodes for high-performance flexible planar micro-supercapacitors, Chem. Eng. J. 353 (2018) 499–506.
[94] H. Wang, B. Zhu, W. Jiang, Y. Yang, W.R. Leow, H. Wang, X. Chen, A mechanically and electrically self-healing supercapacitor, Adv. Mater. 26 (2014) 3638–3643.
[95] Y. Yue, N. Liu, Y. Ma, S. Wang, W. Liu, C. Luo, H. Zhang, F. Cheng, J. Rao, X. Hu, J. Su, Y. Gao, Highly self-healable 3D microsupercapacitor with MXene–graphene composite aerogel, ACS Nano 12 (2018) 4224–4232.
[96] J. Wang, V.K. Bandari, D. Karnaushenko, Y. Li, F. Li, P. Zhang, S. Baunack, D.D. Karnaushenko, C. Becker, M. Faghih, T. Kang, S. Duan, M. Zhu, X. Zhuang, F. Zhu, X. Feng, O.G. Schmidt, Self-assembly of integrated tubular microsupercapacitors with improved electrochemical performance and self-protective function, ACS Nano 13 (2019) 8067–8075.
[97] Y.S. Zhang, A. Khademhosseini, Advances in engineering hydrogels, Science 356 (2017), eaaf3627.
[98] H.H. Hsu, Y. Liu, Y. Wang, B. Li, G. Luo, M. Xing, W. Zhong, Mussel-inspired autonomously self-healable all-in-one supercapacitor with biocompatible hydrogel, ACS Sustain. Chem. Eng. 8 (2020) 6935–6948.
[99] Y. Shi, Y. Wang, Y. Gu, L. Zheng, S. Ma, X. Xu, Self-healable and stretchable ionogels serve as electrolytes and substrates for integrated all-in-one micro-supercapacitors, Chem. Eng. J. 392 (2020) 123645.
[100] Y. Dou, T. Pan, A. Zhou, S. Xu, X. Liu, J. Han, M. Wei, D.G. Evans, X. Duan, Reversible thermally-responsive electrochemical energy storage based on smart LDH@P(NIPAM-co-SPMA) films, Chem. Commun. 49 (2013) 8462.
[101] P. Zhang, J. Wang, W. Sheng, F. Wang, J. Zhang, F. Zhu, X. Zhuang, R. Jordan, O.G. Schmidt, X. Feng, Thermoswitchable on-chip microsupercapacitors: one potential self-protection solution for electronic devices, Energy Environ. Sci. 11 (2018) 1717–1722.

[102] S. Ma, Y. Shi, Y. Zhang, L. Zheng, Q. Zhang, X. Xu, All-printed substrate-versatile microsupercapacitors with thermoreversible self-protection behavior based on safe sol–gel transition electrolytes, ACS Appl. Mater. Interfaces 11 (2019) 29960–29969.
[103] L. Yin, X. Huang, H. Xu, Y. Zhang, J. Lam, J. Cheng, J.A. Rogers, Materials, designs, and operational characteristics for fully biodegradable primary batteries, Adv. Mater. 26 (2014) 3879–3884.
[104] Y. Wu, D. Ye, Y. Shan, S. He, Z. Su, J. Liang, J. Zheng, Z. Yang, H. Yang, W. Xu, H. Jiang, Edible and nutritive electronics: materials, fabrications, components, and applications, Adv. Mater. Technol. 5 (2020) 2000100.
[105] A. Maitra, S.K. Karan, S. Paria, A.K. Das, R. Bera, L. Halder, S.K. Si, A. Bera, B.B. Khatua, Fast charging self-powered wearable and flexible asymmetric supercapacitor power cell with fish swim bladder as an efficient natural bio-piezoelectric separator, Nano Energy 40 (2017) 633–645.
[106] I.M. Mosa, A. Pattammattel, K. Kadimisetty, P. Pande, M.F. El-Kady, G.W. Bishop, M. Novak, R.B. Kaner, A.K. Basu, C.V. Kumar, J.F. Rusling, Ultrathin graphene–protein supercapacitors for miniaturized bioelectronics, Adv. Energy Mater. 7 (2017) 1700358.
[107] Y. Lu, K. Jiang, D. Chen, G. Shen, Wearable sweat monitoring system with integrated micro-supercapacitors, Nano Energy 58 (2019) 624–632.
[108] J. Park, D.B. Ahn, J. Kim, E. Cha, B.-S. Bae, S.Y. Lee, J.U. Park, Printing of wirelessly rechargeable solid-state supercapacitors for soft, smart contact lenses with continuous operations, Sci. Adv. 5 (2019), eaay0764.
[109] X. Wang, W. Xu, P. Chatterjee, C. Lv, J. Popovich, Z. Song, L. Dai, M.Y.S. Kalani, S.E. Haydel, H. Jiang, Food-materials-based edible supercapacitors, Adv. Mater. Technol. 1 (2016) 1600059.
[110] C. Gao, C. Bai, J. Gao, Y. Xiao, Y. Han, A. Shaista, Y. Zhao, L. Qu, A directly swallowable and ingestible micro-supercapacitor, J. Mater. Chem. A 8 (2020) 4055–4061.

Index

Note: Page numbers followed by *f* indicate figures and *t* indicate tables.

G

H

I

K

L

M

N

O

P

R

S

FSC
www.fsc.org
MIX
Papier | Fördert gute Waldnutzung
FSC® C083411

Kolibri 360 GmbH
Ferdinand-Jühlke-Straße 7
99095 Erfurt, Deutschland
produktsicherheit@kolibri360.de